Microwave Diode Control Devices

Robert V. Garver
Harry Diamond Laboratories

Microwave Diode Control Devices

Artech

Copyright © 1976

ARTECH HOUSE, INC.

Printed and bound in the United States of America.

All rights reserved. No part of this book may be reproduced or utilized in any form or by any means, electronic or mechanical, including photocopying, recording, or by any information storage and retrieval system, without permission in writing from the publisher.

Standard Book Number: 0-89006-022-3

Library of Congress Catalog Card Number: 74-82596

*To God who answers so many
of my prayers giving me inspiration,
insight, and patience,
and to Shirley, my wife,
one of God's angels, who provides
a pleasant and loving home
in which to live and work.*

Preface

This book is based on the microwave diode switching development at the Harry Diamond Laboratories, a research project sponsored by the Army Materiel Command. This development embodied many of the basic diode switching concepts and led to further development of attenuators, limiters, and phase shifters. To make the book more useful, related concepts are referenced. I wish to thank my colleagues at the Harry Diamond Laboratories and the IEEE Professional Group on Microwave Theory and Techniques for many fruitful discussions. I especially wish to thank Edward G. Spencer, Robert D. Hatcher, John E. Tompkins, Howard S. Jones, Jr., John W. Seaton, Maurice M. Apstein, Helmut Sommer, Joseph J. Witte, and Billy M. Horton for their guidance and encouragement.

One gracious lady has always been ready to convert my hieroglyphics and arrow games into perfect copy. The speed and perfection of her typing made the task of preparing this manuscript more pleasant. To Thelma Southworth, thank you.

ROBERT V. GARVER

Table of Contents

Chapter 1 Introduction   1
  A. History   1
  B. Diode vs Ferrite   5
  C. Related Techniques   6
  D. Systems Applications   8

2 Basic Concepts   15
  A. In-Line Switches   19
  B. Diode Modes   26
  C. Reflection Switches   39
  D. Variable Attenuation   41

3 Biasing Circuits   45
  A. Designing for Maximum RF Bandwidth   46
  B. Designing for Low VSWR   53
  C. Designing for Fast Switching   63
  D. Suppression of Modulator Transients   72
  E. Suppression of RF Leakage Out of the Bias Port   82

Chapter 4 Basic Limitations 85
    A. Limits Imposed by Diode Q   86
    B. Limits Imposed by f_R 96
    C. Power Limitations   105
    D. Other Limitations   115

5 Control Element Design 123
    A. TEM Transmission Line — Stripline   124
    B. TEM Transmission Line — Microstrip   145
    C. Waveguides   151

6 Multiple Diode Switches 157
    A. Filter Theory (Lumped Circuits)   157
    B. Hybrid Attenuation Equation (Mixed Lumped Circuit and Transmission Line Elements)   160
    C. Series-Shunt — Lumped Elements   178

7 Multiple Throw Switches 183
    A. Use of Coupling Devices   183
    B. Lumped Circuit Element Structure   188
    C. Use of Stubs   193

8 Matched Switches and Attenuators 197
    A. Coupling Devices   197
    B. Multiple Diode Absorbing   200
    C. Lumped Element Variable Attenuators   209

Chapter 9 Limiters 215
    A. Output Characteristics   215
    B. Recovery Time   219
    C. PIN Diode Limiters   222
    D. Useful Circuits   227
    E. Measurement of Limiter Performance   232

10 Phase Shifters 235
    A. Digital Phase Shifters   238
    B. Continuous Phase Shifters   280

11 Other Control Devices 285
    A. Doubly Balanced Mixer/Modulator   285
    B. Vector Generator   287
    C. Sampling Gates   290
    D. Bulk Diode Switches   293

Appendix A Diode Measurement 297

Appendix B Diode Junction Theory 323

Appendix C Filter Theory 333

Appendix D Matrix Relationships 345

Appendix E Transmission Line Topics 349

Appendix F Driver Circuits 359

Author Index 365

Subject Index 367

introduction

Introduction

Chapter 1

In Tolkien's trilogy[1], the Hobbits delight in writing books in which everything they know is elegantly arranged in one place. In the field of engineering, such books, written by those who have worked in a specialty area for a long period of time, are valuable to other engineers who occasionally make use of the special information. These books are most useful if they are structured so that they can be accessed efficiently by a variety of users. The systems engineer needs theory allowing him to predict realizable limits of the various devices and to specify realistic component requirements. The component engineer requires specific design equations and charts as well as a clear connection between theoretical variables and structure. The young engineer on his first engagement with the specialty needs a simple, direct, and precise derivation of the theory as it relates to the component limitations and design structure.

A. History

The subject of microwave diode control devices normally elicits thoughts of the PIN diode and its applications. But, in the beginning there was no PIN diode. Diode switches in their infancy were made with point contact diodes. Although some development had been done earlier[2,3] with mixer diodes, none of the work provided clearly defined switching, i.e., insertion loss less than 1 dB and isolation greater than 20 dB. The birth date of the diode switch is considered to be November 1955 when Edward G. Spencer of the U.S. Army Diamond Ordnance Fuze Laboratories, (now the U.S. Army Harry Diamond Laboratories), had Mary Ann Armistead put a 1N263 germanium point contact mixer in a special X-band (9-10 GHz) mixer mount[4] having two flanges instead of one flange and a short circuit. With a 50 mA forward bias current, it gave 0.7 dB insertion loss, and with reverse bias of 1.0 to 1.5 V, it gave 25 to 35 dB isolation over a bandwidth of 1 GHz. When the same thing was done with a 1N23 silicon mixer diode, the insertion loss was 4 dB and the isolation was 6 dB. Yet, they were both good mixers at X-band. The fragile 1N263 also burned out when it was not handled properly, e.g., touching hand to waveguide before diode touches waveguide or passing diode from person to person without having both persons touch the diode at the same time. While the 1N263 gave good switching performance for 1 mW incident power, the isolation was only 10 dB for 10 mW incident power and even less at higher powers. The diode switching speed had also been observed to be as fast as the fastest oscilloscopes available in that day, 3.5 — 5.0 nsec[5]. A whole field of questions remained unanswered. Why does the 1N263 switch while the 1N23 does not? What property of the 1N263 forces the incident power to be low? How fast is the switch? These questions faced the author when he was assigned the diode switching project in August 1956. The switch was already being built into a high resolution radar that could detect a ball being thrown from a two-story building or sort people walking up and down an aisle. Today a time domain reflectometer can resolve discontinuities less than an inch apart. The switch had important possibilities.

One by one the questions were answered. The germanium die was removed from a burned out 1N263 and put in the 1N23 cartridge. It switched. It was the germanium, not the cartridge that made the difference between the 1N23 and 1N263. John Rosado, who was then a summer employee at DOFL, put the germanium die of the 1N34 into the 1N23 cartridge. It switched and higher power did not deteriorate the isolation. (The insertion loss was about 3 dB.) The power limitation[5] of the 1N263 was caused by its low breakdown voltage (1.5 V). A study revealed that surface traps were making the difference between silicon and germanium[6].

R. Wigington and J. Tippett of the National Security Agency (NSA) were just beginning their studies of high speed logic in October 1956. The transistor was not very fast. It broke 5 nsec late in 1957[7]. The speed of the 1N263 diode switch was measured on the NSA EG&G traveling wave oscilloscope to measure the rise time. The result was 3 nsec, the speed of the X-band detector[8]. NSA redesigned the detector. The diode switch was measured again. The result was 1 nsec, the speed of the traveling wave oscilloscope. NSA decided to have someone build a sampling oscilloscope based on some earlier British work and have another look when it became available. Tektronix, Dumont and other oscilloscope manufacturers were not interested in building a sampling oscilloscope. Finally, NSA found a small company that would do it — Lumitron. Lumitron needed a fast diode for their sampling gate. The fastest known diode was the

INTRODUCTION

1N263 based on the NSA-DOFL data. Eventually, the 1N263 X-band diode switch was measured and found to have a rise time of .2 — .3 nsec which was dominated by the RF filters required in conjunction with the switch[8].

Another interesting facet in the history of diode switching, is the difficulty encountered in convincing industry to manufacture the switches once improvements over the 1N263 were determined[5]. The germanium of the 1N263 had a doping density of 2×10^{18} donors/cc. A more rugged diode switch could be made by using germanium having a doping density of 4×10^{16} donors/cc. This doping provided a higher breakdown voltage, allowed the point contact to be larger with the same capacitance as the 1N263, and still had low enough resistance at forward bias to provide 1 dB or less insertion loss. The diode whisker had to be made larger in diameter to keep the pressure on the larger point contact. The author made the original diodes himself and also made the diodes for other projects at DOFL for which engineers were rapidly designing the switches into systems. Unfortunately, the systems engineers were not familiar with diodes switches and were not building safeguards into their systems to prevent burning them out with turn-on/off or other transients. And any time a system failed that had a diode switch in it, the failure was attributed to the switch. To keep the diode switch in the system, the author had to learn the operation of the system and help the engineer find the cause of failure. Many efforts were made to get Philco, the manufacturer of the 1N263, and others to experiment and make these improved switches but with very little success. It wasn't until 3 or 4 years later that the X-band diodes switches became available commercially. They were very profitable products to the few companies in the business.

No small factor in the development of practical diode switches was the "Crystal Rectifier" (Blue Book) project undertaken at Bell Telephone Laboratories under the sponsorship of the Army Signal Corps. This project continued under various titles (49 Quarterly Reports) from July 1954 to October 1967. It gave birth to the varactor[9] and parametric amplifier; the PIN diode[9] for switching and limiting, and, finally, the IMPATT diode oscillator.

After the early X-band efforts at DOFL in 1956, Bell Labs proposed the PIN diode[9] in 1958. From 1958 to 1959, AEL[10] and Sylvania[11] developed broadband TEM transmission line switches using inexpensive gold bonded diodes. In 1960-1961, DOFL developed the theory for the TEM switches[12] which opened the field for improved bandwidth and power handling. Diode switches became one of the few microwave devices that performance could be accurately predicted from theory. Many companies began selling TEM diode switches.

Activity on limiters began in 1959-1960 at Bell Labs[13] providing the theory for PIN diode limiting and giving very good results in TEM transmission line. Switching occured first at X-band and was followed by switching in TEM transmission line. Limiting was developed first in TEM transmission line followed in 1961 by the development of an X-band diode limiter[14] at DOFL.

Diode phase shifters began their evolution in August 1957 with the development of a switched line phase shifter by Emerson[15]. The reflection shifter[16] first appeared in 1959. The idea for a loaded line phase shifter[17] was begun in 1959. The High-Pass Low-Pass phase shifter[18] didn't appear until June 1970 when microwave integrated circuits were well developed.

Hewlett-Packard, entering the diode field, introduced the variable attenuator[19] in July 1962.

Complete diode control device bibliographies were published by Boudreau and Assaly[20], and by Ryder, Brown and Forest[21].

Diode switching had an unusual development in comparison to other subjects in the field of solid-state microwave devices. Most command a large following for a period of time and then pass into the background as they are overshadowed by new technologies. They behave like fads that have their day. A great number of engineers rush into each new fad giving it quick development and short life. For example, tunnel diodes were very popular as a candidate basic element for amplifiers, oscillators, and logic elements. But, transistor technology caught up to the tunnel diode in operating frequency and transistor circuits now perform most of the same functions with greater simplicity and reliability. Parametric amplifiers, once very popular, have now been largely replaced by better designed mixers using hot carrier diodes. Harmonic generators once commanded significant attention but now are largely replaced by microwave transistors that produce substantial output power. Today, 1975, the fad is diode oscillators and amplifiers stemming from IMPATT and GUNN diodes.

The diode switch never had a zenith. It serves a simple, basic function and nothing seems to take its place. It has become the basic building block for limiters, phase shifters, and attenuators. It takes the place of some devices used before its time. Doppler and amplitude modulation were obtained with chopping wheels. TR and ATR tubes were used for sharing a radar antenna. Oscillator tubes were modulated internally rather than at their output. And, of course, there has always been competition between diodes and ferrites as to which can perform a control function better. More often than not, the technology

INTRODUCTION

selected is the most familiar to the engineer having authority to make the decision. The diode switch endures as an important and current topic.

B. Diode vs Ferrite

The question of which to use, the diode or ferrite, warrants greater discussion. The design information presented in this book should help the engineer give the diode fair consideration. A number of factors influence the decision.

- *System Structure* — When the RF portion of the system is integrated on a ceramic substrate or in stripline, the diode control devices can be added easily and economically using the same methods and materials already being employed to fabricate the subsystem. When the RF circuit is being made of components having connectors (waveguide or coax), this advantage no longer exists and other factors must be considered.

- *Frequency* — In general, diodes have the advantage at lower carrier frequencies while the ferrites, because of their distributed nature, have the advantage at higher carrier frequencies. As both technologies improve, however, the ferrites reach down in frequency (by special doping) while the diodes reach higher in frequency (as their cutoff frequency gets higher). At high modulation frequencies, the diode demonstrates a clear advantage because of its inherent high speed. On the other hand, the latching ferrite has an advantage at low modulation frequencies because it consumes less average system power.

- *Reliability* — Based on World War II radar experiences, some engineers have reservations about microwave diodes. The mixer diodes were always burning out. These engineers feel that diodes are unreliable. On the other hand, engineers familiar with transistor circuits will know that the miniature inductors and transformers used with them are usually the most likely to fail in circuits. Ferrites and diodes are susceptible to poor reliability if not properly engineered. The ferrite winding should not be too fine a wire and high current surges should be avoided. The diode should have a margin of safety in its RF power ratings, and, it too, should be protected from large current spikes.

- *Switching* — In the areas of power rating, insertion loss and isolation both devices can be made equally effective by careful designing in the overlapping carrier frequency range.

- *Costs* — The initial and running costs of both devices also tend to stay competitive, but one is usually forced over the other because the in-house technology may be better developed.

- *Production* — The diode sometimes gets the advantage in production because the circuits can be mass-produced — no trimming required. There are often problems maintaining close ferrite/metal wall interfaces which can cause intolerable deviations in performance.

- *Temperature Sensitivity* — The ferrite also is ten times more sensitive to temperature variations than diodes.

- *Bandwidth* — Both devices can give adequate bandwidth for most systems. The diode can give extreme bandwidth but this is usually not needed.

- *Burnout* — The ferrite stands out best where another very high power radar might force its power down into a lower power radar containing a control device. If the diode is not designed for high power, it can be damaged. The ferrite will simply cool off after the exposure and resume its normal function.

C. Related Technologies

Varactor/Pin Diode — Both developments were introduced by BTL[9] in 1958. Without them, diode switches would have never controlled more than 1-10 watts. The quality of these diodes became measured by their cutoff frequency, f_c, the frequency at which the capacitive junction reactance equals the parasitic series resistance. Later, it became evident[22] that the limits on insertion loss and isolation of diode switches are dictated by f_c.

Filter Theory — The period from 1955 to 1965 included significant advances in the design of microwave filters[23], carried on mostly at Stanford Research Institute. Diode switching began using the technology when the switch designers discovered that broad bandwidth could be obtained by making the diode parasitic capacitance part of a filter structure[24]. These ideas are utilized in Chapters III — Biasing Circuits, VI — Multiple Diode Switches, and VII — Multiple Throw Switches. A brief summary of the basic relationships of filter theory is given in Appendix C to aid the reader in its application to diode switching.

Matrices/The Computer — The analog computer emerged in the late 1950's and the digital computer in the 1960's. Prior to the emer-

INTRODUCTION

gence of the general purpose digital computer (and Fortran), numerical calculations were very tedious. They were normally done only by graduate students and "algebraic technicians." One of the few fields that could afford them was antenna design. With the advent of Fortran computers, and later, desk top computers, it became necessary for the engineer to have a better understanding of the basic equations for his devices and better theoretical tools for setting up his numerical analysis. ABCD matrices became very useful for calculations because elements can be cascaded by a straight matrix multiplication. The S-parameters have the familiar sence of Γ, the voltage (or current) reflection coefficient. S_{21} is the forward voltage transfer coefficient. These matrix methods have been used widely in this book where more simple models become too complex. The relationships of these mathematical models are sketched in Appendices D — Matrix Relationships and E — Transmission Line Topics.

Integrated Circuits — In pure integrated circuits all active and passive components are fabricated on the same semiconductor substrate. In hybrid integrated circuits, the substrate is an insulator (ceramic or ruby) and some components are mounted as chips while others may be printed with the conductor pattern. The object of integrated circuits is to eliminate space normally used up by connectors and covers. And certainly, microwave devices use up lots of space with connectors. The standard method for estimating the cost of a coaxial component is to allow $25.00 for the connectors and then add the cost of the inner workings. Elimination of the connectors also eliminates the need to have everything designed for a 50-ohm characteristic impedance. With microwave integrated circuits, 50 ohms is needed only at the RF connectors to the world outside the substrate. Historically, microwave energy has been conducted along the following sequence of preferred transmission lines: waveguide, coax, stripline, microstrip. Waveguide gave way to coax as the X-band spectrum became crowded and other system factors favoring the lower frequencies came into the foreground. Coax gave way to stripline in system work when low loss teflon-fiberglass board was developed providing significant savings in cost and space. This was really the first transition into microwave integrated circuits (MIC) because complete RF assemblies were printed on a single board from one master pattern. Stripline gave way to microstrip when uniform, low loss, high dielectric substrates were developed, and, when fabrication tools and chip elements became available for making up the circuits.

Diode control devices are especially compatible with microwave integrated circuits. The performance of diode switches is severely limited by the parasitic inductance and capacitance of the diode cartridge.

Stripping this cartridge away allows the diode switch to have a useful bandwidth, almost as wide as the microstrip substrate. Control diodes are implantable semiconductor chips requiring only the standard MIC tools for installation. The other pieces (capacitors, bias stubs) needed in diode control devices are also included in the MIC fabrication technology. Since diode mixers and oscillators are equally as adaptable to MIC, diodes have become almost as plentiful in MIC as transistors have in pure integrated circuits at lower frequencies.

Is the next step to have fully integrated circuits for microwaves? A reverse trend can already be seen. Losses are very important in many microwave applications. Fully integrated circuits made for microwaves have been too lossy[25] because the semiconductor substrate must necessarily be lossy in order to be a semiconductor. Even microstrip is more lossy than teflon-fiberglass stripline. Some designs in microstrip have had to be converted to stripline in order to satisfy low loss requirements. Fully integrated circuits for microwaves may be useful in the future where power levels are moderate, losses are not important, extreme logic speeds are required, and feedback instabilities are subdued.

D. Systems Applications

Radar is an extension of Man's eyes. Electronic communication is an extension of his ears and mouth. Electronic instrumentation is a magnification of his sense of touch. The computer is an extension of his mind.

Radar — Radar permits Man to see the presence of objects when it is dark or cloudy or when the objects are too far away to be seen with the naked eye. Radar permits Man to see precisely how high he is above the ground or to make maps of the ground features.

The simplest radar consists of a high power pulsed oscillator connected to a narrow beam antenna and a video detector connected to another narrow beam antenna pointed in the same direction. A simple oscilloscope connected to the detector will show the distance to the reflecting object when the trace is started with the transmitted pulse. The return pulse is delayed about 10 μsec/mile. To improve sensitivity, a superheterodyne mixer is used instead of a detector. RF power leakage from the transmitter antenna to the receiver antenna can burn out the mixer diode(s). There lies the first need for a protective device which need can be fulfilled by a diode switch or limiter. The limiter has advantages in that connecting wires and drive circuits are not required, and it protects the receiver from the pulses of other radars as well.

INTRODUCTION

The next level of complexity is added to the radar by having the transmitter and receiver share the same antenna. Considerable cost is incurred in making these large maneuverable antennas. A double throw switch is required to switch alternately between transmitter and receiver. In this application, the switch must have low loss because any losses must be overcome by increased transmitter power or an even larger antenna. The switch must be able to control the high power of the transmitter. It must have enough isolation to reduce the undesired transmitter pulse at the mixer to a level that will not burn out the mixer, or on short range radars, cause receiver saturation. The switch must be fast enough to engage the receiver before the return pulse is expected. It is also advantageous if the double throw switching function can be obtained passively with a limiter or limiters.

The next most common use of diode control devices in radar is in the improvement of the antenna maneuverability. Because of the large mass and high inertia of a conventional radar antenna, it is difficult to change back and forth between rotating search scan and a relatively fixed track scan. An electronically controlled array radar can provide inertialess antenna scanning. There are two methods of providing electronic pointing of an array: fixed matrix input control and distributed phase shifter control. A Butler array feed matrix has many input ports each one pointing the radiation of the antenna array in a different direction. Scanning is obtained by sequentially connecting the transmitter and receiver to different ports. When the scan is controlled by binary logic, then one transmitter and receiver are easily controlled by a "fan-out" of double throw switches connecting it to the 2^n input ports. All of the switches must be able to control the transmitter power. Insertion loss must be kept low because the power must go through n ON switches for transmit and through them again for receive.

Distributed phase shifter control of array scan has the advantage that no control element is subjected to the full transmitter power. The transmistter power is distributed among all of the radiating elements and phase shifters. Most systems are controlled by binary logic and have phase shift bits of $180°$, $90°$, $45°$, $22.5°$. Because the bits are in series, it is important that each bit have low insertion loss. No more than 3 or 4 bits are normally used because more bits add insertion loss and do not improve radiation efficiency enough to make up for the added insertion loss, and accumulated phase error and interacting VSWR induced phase error become larger than the smallest bit eliminating any precision added by the smallest bit. It is also important that the phase error of each bit be small because being in series they accumulate, reducing radiation efficiency and increasing sidelobe levels.

With the freedom of inertia allowed by electronic steering, one antenna array can perform a number of functions by time sharing it. It can search, track a number of objects, and even perform communication functions all at the same time.

Another radar function that control diodes are sometimes required to perform is information processing. For example, normally information processing is performed after the mixer, but when multiple RF signals must be processed, then, multiple local oscillators, mixers and IF circuits can be obviated by doing the processing at RF. As another example, a simple diode switch at the receiver input can be used to time-gate out undesired signals.

For every radar, there must be at least one radar tester or target simulator. Furthermore, all discussions thus far have presumed pulse radar. There are also FM and doppler radars. To test these radars — to simulate targets for them — requirements exist for amplituded modulators, balanced modulators, single sideband modulators, phase modulators, and variable attenuators. The requirements for these devices are usually expressed in terms of level of undesired sidebands and precision of transfer function.

Communications — Microwave is a vehicle for point to point communications because of the large bandwidth (in MHz) available at the higher frequencies. A communications system includes a sender, a receiver, and often intermediate stations between them for relay or traffic control. These stations may be buried coaxial systems, the familiar microwave relay towers, or satellites. When the station serves as a relay only, it may have only RF amplifying equipment, or, in the case of satellites, it may change frequency and amplify. Changing frequency involves mixing and modulating. Since the RF spectrum in communications is crowded, the modulator must have low cross modulation products to avoid coupling between channels. Since there may be many relay stations between sender and receiver, the distortion of each station must be low. For maximum utilization of a carrier, the relay components must have wide bandwidth for both the RF and the modulation.

In a traffic control station, RF switches may be used for sorting out channels, but, this function is usually performed after demodulation.

As digital computers are being used more and more as major tools for information and control systems, the information communicated takes on a digital (as opposed to analog) character. Thus, the trend in communications sytems is toward digital modulation. Once the modulation is established as digital then an advantage in using phase

INTRODUCTION

modulation is obtained. For the same spectrum width, the average power can be higher for the same peak power. A zero with amplitude modulation is no power at all while a zero with a binary phase phase modulation is full power 180° out of phase with a one. The average power for the zero is thus radiated rather than absorbed in the modulator. This increased power allows the relay stations to be farther apart requiring fewer stations. In this application, high speed digital phase modulators are required to meet tough specifications for high modulation rate and wide RF bandwidth. Gigabit rates are presently obtainable (see Chapter III — Biasing Circuits, Section D — Designing for Fast Switching Time), and wide RF bandwidth is obtainable (Chapter X — Phase Shifters, Section A — Digital Phase Shifters).

Instrumentation — Diode switches have been used in two general categories in support of microwave instrumentation. In one category, the diode switch is used to modulate cw power so that it may be detected with a video detector and amplified with a narrow band amplifier. The most common use of this method is in the modulation at 1 kHz of conventional signal generators. The Hewlett-Packard PIN line attenuator is widely used in this application. Simpler modulators have been used for chopping the input of spectrum surveillance systems. Network analyzers use balanced modulattion both for modulating the subcarrier in subcarrier modulated systems and in sampling gates for wideband detection systems. Very sensitive detection can be obtained when the modulator and amplifier are phase locked which provides synchronous detection. Another application which is not strictly modulation, but similar to it, is that in which a double throw switch is used to connect a receiver alternately to the power output of a noise source and the power received by a radar astronomy antenna. Considerable precision in the measurement of sky temperature is permitted by the freedom from receiver drift.

The second category of the use of diode switches in instrumentation is in the analysis of transient phenomena in gases and plasmas[26]. As the event is repeated, the diode switch gates a pulse of diagnostic microwaves into the sample after a given delay. The phase and amplitude of the reflected and transmitted pulses reveal the internal state of the gas or plasma.

REFERENCES

1. J.R.R. Tolkien, *The Hobbit*, (no date); *Lord of the Rings*, Houghton, London, 1954-56.

2. D.J. Grace, "A Microwave Switch Employing Germanium Diodes," Tech. Rep. No. 26, *Appl. Electronics Lab.*, Stanford University, Stanford, Calif., Jan. 1955.

3. F.S. Cole, "A Switch Detector Circuit," *IRE Trans. on Microwave Theory and Techniques*, vol. MTT-3, pp. 59-61, Dec. 1955.

4. M.A. Armistead, E.G. Spencer, and R.D. Hatcher, "Microwave Semiconductor Switch," *Proc. IRE*, Vol. 44, p. 1975, Dec. 1956.

5. R.V. Garver, E.G. Spencer, and R.C. LeCraw, "High Speed Microwave Switching of Semiconductors," *J. Appl. Phys.*, vol. 28, pp. 1336-1338, Nov. 1957.

6. R.V. Garver, J.A. Rosado, and E.F. Turner, "Theory of the Germanium Diode Microwave Switch," *IRE Trans. on Microwave Theory and Techniques*, vol. MTT-8, pp. 108-111, Jan. 1960.

7. C.G. Thornton and J.B. Angell, "Technology of Micro-Alloy Diffused Transistors," *Proc. IRE*, vol. 46, pp. 1166-1176, June 1958.

8. R.V. Garver, "Sub-Nanosecond Microwave Switching," *1965 IEEE Military Electronics Conf. Rec.* — Mil-E-Con 9, pp. 143-145.

9. A. Uhlir, "The Potential of Semiconductor Diodes in High-Frequency Communications," *Proc. IRE*, vol. 46, pp. 1099-1115, Jun. 1958.

10. L. Riebman, "Study of Wide-Open Receiver Detector Capabilities," *American Electronics Labs., Inc.*, Philadelphia, Pa., AEL Tech. Rep. 57052-3, Signal Corps Contract, DA-36-039-SC-74813, July-December 1958.

11. M. Bloom, "Microwave Switching with Computer Diodes," *Electronics*, vol. 33, pp. 85-87, Jan. 1960.

12. R.V. Garver, "Theory of TEM Diode Switching," *IRE Trans. on Microwave Theory and Techniques*, vol. MTT-9, pp. 224-238, May 1961.

13. D. Leenov, J.H. Forster, and N.G. Cranna, "PIN Diodes for Protective Limiter Applications," *1961 ISSCC Digest*.

14. R.V. Garver and D.Y. Tseng, "X-Band Diode Limiting," *IRE Trans. on Microwave Theory and Techniques*, vol. MTT-9, p. 202, Mar. 1961.

15. E.M. Rutz and J.E. Dye, "Frequency Translation by Phase Modulation," *1957 IRE WESCON Conv. Rec.*, pt. 1, pp. 201-207.

16. E. Stern, "A Variable Reactance Diode Phase Shifter," *IRE-PGMTT Natl. Symp.*, Harvard Univ., Cambridge, Mass., June 2, 1959.

17. H.N. Dawirs and W.G. Swarner, "A Very Fast Voltage Controlled Microwave Phase Shifter," *Microwave Journal*, vol. 5, pp. 99-107, Jan. 1962.

18. P. Onno and A. Plitkins, "Miniature Multi-Killowatt PIN Diode MIC Digital Phase Shifters," *1971 IEEE-GMTT Int. Microwave Symp. Dig.*, Catalog No. 71 C25-M, pp. 22-23.
19. J.K. Hunton and A.G. Ryals, "Microwave Variable Attenuators and Modulators Using PIN Diodes," *IRE Trans. on Microwave Theory and Techniques*, vol. MTT-10, pp. 262-273, July 1962.
20. C.A. Boudreau and R.N. Assaly, "Electronically Controlled Microwave Devices," *Lincoln Lab. Lib. 30th Reference Bibliography*, 7 Oct. 1966, DDC No. AD 805508.
21. R.M. Ryder, N.J. Brown, and R.G. Forest, "Microwave Diode Control Devices," *Microwave Journal*, pt. 1, pp. 57-64, Feb. 1968, pt. 2, pp. 115-122, Mar. 1968.
22. R.V. Garver, "Fundamental Limitations in RF Switching Using Semiconductor Diodes," *Proc. IEEE*, vol. 52, pp. 1382-1383, Nov. 1964.
23. G.L. Matthaei, L. Young and E.M.T. Jones, *Microwave Filters, Matching Networks and Coupling Structures*, McGraw-Hill Book Co., N.Y., N.Y., 1964.
24. P. Clar, "Optimum Design of Fast Acting Broadband Multithrow Diode Switches," *1963 IEEE-PGMTT National Symposium Digest*, pp. 105-111, May 1963.
25. B.W. Battershall and S.P. Emmons, "Optimization of Diode Structures for Monolithic Integrated Microwave Circuits," *IEEE Trans. on Microwave Theory and Techniques*, vol. MTT-16, pp. 445-450, July 1968.
26. F.L. Tevelow, "Microwave Interferometer Measurements in Shocked Air." *J. Appl. Phys.*, vol. 38, pp. 1765-1780, 15 Mar. 1967.

Basic Concepts

Chapter 2

The understanding of most microwave semiconductor diode control devices is based on the diode switch. The switch is the simple device that serves as basic building block for these other control devices. A diode variable attenuator is a switch that goes continuously and smoothly from its low loss, "on" state to its high loss, "off" state. A limiter is a switch that is normally "on" at low level incident power and turns itself "off" at high level incident power by rectifying its own "off" biasing current. In other words, a limiter is a self-activating switch having the proper switching polarity and activated by high-level incident power. A phase modulator can be made by switching between different phase paths or by switching between different transmission line loading elements. Thus a conceptual derivation that begins with the diode switch will provide a good foundation for understanding the other devices.

A light switch is close to ideal. A knife switch is even closer to ideal. When the knife switch is closed, the high pressure over the large contact area provides a very low resistance, for all practical purposes zero ohms. When the knife switch is open, the resistance is extremely high, practically infinite. The switch is usually placed in series between a voltage source and light bulb so that when it is open no current flows, etc. The house current or battery normally used provides a constant voltage, i.e., the source has a very low output impedance, much lower than the resistance of the load (light bulb).

A switch in a transmission line is different from a light switch in that the voltage source is replaced by a generator having an output impedance equal to that of the load. In order to operate efficiently the source and load impedances are equal to the transmission line impedance. Microwave transmission lines are usually coaxial, stripline, or waveguide, all of which may be represented by the parallel line Letcher model. Note that in a transmission line the alternate scheme

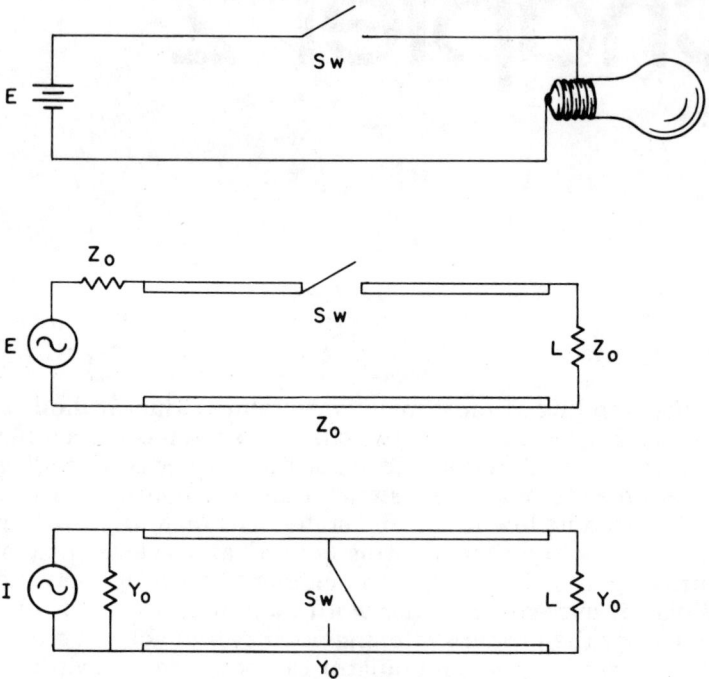

Fig. 2-1 Basic Switch Operation

of having the switch in shunt with the load becomes a possibility. In Fig. 2-1, the first Z_0 (Y_0) is the generator output impedance and the second Z_0 (Y_0), designated by L, is the matched load terminating the transmission line.

Switches like the knife switch can be made in coaxial transmission line by using mercury switches, like those used in delay line nanosecond pulse generators. When the switch is closed the structure looks like the center section of 50 ohm coaxial transmission line. When the switch is open it looks like a series 0.5 pF capacitor.

BASIC CONCEPTS

A useful concept in transmission lines is traveling waves. A well matched transmission line has only a forward traveling wave, going from generator to matched load. This is called incident power and denoted by P_i. When the wave hits a dicontinuity some power is absorbed in the discontinuity, denoted by P_a, some is reflected by the discontinuity, denoted by P_r, and some is transmitted past it, denoted by P_t. According to convention, the reflected wave travels to the left and is absorbed by the generator. Thus when a switch in the "Off" state is put in a transmission line, the incident power is largely reflected by the switch and returned to the generator.

An ideal diode switch will either transmit or reflect all incident power. None will be absorbed in the diode in either switching state. This is quite different from the performance of a conventional variable attenuator which will either transmit or absorb all incident power. The conventional variable attenuator thus has a low VSWR in all states while the basic diode switch has a low VSWR for the "On" state and a very high VSWR for the "Off" state. When the reflected power causes problems, the diode switch must be preceded in the transmission line by an attenuator, isolator, or some other combination of components to provide low VSWR (Chapters IV and VI). In many cases the switch is used to direct the flow of power wherein the high reflectivity of the "Off" diode is used to conserve RF power.

The open circuit and short circuit conditions of a knife switch are approximated by a semiconductor diode. The DC characteristics of a diode are shown in Fig 2-2. Depending on the semiconductor and dopants, the diode begins forward conduction, V_F, at 0.3V to 1.1V. For large enough reverse bias the diode again begins conducting. V_B is the voltage at which the reverse current reaches 10 μA. When the diode is forward biased to point 1, a small AC signal will see the low resistance calculated from the slope $\frac{dV}{dI}$. At reverse bias point 2, practically no current flows for all parts of the AC cycle approximating the open position of the knife switch. (This simple model is correct for hot carrier diodes at all frequencies and the PIN diodes only at low frequencies.)

From a practical point of view an equivalent circuit can be defined for a diode which has been quite useful in predicting the performance of conventional diodes when used as switches up to 10 GHz. The most common glass package and pill package diodes are shown in Fig. 2-3. The equivalent circuit shown in the figure represents both packages.[1] C_D represents the diode reverse bias depletion layer capa-

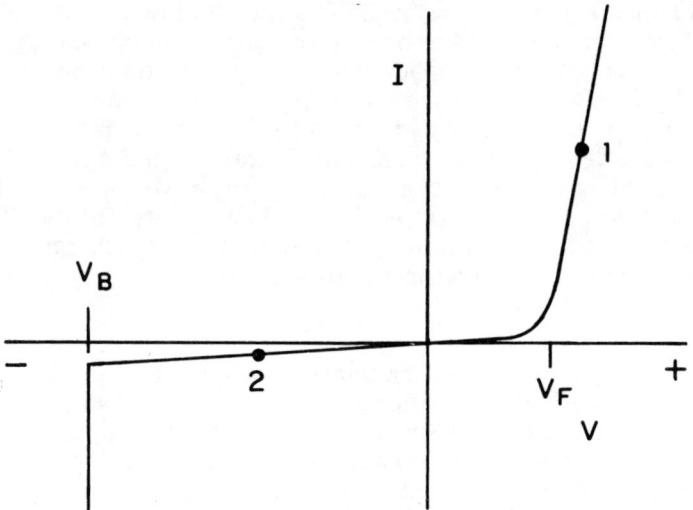

Fig. 2-2   Diode V-I Characteristic

citance. At forward bias this capacitance is shunted by conduction current, effectively closing the switch S_w. R_s is the series resistance of the diode attributed mainly to the finite conductivity of bulk semiconductor beyond the junction. The remainder of the equivalent circuit is caused by the diode package. L_w results from the fine wire required to go from the tiny junction to the mechanically convenient dimensions of the package. C_c is the capacitance between the end terminals of the package, having a significant capacitance because of the higher than air dielectric constant of the ceramic holding the package together. Typical values are as follows:

	R_s	C_D	L_w	C_c
Glass	1 − 2 ohms	.05 − .2 pF	3 nH	.05 pF
Pill	1 − 2 ohms	.05 − .2 pF	0.3 nH	.1 pF

The difference between forward conduction and reverse bias is approximated in the circuit by the perfect switch in parallel with C_D, being closed and opened respectively. In reality, the impedance of C_D becomes a short circuit in a varactor diode by having C_D become infinite for forward bias. In a PIN diode or hot carrier diode a resistance in parallel with C_D, here infinite, decreases to practically zero ohms for forward conduction when the depletion layer is flooded with injected carriers.

BASIC CONCEPTS

A. In-Line Switches

The concepts of the transmission line and diode can now be combined to predict the switching properties of diodes. As long as a diode is mounted in a bilaterally matched lossless transmission line, the length of line on each side of it will not alter its switching performance. Furthermore as long as the diode is small compared to a wavelength ($< 0.1\lambda$) it can be represented by the lumped circuit equivalent circuit shown in Fig 2-3. Thus by setting line lengths equal to zero the equivalent circuits shown in Fig. 2-4 can be drawn.[2] The Z_0 (Y_0) and generator are the same as in Fig. 2-1. The diode is represented by Z and Y.

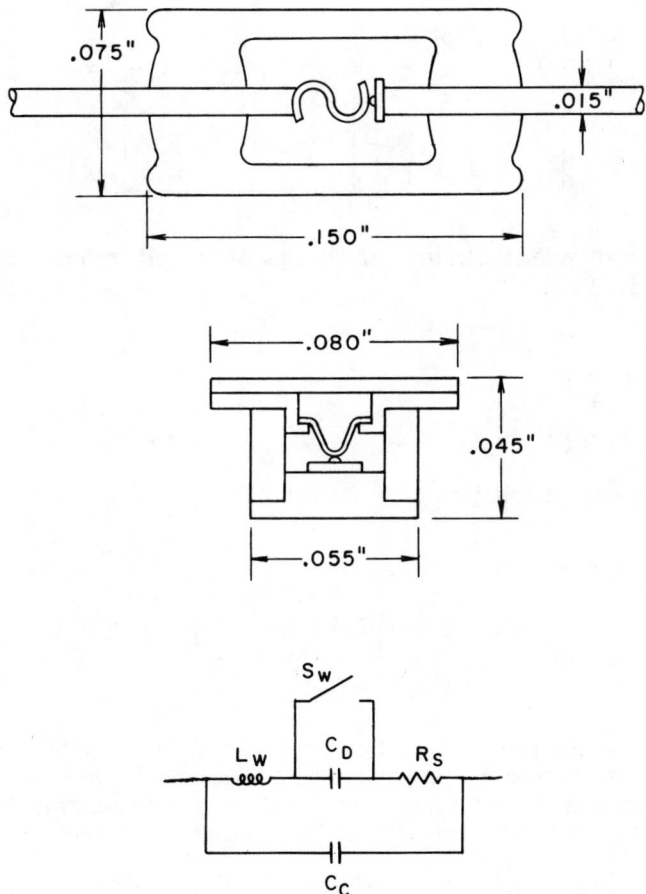

Fig. 2-3 Diodes and their Equivalent Circuit

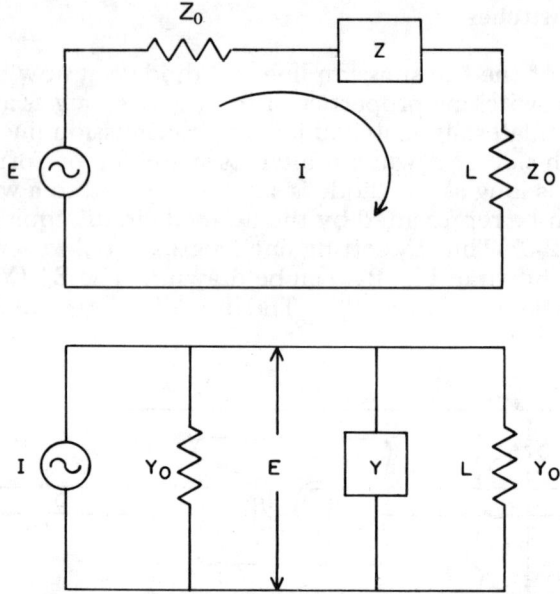

Fig. 2-4 Equivalent Circuits of Diodes Mounted in Transmission Lines

$$Z = R + jX \tag{2-1}$$

$$Y = G + jB \tag{2-2}$$

Using the series circuit first

$$I = \frac{E}{2Z_0 + Z} \tag{2-3}$$

$$P_L = \frac{1}{2} I I^* Z_0 = \frac{\frac{1}{2} E^2 Z_0}{(R + 2Z_0)^2 + X^2} \tag{2-4}$$

in which P_L is the power absorbed by L and I^* is the complex conjugate of I (peak current). The attenuation, α, of a diode inserted in a transmission line is defined as the ratio in decibels of power incident to the diode, P_i, to power transmitted past the diode P_t ($P_t = P_L$).

$$\alpha = 10 \log \frac{P_i}{P_t} \tag{2-5}$$

BASIC CONCEPTS

When the series diode is zero ohms impedance all of the incident power is transmitted past the diode and absorbed in L.

$$P_i = P_L \text{ for } Z = 0. \tag{2-6}$$

Using Eq. (2-4)

$$P_i = \frac{E^2}{8 Z_0} \tag{2-7}$$

Normally the diode impedance will not be zero ohms, but some finite value, and will cause the power transmitted past it to be some reduced value $P_t = P_L$ for $Z = Z$.

Using Eqs. (2-4, 2-5 and 2-7)

$$\alpha = 10 \log \left[\left(\frac{R}{2 Z_0} + 1 \right)^2 + \left(\frac{X}{2 Z_0} \right)^2 \right]. \tag{2-8}$$

A similar equation is derived for shunt diodes interchanging I and E, R and G, X and B, and Y and Z. Thus for shunting diodes

$$\alpha = 10 \log \left[\left(\frac{G}{2 Y_0} + 1 \right)^2 + \left(\frac{B}{2 Y_0} \right)^2 \right] \tag{2-9}$$

The same equations may be derived using matrices (Appendix D). Given a matrix of the form

$$\begin{bmatrix} A & B \\ C & D \end{bmatrix} \tag{2-10}$$

a series impedance in a transmission line is represented by

$$\begin{bmatrix} 1 & Z \\ 0 & 1 \end{bmatrix} \tag{2-11}$$

a shunt admittance, by

$$\begin{bmatrix} 1 & 0 \\ Y & 1 \end{bmatrix}$$

and the equation for attenuation in a transmission line is

$$\alpha = 10 \log \frac{1}{4} \left| A + \frac{B}{Z_0} + Z_0 C + D \right|^2. \qquad (2\text{-}12)$$

Substituting 2-11 in 2-12 gives

$$\alpha = 10 \log \frac{1}{4} \left| 2 + \frac{Z}{Z_0} \right|^2 \qquad (2\text{-}13)$$

which reduces to Eq. (2-8), etc.

Eq. (2-8 and 2-9) give circular arcs of eqi-attenuation curves on a rectangular grid plot of normalized impedance or admittance.

Circle centers are at $-2 + j\,0$ and radii are $2 \times 10^{\alpha/20}$. The eqi-attenuation curves are also circles on the Smith Chart as shown in Fig. 2-5.

Plotting the impedance or admittance of a diode at the frequency of interest normalized to transmission line impedance on Fig. 2-5 indicates directly the switching properties to be expected from the diode.

The phase shift of a network represented by an ABCD matrix (ref. Appendix D) is given by

$$\phi = \tan^{-1} \left[\frac{Im\,(A + B/Z_0 + C\,Z_0 + D)}{Re\,(A + B/Z_0 + C\,Z_0 + D)} \right] \qquad (2\text{-}14)$$

Substituting 2-11 in 2-14 the phase shift from an impedance in series in a transmission line given by

$$\phi = \tan^{-1} \left[\frac{X/Z_0}{2 + R/Z_0} \right] \qquad (2\text{-}15)$$

These are shown as the eqi-phase arcs on the Smith Chart of Fig. 2-5.

The utility of these curves may be demonstrated by viewing the normalized impedance and admittance of a typical PIN diode.

Consider an HP 3001 PIN junction diode measured at 1 GHz and normalized to 50 ohm transmission line characteristic impedance superimposed on the attenuation Smith Chart as shown in Fig. 2-6. Note that for this diode $L_W = 3.7$ nH $R_s = 2\Omega$ $C_c + C_D = 0.2$ pF. As illustrated in Figs. 2-6 and 2-7 placing this diode in series in a 50 ohm transmission line would provide $< \frac{1}{2}$ dB insertion loss (δ) and 18

BASIC CONCEPTS

dB isolation (η). Shunting a 50 ohm transmission line it would provide 0.12 dB insertion loss and 4 dB isolation.

Note that this diode functions better in series than in shunt. The required impedance range of a diode giving 1 dB to 20 dB switching is calculated using Eq. (2-13).

$$\alpha = 10 \log \left| 1 + \frac{Z}{2 Z_0} \right|^2 \qquad (2\text{-}16)$$

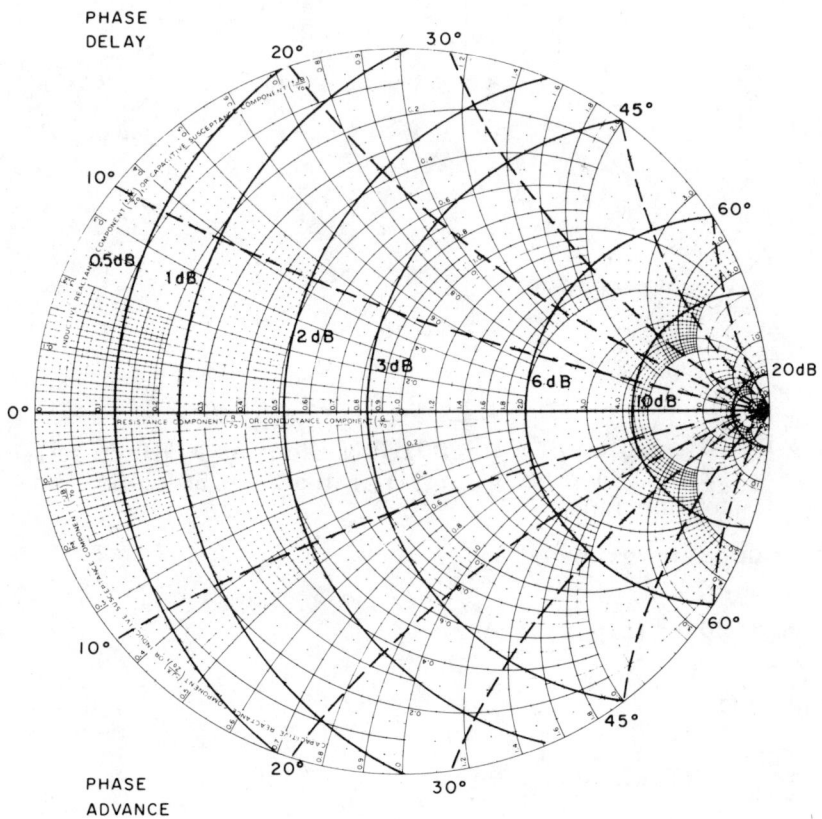

Fig. 2-5 Transmission Loss and Phase from Normalized Diode Impedance or Admittance

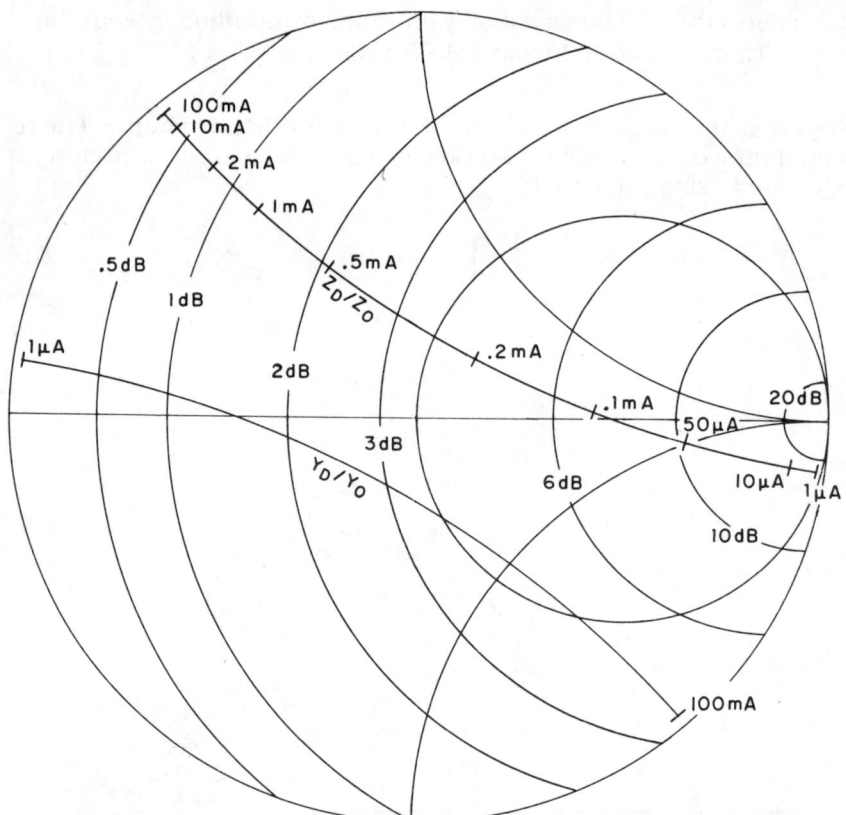

Fig. 2-6 Attenuation of a HP3001 PIN Diode at 1 GHz in series (Z_D/Z_0) and Shunt (Y_D/Y_0) in a $Z_0 = 50$ ohms Transmission Line

which can be approximated by

$$\alpha \approx 20 \log\left(1 + \frac{|Z|}{2 Z_0}\right) \tag{2-17}$$

and provide

$$\frac{|Z|}{2 Z_0} = 9 \text{ for 20 dB}$$

or approx.

$$\frac{|Z|}{Z_0} = 20. \tag{2-18}$$

BASIC CONCEPTS

Similarly for a shunt diode

$$\frac{|Y|}{Y_0} \approx 20 \approx \left(\frac{|Z|}{Z_0}\right)^{-1} \qquad (2\text{-}19)$$

$$\frac{|Z|}{Z_0} \approx .05 \qquad (2\text{-}20)$$

For the low loss state (1 dB).

$$\frac{|Z|}{2 Z_0} = .13$$

$$\frac{|Z|}{Z_0} \approx .25 \text{ for series} \qquad (2\text{-}12)$$

and

$$\frac{|Z|}{Z_0} \approx 4 \text{ for shunt.} \qquad (2\text{-}22)$$

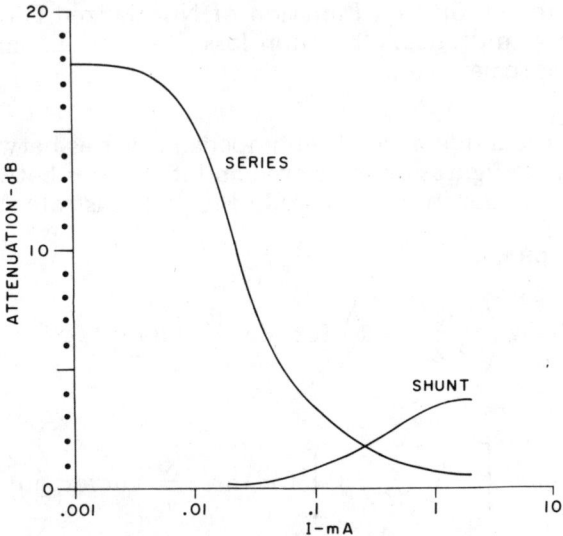

Fig. 2-7 Attenuation as a Function of Current of PIN Diode Switch of Fig. 2-6

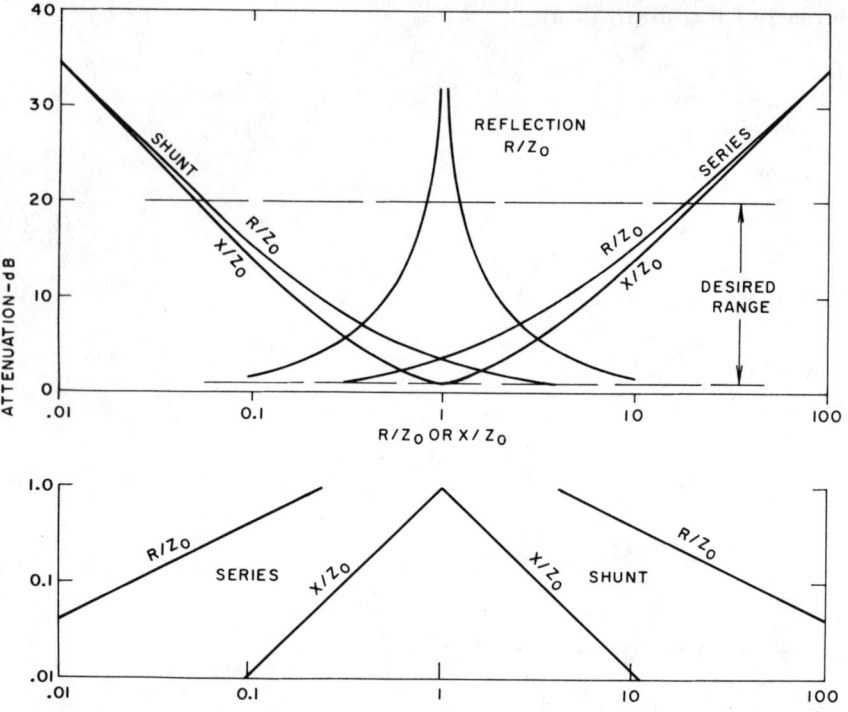

Fig. 2-8 Attenuation as a Function of Normalized Diode Impedance or Admittance. Insertion loss is shown expanded on a Log Log scale.

Therefore if the diode normalized impedance varies between .25 and 20 the series configuration is favored and if it varies between .05 and 4 the shunt configuration is favored. Fig. 2-8 illustrates this point.

Using Eqs. (2-8 and 2-9)

$$\alpha = 20 \log \left(\frac{R}{2 Z_0} + 1 \right) \text{ for the } \frac{R}{Z_0} \text{ curves.} \quad (2\text{-}23)$$

and

$$\alpha = 10 \log \left[1 + \left(\frac{X}{2 Z_0} \right)^2 \right] \text{ for the } \frac{X}{Z_0} \text{ curves, etc.} \quad (2\text{-}24)$$

B. Diode Modes

Due to the reactive elements in the diode equivalent circuit shown in

BASIC CONCEPTS

Fig. 2-3 the diode impedance goes through various resonances as frequency is changed. These resonances when controlled can provide useful switching at higher frequencies.[2] Most waveguide switches in the 10 to 20 GHz frequency range use one of these resonances while most low power TEM transmission line switches use an extension of the low frequency performance, avoiding resonances.

The computed attenuation of a series diode in 50 ohm transmission line is shown in Fig. 2-9 as a function of frequency. The equivalent circuit parameters are C_c = 0.2 pF, L_w = 5nH, C_d = 0.2 pF, and R_s = 5 ohms. It is assumed that these parameters do not change drastically with frequency. From Fig. 2-9 it is seen that high isolation is available at three frequencies which are labeled Modes 1, 2 and 3.

Mode 1 is the baseband mode; Mode 2 is the resonant diode mode; and Mode 3, the higher order diode resonance mode. Referring to Fig. 2-10, the high isolation of Mode 1 is due to the high series reactance of the reverse biased diode capacitance at low frequencies which impedes the flow of current in the load L. The low insertion loss of Mode 1 is attributed to the low series impedance of R_s and L_w at low frequencies. As frequency is increased, the parallel resonance between L_w and C_c is encountered at forward bias which presents a high series impedance again impeding the flow of current to L and providing the isolation peak labeled Mode 2. As frequency is increased on the reverse biased equivalent circuit the series resonance

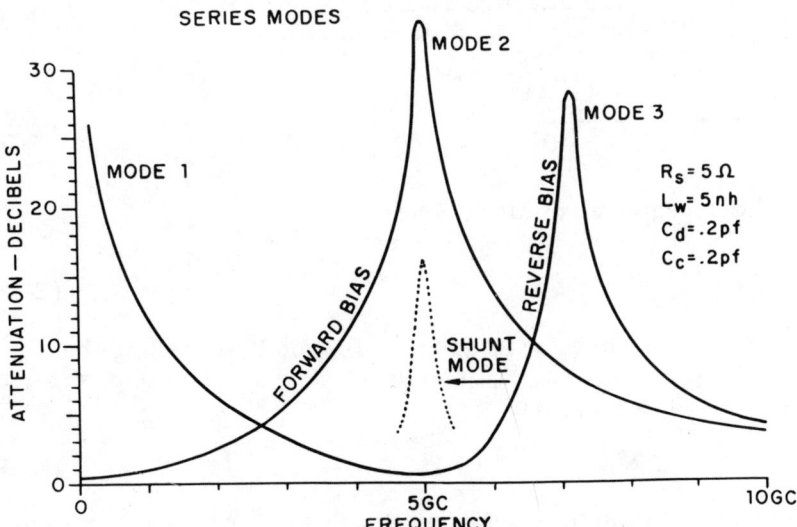

Fig. 2-9 Attenuation of Series Diode Switching Modes

between L_w and C_D is approached which provides a low impedance current path for the low insertion loss shown in Fig. 2-9 at 5 GHz. As frequency is increased further on the reverse biased diode the series $L_w - C_D$ combination becomes inductive and soon parallel resonates with C_c giving the high isolation peak labeled Mode 3. The forward biased diode is capacitive above Mode 2 which by itself does not present a low insertion loss for Mode 3 but, with the addition of a series inductance L_T, could be made to provide low insertion loss, see Fig. 2-10. L_T may be realized by including a short section of high impedance transmission line adjacent to the diode.

Note that the high isolation and low insertion loss occur at the same frequency in Fig. 2-9 for Mode 2 because C_c and C_D are equal. At forward bias L_w parallel resonates with C_c and at reverse bias it series resonates with C_D, therefore when $C_c = C_D$ these occur at the same frequency. Note also that the center frequency of Mode 3 is $\sqrt{2}$ times the frequency of Mode 2.

In Fig. 2-10 the terms L_w, R_s, C_c and C_D are defined in Fig. 2-3. The other terms are simply derived and are given by

$$C' = \frac{(1 - \omega^2 L_w C_c)^2 + (\omega R_s C_c)^2}{\omega^2 [C_c(R_s^2 + \omega^2 L_w^2) - L_w]} \tag{2-25}$$

$$R' = \frac{R_s}{(1 - \omega^2 L_w C_c)^2 + (\omega R_s C_c)^2} \tag{2-26}$$

$$L' = L_w - \frac{1}{\omega^2 C_D} \tag{2-27}$$

To find the frequency of Mode 3 set

$$\omega_3 L' = \frac{1}{\omega_3 C_c} \tag{2-28}$$

in which ω_3 is the angular frequency of Mode 3. Recalling that $C_c = C_D$, defining $\omega_2 = (C_c L_w)^{-\frac{1}{2}}$, and using Eq. (2-27) with $\omega = \omega_3$ the expression can be obtained

$$\omega_3 = \sqrt{2}\ \omega_2 \tag{2-29}$$

Modes similar to the three series modes also occur for shunt diodes as shown in Fig. 2-11. A very efficient diode switch[2,4] can be made at 2.5 or 5.0 GHz using a reverse biased varactor (as shown in Fig. 2-11).

BASIC CONCEPTS

In the absence of L_T, Mode 3 is not evident. As mentioned earlier, Shunt Mode 2 or modifications of it are most common in waveguide, while Shunt and Series Mode 1 are most common in TEM transmission line, even up to 20 GHz in integrated circuits.

1. Mode 1

The reason for the wide use of Mode 1 is that it is possible to obtain extreme bandwidths with it. As can be seen from Fig. 2-9, even a simple conventional diode gives quite satisfactory performance from 0 GHz to 1 GHz. Lowering the inductance and capacitance of the diode raises the upper frequency of good performance and improves performance at lower frequencies. Therefore efforts to improve the high frequency performance of these switches also improve low frequency performance. The earliest versions of these switches used a gold-bonded germanium diode in series in 50 ohm coaxial transmission line. These devices provided less than 1 dB insertion loss and more than 20 dB isolation up to about 4 GHz. As semiconductor technology advanced, PIN junction diodes have become the most advantageous element for broad-band diode switching. With one of these diodes shunting miniature stripline it is possible to obtain less than 1 dB insertion loss and greater than 20 dB isolation up to 20 GHz.

A better appreciation of this mode may be obtained from Fig. 2-12. Using the equivalent circuit for the diode shown in Fig. 2-10 for Mode 1, it is inductive at forward bias and capacitive at reverse bias. The isolation of a series diode will be dominated by its reactance which is the reverse bias capacitance. Thus

$$\eta_{series} = 10 \log \left[1 + \left(\frac{1}{2\omega C Z_0} \right)^2 \right] \quad (2\text{-}30)$$

from Eq. (2-8). The isolation of a shunt diode is dominated by its reactance which is its forward bias inductance. Thus

$$\eta_{shunt} = 10 \log \left[1 + \left(\frac{1}{2\omega L Y_0} \right)^2 \right] \quad (2\text{-}31)$$

These equations provide the isolation curves shown in Fig. 2-12. Note the equivalance between L and C in the equations.

$$\omega C Z_0 = \omega L Y_0 \quad (2\text{-}32)$$

$$L = Z_0^2 C \quad (2\text{-}33)$$

For

$$Z_0 = 50$$

$$L = 2500 C \quad (2\text{-}34)$$

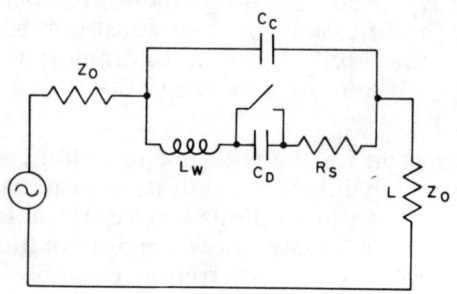

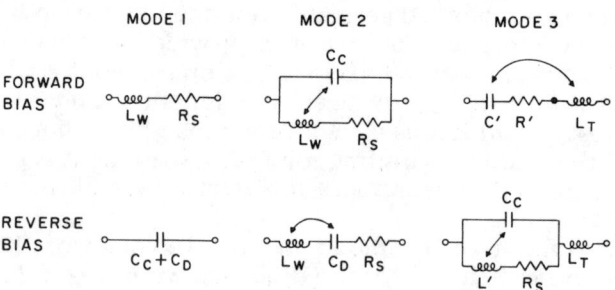

Fig. 2-10   Diode Elements Producing the Switching Modes

Thus the shunt inductance that will provide the same isolation as a series capacitance of 0.1 pF (10^{-13} F) is .25 nH (2.5 x 10^{-10} H).

The same equivalence can be derived from the insertion loss states.

$$\delta_{series} = 10 \log \left[1 + \left(\frac{\omega L}{2 Z_0}\right)^2\right] \qquad (2\text{-}35)$$

$$\delta_{shunt} = 10 \log \left[1 + \left(\frac{\omega C}{2 Y_0}\right)^2\right] \qquad (2\text{-}36)$$

$$\frac{\omega L}{Z_0} = \frac{\omega C}{Y_0} \qquad L = Z_0^2 C \qquad (2\text{-}37)$$

Later chapters will deal with techniques for reducing effective L and C (IV) and the use of multiple diodes for expanding bandwidth (V).

BASIC CONCEPTS

2. Mode 2

A Mode 2 switch is shown in Fig. 2-13.[3] When the diode junction is reverse biased, the junction capacitance series resonates with the "whisker" inductance. This series resonance provides a low impedance across the waveguide which reflects incident power. At forward bias the "whisker" inductance parallel resonates with the cartridge capacitance to provide high impedance across the waveguide which permits incident power to travel past the diode with very little loss.

Another way of visualizing the performance of the Mode 2 switch is to consider the junction capacitance and diode cartridge as a section of transmission line which has an effective length and characteristic impedance. This length of line connects the two terminals of the tiny diode junction to the two terminals comprised of the upper and lower waveguide walls. At the center frequency of operation it is a quarter wavelength long. Thus at reverse bias the high impedance of the junction is transformed by the quarter wavelength to a short circuit across the waveguide; conversely, the forward biased low impedance of the diode junction is transformed to an open circuit across the waveguide.

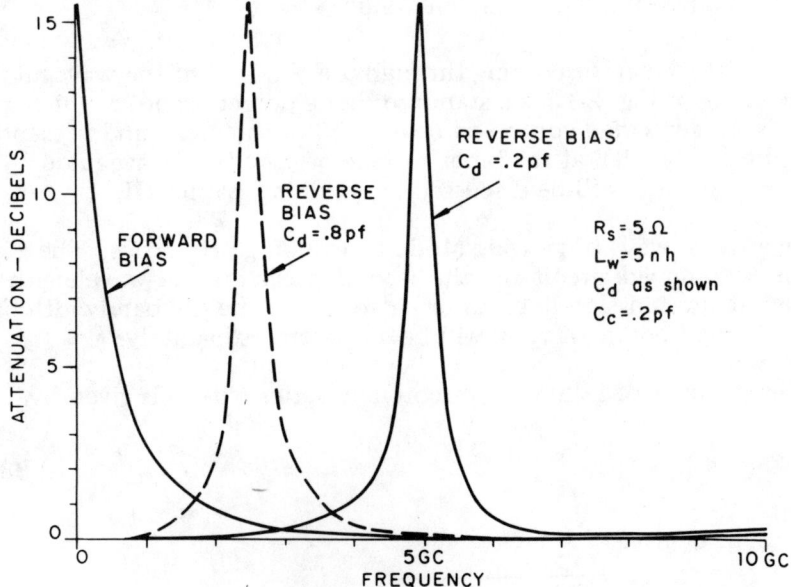

Fig. 2-11 Attenuation of Shunt Diodes Modes

32   MICROWAVE DIODE CONTROL DEVICES

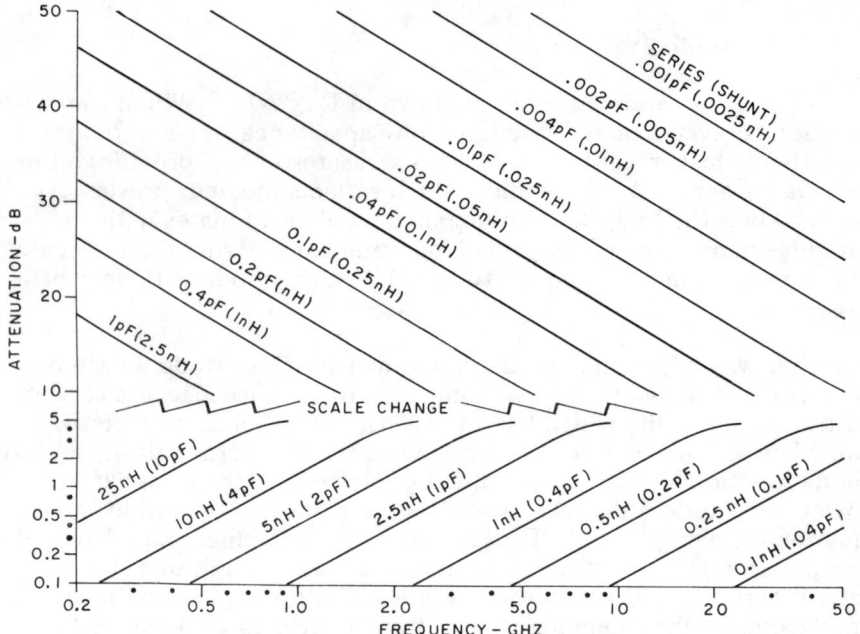

Fig. 2-12 Expanded Characterization of Mode 1 Switching in a 50-ohm Transmission Line.

The coaxial connector coming through the top wall of the waveguide to the diode in Fig. 2-13 is a standard diode detector choke which permits the switching voltage to be applied to the diode and presents zero ohms to the RF at the inner surface of the upper waveguide wall. Bias circuits will be discussed in detail in Chapter III.

There are two ways of viewing Mode 2 switching properties. The first is as a lumped circuit and the second is as a quarter-wavelength section of transmission line. In order to determine the bandwidth of Mode 2, these points of view will be considered separately.

The reactance X of a lumped circuit C in series with L is given by

$$X = \omega L - \frac{1}{\omega C} \tag{2-38}$$

Solving for ω gives

$$\omega = \frac{X}{2L} \pm \sqrt{\left(\frac{X}{2L}\right)^2 + \left(\frac{1}{LC}\right)} \tag{2-39}$$

BASIC CONCEPTS
33

The positive solution gives positive ω, and

Substituting
$$\omega_c = \frac{1}{\sqrt{LC}} \tag{2-40}$$
and
$$X_{RES} = \omega_0 L \tag{2-41}$$
into Eq. (2-39) gives

$$\frac{\omega}{\omega_0} = \left(\frac{1}{2}\frac{X}{X_{RES}}\right) + \sqrt{\left(\frac{1}{2}\frac{X}{X_{RES}}\right)^2 + 1} \; . \tag{2-42}$$

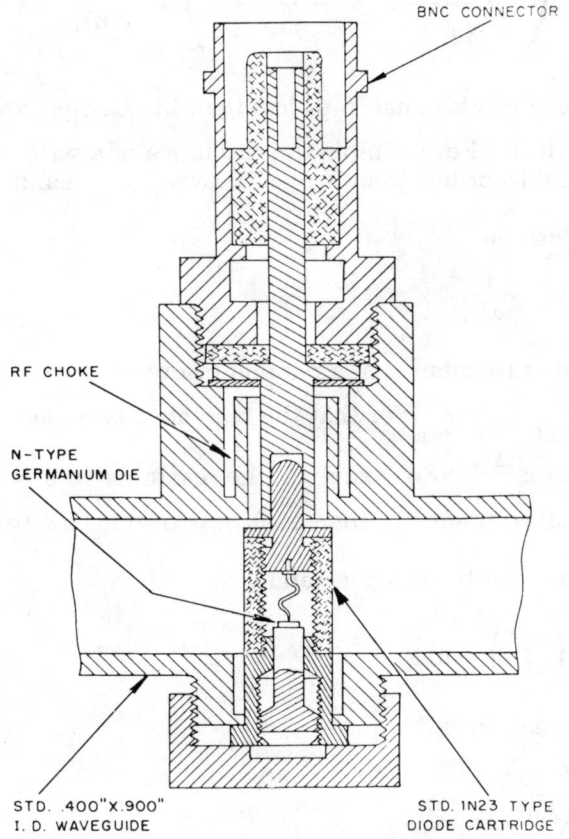

Fig. 2-13 X-Band Waveguide Diode Switch Working in Mode 2

Taking bandwidth as $\Delta\omega$ then at

$$\omega = \omega_0 + \frac{\Delta\omega}{2}$$

X is positive and at

$$\omega = \omega_0 - \frac{\Delta\omega}{2}$$

X is negative, therefore,

$$\frac{\Delta\omega}{\omega_0} = \left(\frac{\omega_0 + \frac{\Delta\omega}{2}}{\omega_0} - \frac{\omega_0 - \frac{\Delta\omega}{2}}{\omega_0}\right) = \frac{|X|}{X_{RES}} \qquad (2\text{-}43)$$

The reactance ratio is equal to or less than $|X|/X_{RES}$ over $\frac{\Delta\omega}{\omega_0}$ fractional bandwidth. For example the reactance of a series resonant circuit is equal to or less than $X_{RES}/10$ over a 10% bandwidth.

In a similar fashion

$$B = \omega C - \frac{1}{\omega L} \qquad (2\text{-}44)$$

yields that the susceptance of a parallel resonant circuit is equal to or less than $\frac{|B|}{B_{RES}}$ over $\frac{\Delta\omega}{\omega_0}$ or that the reactance is equal to or greater than $\frac{|X|}{X_{RES}}$ over $\frac{\Delta\omega}{\omega_0}$. For example, the reactance of a parallel resonant circuit is equal to or greater than $10\, X_{RES}$ over a 10% bandwidth.

These relations can be stated generally as

$$|X| \leq \left(\frac{\Delta\omega}{\omega_0}\right) X_{RES} \qquad (2\text{-}45)$$

for series resonance, and

$$|X| \geq \left(\frac{\omega_0}{\Delta\omega}\right) X_{RES} \qquad (2\text{-}46)$$

for parallel resonance.

BASIC CONCEPTS

Since Mode 2 is used most in waveguide the bandwidth of insertion loss and isolation will be calculated for shunt switching. Recalling from Eq. (2-9) for reactance limited shunt switching

$$\alpha = 10 \log \left[1 + \left(\frac{B}{2 Y_0} \right)^2 \right] \qquad (2\text{-}47)$$

$$= 10 \log \left[1 + \left(\frac{Z_0}{2 X} \right)^2 \right] \qquad (2\text{-}48)$$

Parallel resonance provides high reactance and thus low insertion loss giving

$$\delta \leqslant 10 \log \left[1 + \left(\frac{Z_0}{2 \left(\frac{\omega_0}{\Delta \omega} \right) X_{RES}} \right)^2 \right] \qquad (2\text{-}49)$$

$$\leqslant 10 \log \left[1 + \left(\frac{Z_0}{2 X_{RES}} \right)^2 \left(\frac{\Delta \omega}{\omega_0} \right)^2 \right] \qquad (2\text{-}50)$$

Series resonance provides the isolation

$$\eta \geqslant 10 \log \left[1 + \left(\frac{Z_0}{2 \left(\frac{\Delta \omega}{\omega_0} \right) X_{RES}} \right)^2 \right] \qquad (2\text{-}51)$$

$$\geqslant 10 \log \left[1 + \left(\frac{Z_0}{2 X_{RES}} \right)^2 \left(\frac{\omega_0}{\Delta \omega} \right)^2 \right] \qquad (2\text{-}52)$$

Eqs. (2-50) and (2-52) were used to generate the curves shown in Fig. 2-14. The same curves apply to series Mode 2 when B_{RES} is substituted for X_{RES} and Y_0 for Z_0.

The waveguide diode switch shown in Fig. 2-13 should thus provide less than 1 dB insertion loss and greater than 22 dB isolation over the 20% bandwidth from 9 GHz to 11 GHz in X-band waveguide when $X_{RES} = \frac{1}{10} Z_0 = 40$ ohms.

$$L_w = \frac{40}{\omega_0} = \frac{40}{2\pi \times 10^{10}} = 6.4 \times 10^{-10} = .64 \text{ nH}.$$

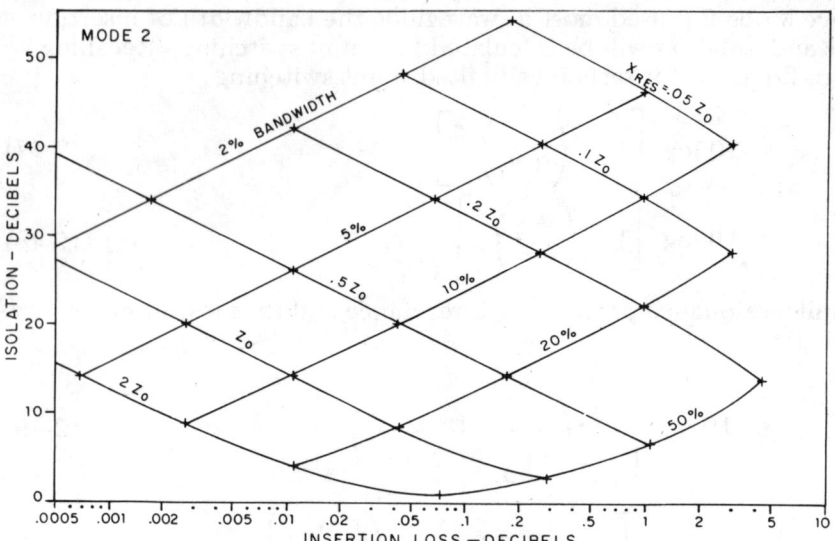

Fig. 2-14 Mode 2 Bandwidth from a Lumped Element Shunt Diode (For a Series Diode Substitute Y_0 for Z_0 and B_{RES} for X_{RES}.)

For example, assume it is desired to make a 1 dB to 30 dB switch over a 10% bandwidth at 10 GHz with a 0.4 nH diode (X_{RES} = 25 ohms). Fig. 2-14 indicates B_{RES} = $10Y_0$ is required (X_{RES} = $Z_0/10$). A 250 Ohm Z_0 will satisfy the requirement. Rectangular waveguide (9." w $\times \frac{250}{400}$ .4" h)= (.9" w $\times$ .25"h) will give the desired switching.

It should be pointed out, however, that the equivalent circuit of the diode is only approximate, especially when the dimensions exceed $\lambda/10 = \frac{3}{10}$ cm. $\approx$ 0.1 in. Since the waveguide is 0.4" high the diode is much too big to use the equivalent circuit with any precision. Furthermore the waveguide characteristic impedance in not uniquely defined and varies with frequency.

Returning to the transmission line length representation of the cartridge, it is interesting to note that when the tiny junction is located midway between the top and bottom waveguide walls, it is actually .17 wavelengths from the waveguide walls. A relative dielectric constant of 2.16 in the ceramic of the cartridge would convert this to .25 wavelengths.

BASIC CONCEPTS 37

The calculation is now made for a quarter-wavelength stub as it would represent the diode cartridge in describing the Mode 2 type performance of a shunt diode. The stub has characteristic impedance Z_Q. When the diode junction is a short circuit the stub has reactance.

$$X = Z_Q \tan(2\pi \ell/\lambda) \tag{2-53}$$

Note that

$$\ell = \lambda_0/4$$

thus

$$2\pi \frac{\ell}{\lambda} = \frac{2\pi}{4} \frac{\lambda_0}{\lambda} = \frac{\pi}{2} \frac{\omega}{\omega_0} \tag{2-54}$$

and that

$$X \geqslant Z_Q \tan \frac{\pi}{2} \left(\frac{\omega_0 + \frac{\Delta\omega}{2}}{\omega_0} \right) = Z_Q \cot \left(\frac{\pi}{4} \frac{\Delta\omega}{\omega_0} \right) \tag{2-55}$$

using Eq. (2-48) this provides

$$\delta \leqslant 10 \log \left[1 + \left(\frac{Z_0}{2 Z_Q} \tan \frac{\pi}{4} \frac{\Delta\omega}{\omega_0} \right)^2 \right] \tag{2-56}$$

In similar fashion an open circuit junction provides

$$X \leqslant Z_Q \tan \left(\frac{\pi}{4} \frac{\Delta\omega}{\omega_0} \right) \tag{2-57}$$

which gives

$$\eta \geqslant 10 \log \left[1 + \left(\frac{Z_0}{2 Z_Q} \cot \frac{\pi}{4} \frac{\Delta\omega}{\omega_0} \right)^2 \right] \tag{2-58}$$

These equations give the curves shown in Fig. 2-15.

When both circuits are constrained to give 1 dB maximum insertion loss and 20% bandwidth, the lumped circuit gives a minimum of 22.1 dB isolation while the quarter wavelength stub circuit gives a minimum isolation of 26.2 dB. Fixing the maximum insertion loss of each at 1 dB and minimum isolation at 20 dB the lumped circuit provides a 22.6% bandwidth while the quarter-wavelength stub circuit gives 28.3% bandwidth.

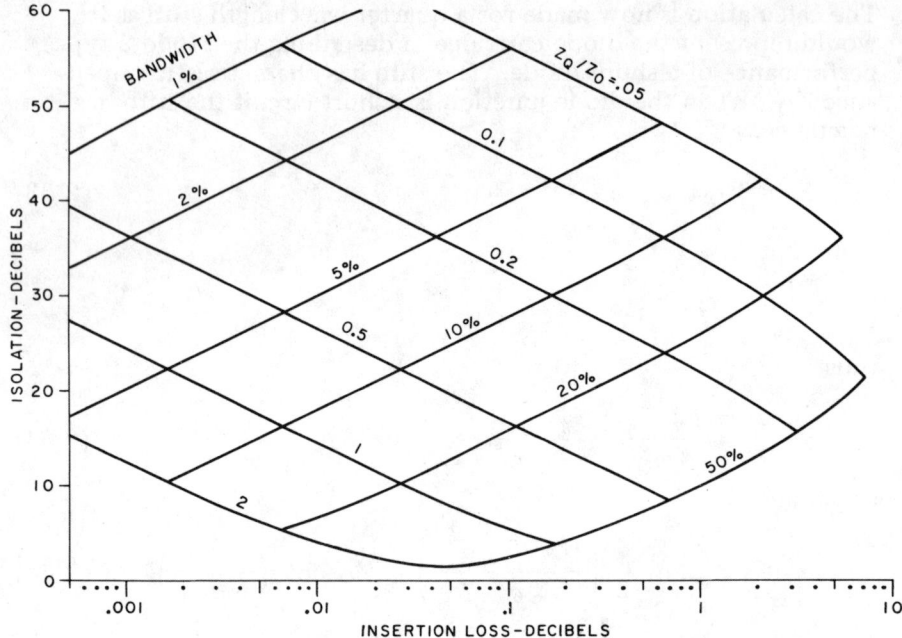

Fig. 2-15 Mode 2 Bandwidth from a Distributed Element Shunt Diode

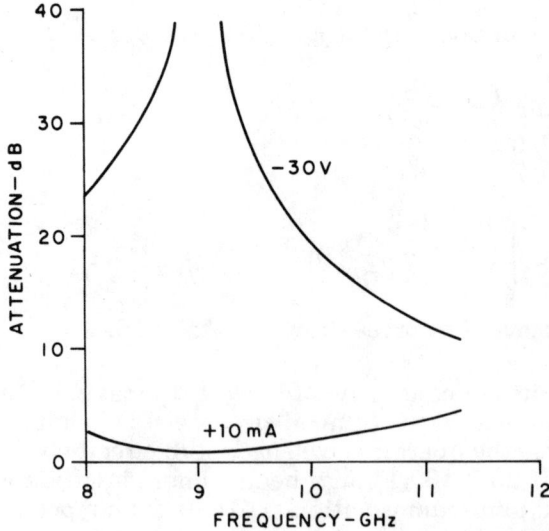

Fig. 2-16 Attenuation of a Waveguide Mode 2 Diode Switch

BASIC CONCEPTS 39

The attenuation of a waveguide diode switch is shown in Fig. 2-16. The data shows $\delta = 1.7$ dB and $\eta = 25$ dB giving 17% bandwidth. Fig. 2-14 would give 19% bandwidth and Fig. 2-15, 26% bandwidth.

C. Reflection Switches

The possible circuits for making switches discussed thus far have included mounting the diode in series or shunting the transmission line and operating in Mode 1 or Mode 2. Mode 3 has never been used practically. Another basic method of making switches is to use the absorbing properties of the diode at intermediate bias.[5] For example, a diode biased to 50 ohms and terminating a 50 ohm transmission line would absorb all power incident upon it. When this diode is connected to a circulator or when two of them are connected to a 90° 3-dB coupler as shown in Fig. 2-17 the combination becomes an absorption diode switch. Power entering port 1 of the circulator is directed to port 2 by circulator action. When the impedance connected to port 2 is a perfect match then no power is reflected to reenter port 2 and emerge from port 3. The result is high isolation from port 1 to port 3. When the termination on port 2 has a high VSWR, as an impedance does which is near the outside edge

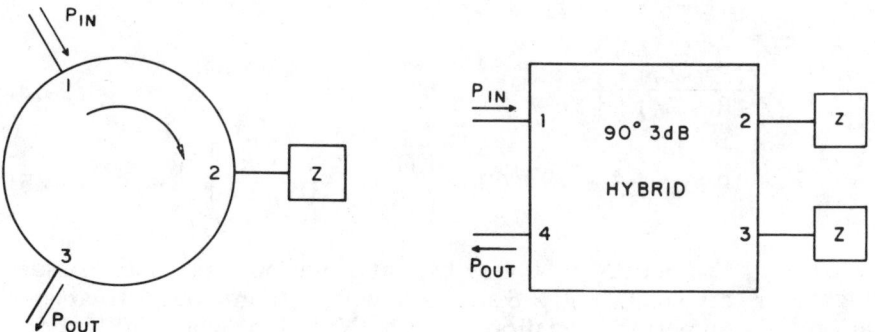

Fig. 2-17   Absorption Diode Switches

of the Smith Chart, then most power emerging from port 2 of the circulator is reflected by the impedance on port 2 only to reenter the circulator at port 2 and be directed to port 3 having a low insertion loss from ports 1 to 3. That power which is reflected from the termination on port 2 emerges from port 3. It can be seen from Fig. 2-6 that a bias of 0.2 mA on the HP3001 PIN Diode will produce a high attenuation when the diode is connected to port 2 of a circulator and a low insertion loss at 0 mA or 100 mA bias. The VSWR (ρ) at 0.2 mA is 1.2. This produces a voltage reflection coefficient given by

$$|\Gamma| = \frac{\rho - 1}{\rho + 1} = \frac{.2}{2.2} = .0909 \qquad (2\text{-}59)$$

The power reflection coefficient is given by

$$\frac{P_R}{P_I} = |\Gamma|^2 = .008263 \qquad (2\text{-}60)$$

The attenuation is given by

$$\alpha = 10 \log \frac{P_I}{P_T} = 10 \log \frac{P_I}{P_R} = 10 \log \frac{1}{|\Gamma|^2} \qquad (2\text{-}61)$$

$$= 20.8 \text{ dB}$$

In general when Z or Y terminate a transmission line having characteristic impedance Z_0 or admittance Y_0 the voltage reflection coefficient is given by

$$\Gamma = \frac{Z_T - Z_0}{Z_T + Z_0} = \frac{Y_0 - Y_T}{Y_0 + Y_T} \qquad (2\text{-}62)$$

$$\Gamma = \frac{(R - Z_0) + jX}{(R + Z_0) + jX} \qquad (2\text{-}63)$$

$$\alpha = 10 \log \frac{1}{|\Gamma|^2} = 10 \log \left[\frac{(R + Z_0)^2 + X^2}{(R - Z_0)^2 + X^2} \right] \qquad (2\text{-}65)$$

When $R \approx Z_0$ and $X \approx 0$ very high attenuation is possible. Assuming X_T is very small, Fig. 2-8 shows how R/Z_0 contributes to attenuation for a reflection type diode switch. Note that when a PIN diode is used to absorb the power then all of the incident power is absorbed by the diode. The maximum average power a PIN diode can absorb is usually in the 1 watt to 10 watt range, thus incident power levels must be below this limit. Power handling capability may be increased by putting a power absorbing resistance in series or shunt with the switching diode.

Also shown in Fig. 2-17 is a 90° 3-dB hybrid with two identical impedances connected to it. These couplers have the property that power into port 1 is split equally between ports 2 and 3 with none

BASIC CONCEPTS   41

emerging from port 4. When perfect identical reflectors are attached to ports 2 and 3 the power reflected by them recombines in the coupler in such a manner that their phases add at port 4 and cancel at port 1. Thus the power reflected by the impedances emerges from port 4 just as it emerged from port 3 of the circulator. This form of matched switch has advantages over the circulator in that (1) it is reciprocal (the circulator gives no variable attenuation for the wave traveling from port 3 to port 1); (2) these hybrid junctions can be made over extreme bandwidth (10 to 1); (3) they are less expensive and smaller in volume and weight than circulators; and (4) the power is divided between two diodes, increasing power capability by a factor of two. Their disadvantage is that they require twice as many diodes as the circulator type.

D. Variable Attenuation

As described in Appendix B, all diodes have voltage current relationship as follows:

$$I = I_0 (\epsilon^{\alpha_{DC} V} - 1) \qquad (2\text{-}66)$$

where I is diode current and V is the voltage across the diode junction. The voltage drop across the series resistance R_s is not included in this equation. The small signal conductance of the diode is given by

$$g = \frac{\partial I}{\partial V} = \alpha_{DC} I_0 \epsilon^{\alpha_{DC} V} \approx \alpha_{DC} I \qquad (2\text{-}67)$$

For a shunting diode

$$\alpha = 10 \log \left[\left(\frac{G}{2 Y_0} + 1 \right)^2 + \left(\frac{B}{2 Y_0} \right)^2 \right] = 20 \log \left(1 + \frac{\alpha_{DC} I}{2 Y_0} \right) \qquad (2\text{-}68)$$

when susceptance is neglected. For hot carrier diodes α_{DC} is $q/kT = 40 V^{-1}$ from dc to above 20 GHz. For PIN diodes the relationship given in Eq. (2-67) is correct only at low frequencies and $\alpha_{DC} = 20 V^{-1}$. At frequencies higher than about 10MHz Eq. (2-67) is only an approximation for PIN diodes. At high frequencies α_{DC} is changed to α_{RF} and is relatively independent of frequency. α_{RF} is dictated by carrier lifetime and junction geometry having a typical value of $80 V^{-1}$. (Refer to Appendix B.)

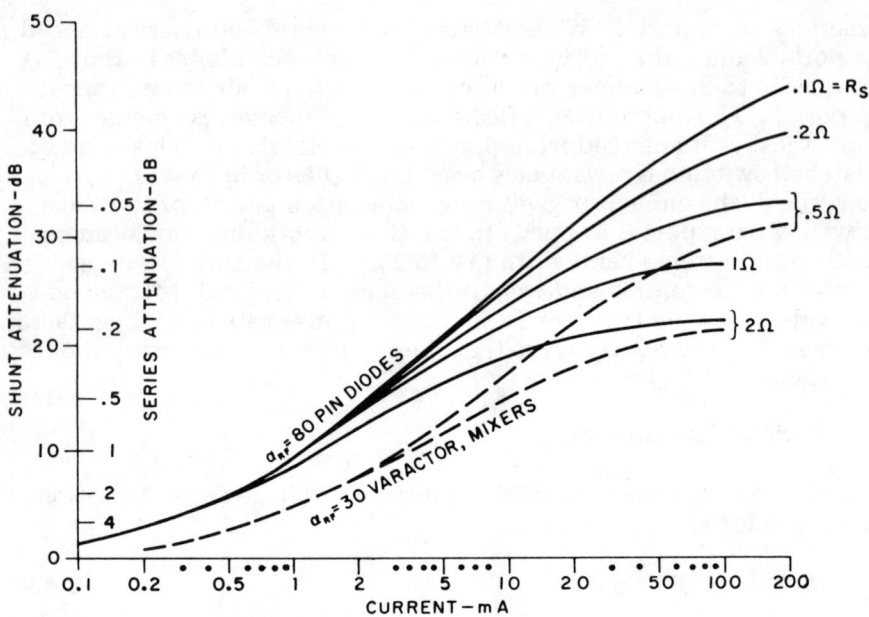

Fig. 2-18 Attenuation of Typical Diodes in 50-ohm Transmission Line

Taking R_s into account total resistance becomes

$$R = R_s + \frac{1}{g} \tag{2-69}$$

and Eq. (2-68) becomes (for a shunt diode)

$$\alpha = 20 \log \left[1 + \frac{1}{2 Y_0 \left(R_s + \frac{1}{\alpha_{RF} I} \right)} \right] \tag{2-70}$$

As current is varied, attenuation varies as shown in Fig. 2-18 for 50 ohm transmission line. It is interesting to note that to obtain 20 dB attenuation from a single PIN diode having R_s = 0 and shunting 50 ohm transmission line 5 mA of bias current is required. (putting more diodes in parallel (in the same electrical plane) will not alter the current requirement for 20 dB.) The total diode current will have to be 5 mA. Techniques will be discussed in Chapter VI for obtaining higher isolations from shunting PIN diodes (with modest bias currents).

BASIC CONCEPTS

The relationships described in this chapter are based on the assumption that diodes can be represented by lumped circuit elements. Measurements up to 4 GHz in stripline and in coaxial transmission line and up to 20 GHz in high dielectric microstrip transmission lines have shown these assumptions to be very good. As a matter of fact, measuring the switching parameters of diodes has become the most accurate means used for measuring diode parameters. Once a diode is characterized, its switching properties may be precisely predicted using the equations contained in this chapter.

REFERENCES

1. H.C. Torrey and C.A. Whitmer, "Crystal Rectifiers," *M.I.T., Rad. Lab. Ser.* McGraw-Hill Book Co., Inc., New York, N.Y., Vol. 15, p. 340 1948.

2. R.V. Garver, "Theory of TEM Diode Switching," *IRE Trans. on Microwave Theory and Techniques*, Vol. MTT-9, pp. 224-238, May 1961.

3. R.V. Garver, E.G. Spencer, and M.A. Harper, "Microwave Semiconductor Switching Techniques," *IRE Trans. on Microwave Theory and Techniques*, Vol. MTT-6, pp. 278-383, October 1958.

4. E.M. Rutz and E. Kramer, "Microwave Modulator Requiring Minimum Modulation Power," *IRE Trans. on Microwave Theory and Techniques*, Vol. MTT-10, pp. 605-610, November 1962.

5. D.A.E. Roberts and S.J. Robinson, "A PIN Diode Modulator for the Frequency Band 2.5-7.5 Gc/s," *The Microwave Journal*, pp. 74-78, December 1963.

Biasing Circuits

Chapter 3

To produce switching action with a diode, it is necessary to supply a switching voltage to the diode. First it is desired to prevent RF energy from leaking out the bias port. Take for example a radar in which a diode switch is used to prevent transmitter power from damaging the mixer diodes. A large RF voltage will be developed across the two diode terminals. These same two terminals are connected to the bias circuit and when not enough RF attenuation is built into the bias circuit, this RF power can be easily coupled through the bias circuit to sensitive parts of the system and significantly reduce system sensitivity. Therefore, RF leakage out the bias port must be kept small.

Another desired characteristic of the bias circuit is the result of the extreme bandwidth of Mode 1 switches. It is often desirable to use a very broad-band biasing circuit to obtain a complete broad-band diode switch. Then one diode switch may be used in any of a variety of applications within a wide frequency range.

A third desired characteristic of bias circuits stems from the fast switching speed of some diodes (for example, hot carrier diodes). Because of their high speed, diode switches have become widely used for the gates on sampling oscilloscopes. While a short discussion will be given on these gates, most interest centers around their use in more conventional control applications. For example, in a close range radar the time between transmit and receive becomes very short and the switch protecting the receiver must be able to go from the "off" state to the "on" state in a matter of nanoseconds. Switches have also been used to generate very short RF pulses having, for example, a spectrum that covers all of X-band with only one dB less average power at the upper and lower bounds as compared to the power at midband. When switches are used to scan a phased array radar the switching speed will limit the rate of scan. Therefore it is frequently desirable to have high-speed biasing circuits to take advantage of the high speed of the diodes and prevent the bias circuit from slowing the switching rate.

A final desired characteristic of the bias circuit is to prevent switching transients from being propagated on the RF transmission line. When a high speed switch is placed in front of a very sensitive mixer-amplifier the high frequency components of the fast switching pulse may be coupled into the mixer as RF and cause the receiver to saturate. The apparent signal causes AGC action in the amplifier which reduces receiver sensitivity until the AGC voltage can decay.

The biasing circuit must keep RF out of the modulator driver circuits, operate over wide RF bandwidths without deteriorating switching performance, permit the rapid change of modulation voltage on the diode, and/or prevent modulating transients from being propagated on the RF transmission line. These desired characteristics of bias circuits will be discussed separately at first and then in combination.

A. Designing for Maximum RF Bandwidth

Waveguide bias connectors were widely used during World War II in mixers and termed "bias chokes."[1] They were quite broad-band and will be discussed later under high-speed switching. The most significant bandwidths will be obtained in TEM transmission lines. The circuits required for biasing series and shunt diodes are shown in Fig. 3-1. The series capacitors are easily made in coaxial transmission line by using a high dielectric constant tube metal plated on the inside and outside. The outer diameter of the cylinder is made the same size as the center conductor of the coaxial transmission line or it is embedded in a metal container having the proper diameter. Capacitance as high as 10,000 pf have been obtained in this manner which are quite good at high frequencies and go very low in frequency. For example, suppose the capacitor were used in series in a 50 ohm transmission line and 0.1 dB insertion loss could be tolerated from it. From the series attenuation equation

$$\alpha = 10 \log \left[1 + \left(\frac{X}{2 Z_0} \right)^2 \right] \qquad (3\text{-}1)$$

$$0.1 = 10 \log \left[1 + \left(\frac{1}{4\pi f C (50)} \right)^2 \right] \qquad (3\text{-}2)$$

$$f \approx 1 \text{ MHz}$$

BIASING CIRCUITS

Thus it is easy to make a broad-band series capacitor in coaxial transmission line.

A series capacitor is made simply in 50 ohm stripline by causing two center strip pieces to overlap with a .001" mylar between them. An overlap of 1" would correspond to a quarter-wavelength at about 2.0 GHz. The capacitance would be

$$C = .225 \, \epsilon_r \, \frac{A}{d} \, pF \qquad (3\text{-}3)$$

where ϵ_r is 2.2 for mylar and A and d are in inches.

$$C = .225 \, (2.2) \, \frac{(1.0)(.090)}{(.001)} = 44.5 \, pF$$

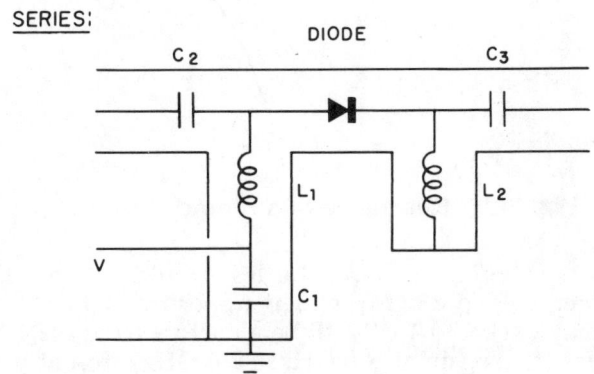

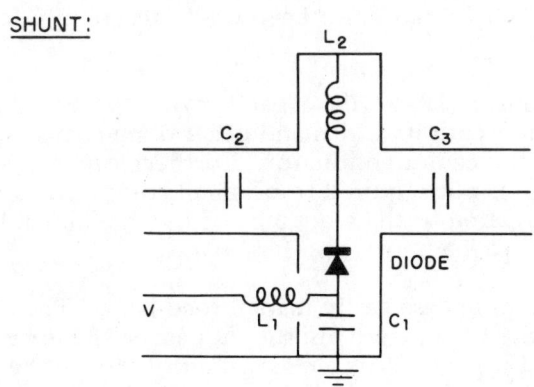

Fig. 3-1   Bias Circuits for TEM Mounted Diodes

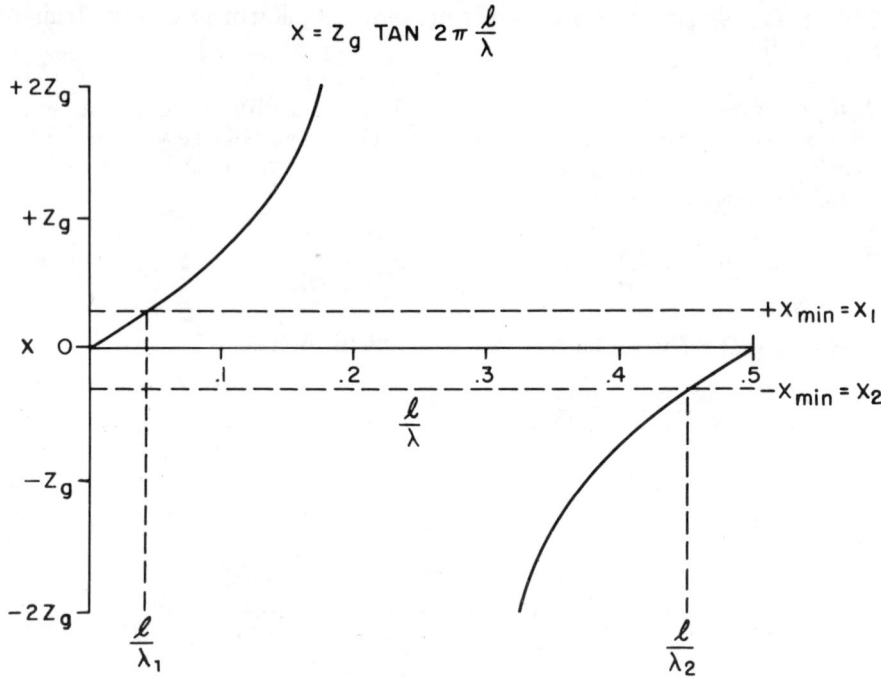

Fig. 3-2 Reactance of a Shunt Stub

Solving Eq. (3-2) as before the insertion loss would rise to 0.1 dB as frequency decreases to 232 MHz. The upper frequency at which insertion loss may exceed 0.1 dB will be a narrow band (464 MHz around 4.0 GHz) (and again at 8.0 GHz, 12.0 GHz, etc.) at which requency the overlapping section is $\lambda_g/2$ long (and multiples thereof). Very small chip capacitors have also been used successfully for stripline and microstrip.

Even more freedom is possible with capacitor C_1. Since it is out of the main transmission path its width and other dimensions are not limited to those of the center conductor. Furthermore, for series diodes the reactance is not required to be small compared to 50 ohms but small compared to the reactance of L, which must be larger than 50 ohms to keep insertion loss low.

The diodes and capacitors are easily made broad-band. The elements remaining to be made broad-band are the inductors. At lower frequencies pre-wound RF inductors are satisfactory. Since the reactance of an inductor decreases with decreasing frequency, a sufficiently large inductance must be selected for a given minimum

BIASING CIRCUITS

operating frequency. For higher frequencies, however, the self-resonance in a large pre-wound RF inductor may cause trouble. Also for high-speed switching a large inductor is undersirable. Therefore, another approach to the design of the biasing lead is needed. First the impedance of a single stub will be discussed, then the use of filter techniques will be considered.

In the early days of radar, the center conductors in coaxial waveguides were supported by shorted quarter-wavelength coaxial sections. These supports had little effect on the microwave transmission over considerable bandwidths. By substituting a capacitance for the short and making the characteristic impedance of the quarter-wavelength line section equal to that of the main guide (Z_0), a bias can be applied to this lead with an attenuation of ¼ dB or less over a 2-to-1 frequency range. Greater bandwidth can be obtained by using a shorted quarter-wavelength biasing lead having higher characteristic impedance as shown below.

The impedance of a shorted length of lossless transmission line of characteristic impedance, Z_g, is

$$Z = jX = j\, Z_g \tan\left(2\pi \frac{\ell}{\lambda}\right) \tag{3-4}$$

which is plotted in Fig. 3-2. For a lossless shunting reactance Eq. (2-9) gives

$$\alpha = 10 \log \left[1 + \left(\frac{Z_0}{2X}\right)^2\right] \tag{3-5}$$

As long as the reactance is greater than some minimum value, the attenuation will be lower than some maximum value. From Fig. 3-2 this is possible from ℓ/λ_1 to ℓ/λ_2. Letting λ_1 correspond to f_1, λ_2 correspond to f_2, and $f_2 = Nf_1$ (where N is the bandwidth factor), then $\lambda_2 = \lambda_1/N$. To find N, set $|X_1| = |X_2|$; then

$$N = \frac{\lambda_1}{2\ell} - 1 \tag{3-6}$$

Combining Eqs. (3-4), (3-5), and (3-6) gives

$$N = \pi \left[\arctan\left(2\,\frac{Z_g}{Z_0}\sqrt{10^{\alpha/10}-1}\right)^{-1}\right]^{-1} - 1 \tag{3-7}$$

The equation is plotted in Fig. 3-3. For $Z_g/Z_0 > 5$ and $\alpha \leq 1$

$$N \approx 3\,\frac{Z_g}{Z_0}\sqrt{\alpha}\,. \tag{3-8}$$

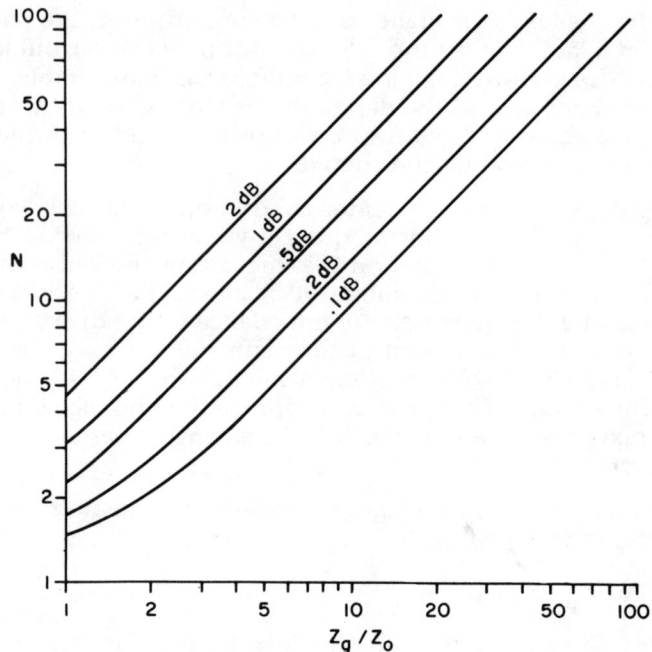

Fig. 3-3 Insertion Loss Bandwidth of Biasing Stubs

Modifying a Z_0 = 50 ohm coaxial T so that one arm has a 42 AWG wire for a center conductor, a Z_g of about 325 ohms is possible, which gives Z_g/Z_0 = 6.5 and N = 20 for α = 1 dB. Thus one biasing lead can be made to work from 4 GHz down to 0.2 GHz. It is pointed out that the 1-dB loss from reflection will exist only at the upper and lower limits. Between these limits, the insertion loss of the biasing lead will be less. For example, the loss will be less than 0.1 dB over a 6-to-1 frequency range. As another example a .002" diameter wire on 1/16" thick teflon-fiberglass stripline boards produces a Z_g of 160 ohms giving Z_g/Z_0 = 3.2. This gives a 1-dB bandwidth of 9.6-to-1 and a 0.1 dB bandwidth of 3-to-1. Knowing N, ℓ may be determined using a modification of Eq. (3-7). Remembering that $\lambda_1 = N\lambda_2$, substituting it in Eq. (3-7), and solving for ℓ

$$\ell = \frac{\lambda_2}{2}\left(\frac{N}{N+1}\right). \qquad (3\text{-}9)$$

The bandwidth attainable by using a straight center conductor is limited by the smallest diameter wire that can be tolerated. Greater bandwidth can be attained with a helical center conductor coaxial transmission line because of the higher characteristic impedance. From Reference 2 the characteristic impedance is given by

BIASING CIRCUITS

$$Z_g = 202nd \sqrt{[1-(d/D)^2]} \log(D/d) \qquad (3\text{-}10)$$

The numbers are taken from Fig. 3-4 in inches. The propagation constant is given by

$$T = 1.49 \, nd \sqrt{\frac{1-(d/D)^2}{\log(D/d)}} \times 10^{-9} \qquad (3\text{-}11)$$

The electrical length K in wavelengths at the frequency f in gigahertz is defined by

$$K = \frac{fT\ell}{12} \times 10^9 \qquad (3\text{-}12)$$

Assuming $\ell/d = 1$ and $K = \frac{1}{2}$ and solving by substituting Eqs. (3-11) and (3-12) into (3-10)

$$Z_g = \frac{813}{fD} \left(\frac{D}{d}\right) \log\left(\frac{D}{d}\right). \qquad (3\text{-}13)$$

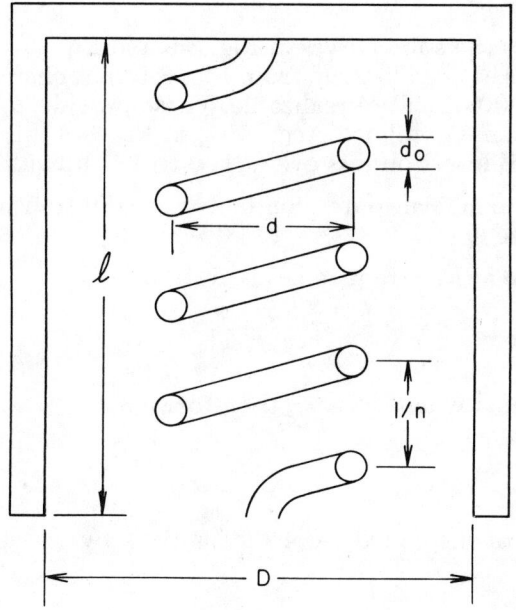

Fig. 3-4   Helix Bias Choke

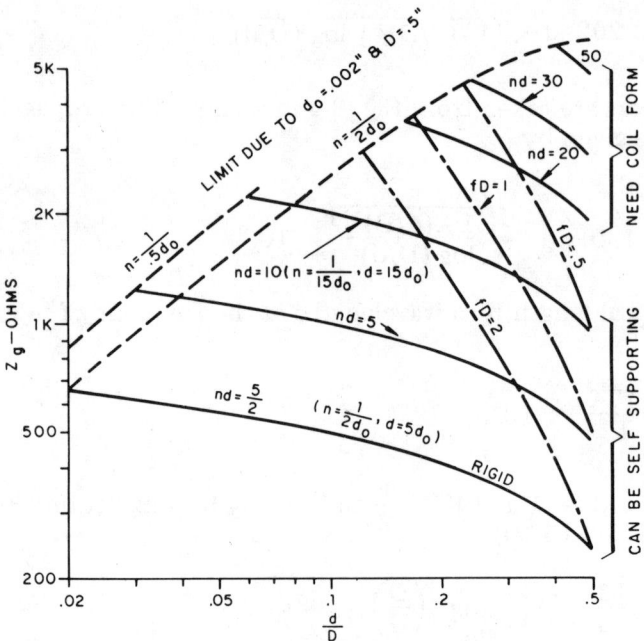

Fig. 3-5   Helix Design Chart

Eq. (3-10) gives the solid curves in Fig. 3-5 and Eq. (3-13) gives the dot-dash curves. It can be seen from Fig. 3-5 that characteristic impedances of 2000 ohms are realizable. These provide $Z_g/Z_0 = 40$ in 50 ohm transmission line. According to Fig. 3-3 this will provide less than 0.4 dB insertion loss over a 100-to-1 bandwidth.

Furthermore the helical center conductor coaxial transmission line will have a finite Q.

The inductance at low frequencies is given by

$$\omega L = Z_g \tan 2\pi \frac{\ell}{\lambda} \approx Z_g \, 2\pi \frac{\ell}{\lambda} \tag{3-14}$$

Resonance occurs when $\frac{\ell}{\lambda} = \frac{1}{2}$; therefore

$$\omega_0 L = \pi Z_g \tag{3-15}$$

The resistance of the coil at resonance is given by

$$R = \frac{\omega_0 L}{Q} = \frac{\pi Z_g}{Q} \tag{3-16}$$

BIASING CIRCUITS

According to the shunt attenuation equation a resistance of 415 ohms shunting a 50-ohm transmission line will cause $\frac{1}{2}$ dB insertion loss. Therefore when Q <15, a Z_g = 2000-ohm line will give $\frac{1}{2}$ dB insertion loss at its half-wave resonance. When the Q is deteriorated in this fashion the upper bound on the helical center conductor is lifted and the bandwidth would seem to be unlimited. However at higher frequencies other modes can be set up on helical center conductors and these can seriously influence insertion loss.

Another useful bias structure is the lossy watchspring built into coaxial connectors. Because of its low Q, resonances are seldom observed and loss remains in the 0.1 dB range.

B. Designing for Low VSWR

So far the biasing elements and the "ON" diode have been considered one at a time. When they are combined in a switch their insertion losses, which are mostly the result of reactive mismatches, do not add directly in dB. These mismatches can at best cancel each other out and at worst add up so that two equal mismatches give approximately four times the insertion loss of one mismatch. Thus when low VSWR is required, the design procedure should take into account all interactions between bias T, diode, and other DC blocking and ground return elements.

The VSWR (compared to insertion loss) is also a much more difficult factor to reduce over a broad bandwidth. The insertion loss δ of a reactive discontinuity having a VSWR ρ is given by

$$\delta = 20 \log \left(\frac{\rho + 1}{2\sqrt{\rho}} \right) \approx (\rho - 1)^2 \qquad (3\text{-}17)$$

which is shown in Fig. 3-6. A typical maximum VSWR for a transmission line component is ρ = 1.2 This VSWR corresponds to an insertion loss of only .036 dB. According to Eq. (3-7) this maximum insertion loss can be obtained over a 1.46 to a 1 bandwidth with a 100 ohm bias lead in a 50 ohm transmission line system. The .5 dB bandwidth for this same bias lead is 4.0 to 1. Thus the .5 dB insertion loss bandwidth is about 2 octaves while the bandwidth for a 1.2 maximum VSWR is about a half octave.

A series diode at forward bias is normally represented by an inductance but really has a finite length and therefore can be more accurately represented by a short length of high characteristic im-

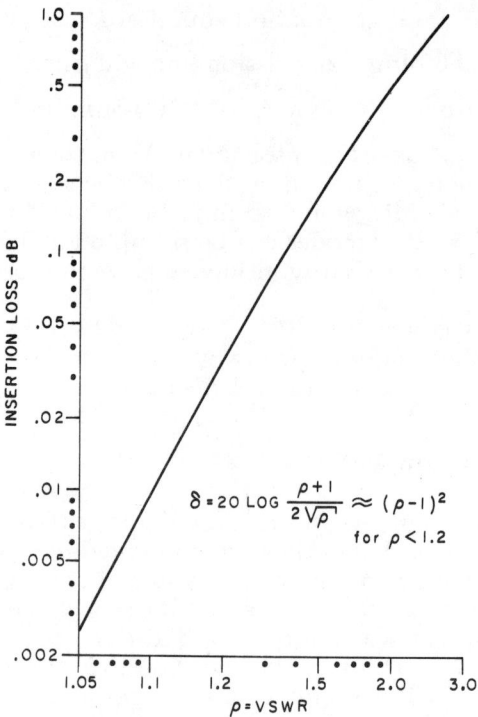

Fig. 3-6 Insertion Loss Induced by a Single Mismatch

pedance transmission line. When the diode is mounted in series in a transmission line that has the same high characteristic impedance, then there will be no increase of insertion loss due to series inductance. A shunt diode at reverse bias is capacitive and by removing some of the distributed capacitance of the transmission line in the plane of the diode, then there will be no increase in insertion loss from diode capacitance. Thus to obtain low VSWR switches (for the "ON" state) it is necessary to find filter structures that will accommodate the diode in them and provide the necessary bias and ground return leads.

Fig. 3-7 shows some filter structures developed for low VSWR bias "T's". All line sections are $\lambda/4$ at the center frequency. Circuit (a) was developed by Muehe and Young.[3] The design equations are

$$\phi = \frac{\pi}{2} \frac{N-1}{N+1} \qquad (3\text{-}18)$$

BIASING CIRCUITS

$$\frac{\rho - 1}{\sqrt{\rho}} Z_Q = \tan \phi \left(\frac{1 - \cos \phi}{1 + \cos \phi} \right) \qquad (3\text{-}19)$$

$$\frac{2 Z_Q}{Z_0'} - 2 Z_0' Z_Q - Z_0'^2 + 1 = \frac{2}{\cos^2 \phi + \cos \phi} \qquad (3\text{-}20)$$

in which N is bandwidth ratio and ρ is maximum VSWR. Ideally, ρ and N should be the independent variables and Z_Q and Z_0' the dependent variables. Then any value of ρ and N would produce the desired characteristic impedances directly. Unfortunately Eq. (3-20) is third degree in Z_0' and not easily solved analytically. Thus when Z_0' and N are made independent variables, Z_Q and ρ may be easily calculated using respectively Eqs. (3-20) and (3-19). Using this procedure Fig. 3-8 is constructed which gives design information directly.

When $Z_Q = Z_0$ and $Z_0' = .8 Z_0$ the structure will provide a maximum VSWR of 1.05 over an octave bandwidth and the combination of $Z_Q = 2Z_0$ and $Z_0' = .8Z_0$ will give a 1.2 maximum VSWR over 2 octaves. Fig. 3-9 shows how this structure can be used to make a diode switch in stripline with a shunt pill diode. The Z_{01}

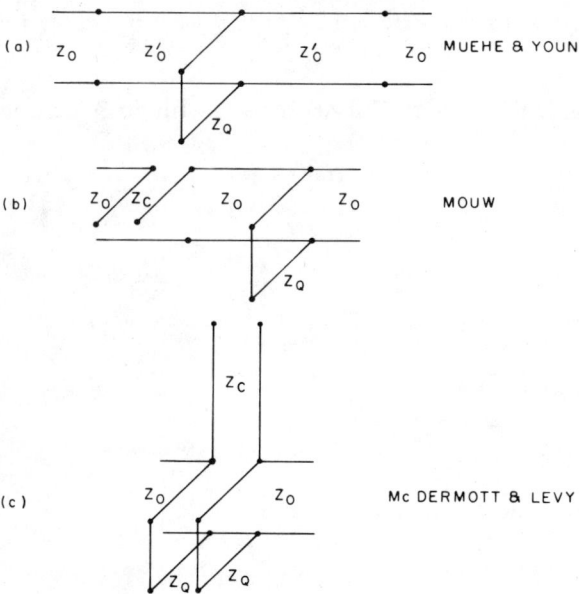

Fig. 3-7   Filter Circuits for Diode Biasing

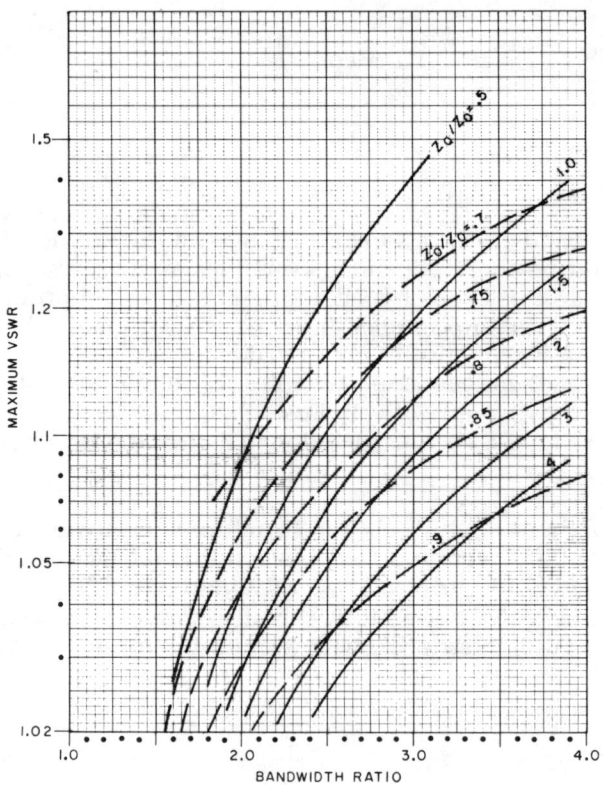

Fig. 3-8 Design Curves for the Muehe & Young Bias T

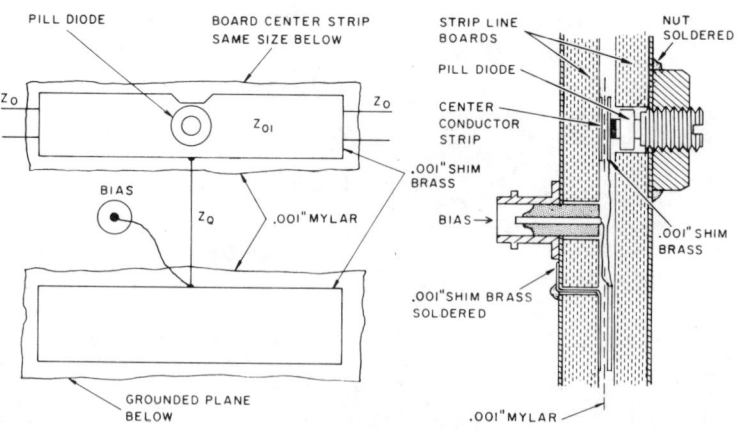

Fig. 3-9 Shunt Diode Bias T Using Muehe & Young Structure

BIASING CIRCUITS

section is printed with the main transmission line on the bottom board. Mylar .001" thick is placed over the Z_{01} line section. A strip of .001" shim brass is cut to exactly cover the Z_{01} line section and another strip or fine wire to give Z_Q is attached to it to provide bias. Then for RF bypass at the bias point use is made of a very small capacitor or two parallel sheets of shim brass separated by mylar (as shown) in which the bottom sheet is well grounded through the stripline board. A small piece of the Z_{01} section is removed in the plane of the diode to compensate for reverse bias diode capacitance. Thus circuit (a) of Fig. 3-7 may be used for making a low VSWR shunt diode switch.

The area of center conductor to be removed may be calculated approximately from the simple parallel plate capacitance formula.

$$C = .225 \, \epsilon_r \, A/d \text{ pF} \tag{3-21}$$

where A is the plate area and d is the plate separation all in inches. For 1/16" teflon-fiberglass stripline ϵ_r = 2.65 and d = .0625"

$$C_\epsilon = 9.55 \text{ pF/in}^2 \text{ for one side.} \tag{3-22}$$

Removal of the dielectric gives

$$C_0 = 3.60 \text{ pF/in}^2 \tag{3-23}$$

for a difference of 5.95 pF/in².

Thus a diode having 0.5 pF reverse bias capacitance can be compensated by removing about 0.1 in² of dielectric around the diode (or between the center conductor and the other ground plane of the stripline). As an alternative, a piece of the center conductor could be removed as shown in Fig. 3-9. Since the center conductor has capacitance to both ground planes the total approximate capacitance C_T is given by

$$C_T = 2 \, C_\epsilon = 19.1 \text{ pF/in}^2 \tag{3-24}$$

and for the 0.5 pF diode about 1/40 in² of center conductor would have to be removed to compensate for reverse biased diode capacitance.

Circuit (b) of Fig. 3-7 was developed by Mouw[4] and is quite satisfactory as a bias T but does not contain all of the bias elements needed for a diode switch. Circuit (c) of Fig. 3-7 was developed by McDermott and Levy[5] but it too does not contain all of the elements needed to make a diode switch. This last bias T was synthesized using

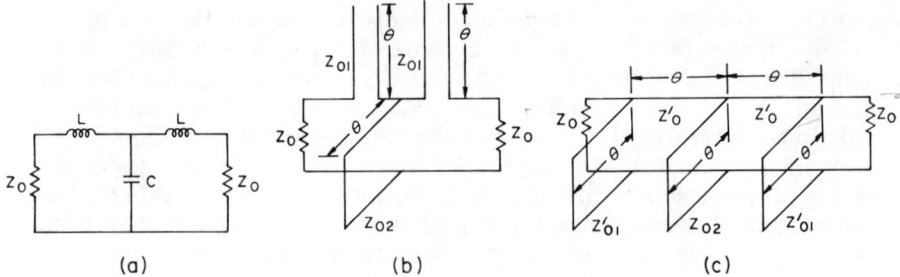

Fig. 3-10   Filter for Low VSWR Series Diode Switch

filter theory which is summarized in Appendix C. Another synthesis is needed, however, to make a low VSWR series diode switch.

The evolution shown in Fig. 3-10 will give the needed circuit for a low VSWR series diode switch. Given a maximum VSWR of 1.1, then for three elements $g_1 = g_3 = .627$ and $g_2 = .969$. (Ref. Appendix C, Table C-1.) For octave bandwidth the pass band should be from $\frac{2}{3}\omega_0$ to $\frac{4}{3}\omega_0$ thus giving a half-bandwidth of $\omega_0/3 = \omega_1$. Using Eqs. (C-21) and (C-22)

$$Z_{01} = g_1 Z_0 \cot\left(\frac{\pi}{2}\frac{\omega_1}{\omega_0}\right) = 54.3\ \Omega \tag{3-25}$$

and

$$Y_{02} = g_2 Y_0 \cot\left(\frac{\pi}{2}\frac{\omega_1}{\omega_0}\right) = \frac{1}{89.5\Omega} \tag{3-26}$$

for $Z_0 = 50$ ohm. The direct transformation from the circuit of Fig. 3-10 (a) to transmission line would be like circuit (b) except open circuits are where short circuits should be and visa-versa. The direct transformed circuit would be a low pass filter having a bandpass to ω_1. Interchanging opens and shorts change the filter to a bandpass filter centered at ω_0 with a pass bandwidth of $2\omega_1$. The application of Kuroda's identify (Ref. Appendix C) converts circuit (b) to circuit (c):

$$Z_0' = Z_0 + Z_{01} = 104.3\Omega \tag{3-27}$$

$$Z_{01}' = Z_0 + \frac{Z_0^2}{Z_{01}} = 96.0\Omega \tag{3-28}$$

using the equations from Appendix C, Fig. C-5.

BIASING CIRCUITS

Similar calculations for other values of maximum VSWR and for octave bandwidth give the curves shown in Fig. 3-11. The filters can be made at any frequency by making the line lengths all $\lambda/4$ at the center frequency. Note that Z_0' is the through line in which the diodes will be mounted and that the characteristic impedance may may be up to 120 ohms for a maximum VSWR of 1.2.

The forward biased diode will appear to be a short length of transmission line having a characteristic impedance in the neighborhood of 100 ohms. The effective characteristic impedance of a diode may be estimated from its forward biased inductance as demonstrated by its insertion loss curve as in Fig. 2-12. Assume the diode is measured in teflon-fiberglass stripline. The propagation velocity v in this stripline is about 2×10^{10} cm/sec. Assume the diode insertion loss as in Fig. 2-12 follows the 2 nH curve and that the diode is physically 1 cm long. The distributed inductance of a transmission line may be calculated by considering the impedance of a short-circuited short length

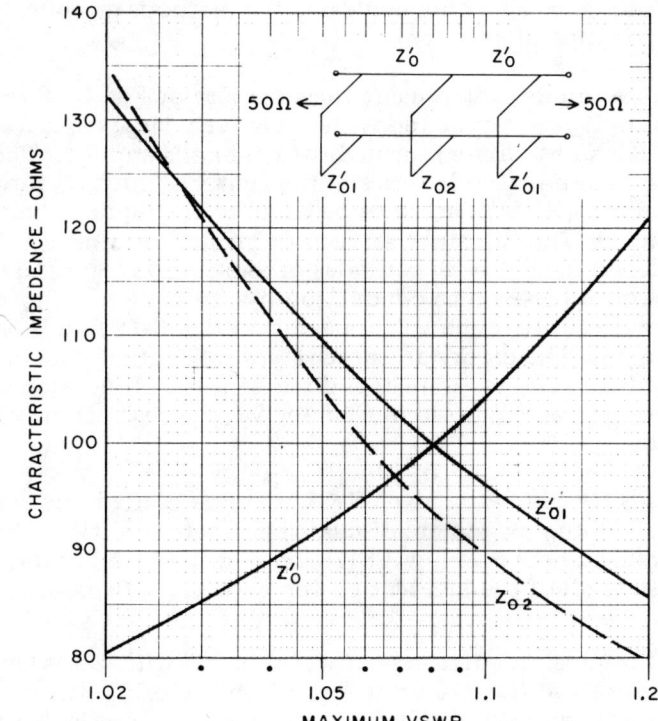

Fig. 3-11 Design Curves for Octave Bandwidth Low VSWR Series Diode Switches

$$X = \omega L = Z_0 \tan 2\pi \frac{\ell}{\lambda} \approx Z_0 \; 2\pi \frac{\ell}{\lambda} = Z_0 \; 2\pi \frac{\ell f}{\nu} = Z_0 \frac{\ell \omega}{\nu} \quad (3\text{-}29)$$

$$L = Z_0 \frac{\ell}{\nu} \quad (3\text{-}30)$$

For one centimeter of 50 ohm transmission line L = 2.5 nH. The 2 nH extra contributed by the diode makes the total for the 1 cm section containing the diode 4.5 nH. Solving Eq. (3-30) for Z_0 and using the 4.5 nH total inductance gives the effective characteristic impedance of the forward biased diode of 90 ohms for the conditions assumed. A more precise measurement method would also take into account the change in propagation velocity in the region of the diode which may be 1 to 2×10^{10} cm/sec due to the higher dielectric constant of the diode glass and the curvature of the diode whisker. Even the transmission line representation of the diode is not exact because the structure is not uniform from end to end. Discontinuities exist at both ends of the whisker and where the filling factor of the glass changes (ref. Fig. 2-3). A more precise measurement of the diode may be obtained using the methods given in Chapter V and Appendix A.

To make a series diode switch using the structure of Fig. 3-10 (c) the middle stub Z_{02} is terminated with a relatively large capacitor (which is an RF short circuit) through which bias is applied. The switching diodes are placed in series in the center of each Z_0' thru transmission line. (Z_0' is selected to be equal to the diode characteristic impedance.) The Z_{01}' stubs serve as dc ground returns. The diodes are placed so that both cathodes or anodes go to ground (so that they are both biased into conduction by the same polarity current). The circuit in stripline will appear as in Fig. 3-12 and is used later in Fig. 5-17. The diodes are a quarter-wavelength apart causing slight residual series resistances and reactances to tune each other out. Furthermore, as discussed in Chapter VI, their isolations will more than add.

Other methods for obtaining low VSWR with the diode in both bias states are more complex and are discussed in Chapter VIII. The methods discussed here give low VSWR only in the "ON" state. In the "OFF" state all of the incident power is ideally reflected.

As a point of interest, an octave bandwidth bias T with series capacitor can be made from the circuit of Fig. 3-7 (a) by using Kuroda's identity to transform part of the Z_Q shorted stub to a series open circuit stub. For a maximum VSWR of 1.05 and octave bandwidth, Fig. 3-8 indicates that Z_Q = 50 ohms and Z_0' = 40 ohms. The Z_Q

BIASING CIRCUITS

stub is equivalent to two 100 ohm stubs in parallel. A transformation using Kuroda's identity of one of the 100 ohm shunt stubs through one 40 ohm section provides a new section of 28.6 ohms = Z_0'' and a series open circuit stub of 11.4 ohms = Z_c as shown in Fig. 3-13. Using the circuit of Fig. 3-7 (c) for a maximum VSWR of 1.05 over an octave bandwidth then Z_Q = 178 ohms and Z_c = 23.8 ohms must be used. The new circuit of Fig. 3-13 is much easier to realize than circuit (c) of Fig. 3-7. An even greater degree of flexibility is possible with the new circuit since taking even less of Z_Q than taken above for the series capacitor leads to any desired lower characteristic impedance for the series open circuit stub. E.g., dividing Z_Q into two parallel stubs, one 52.6 ohms and the other 1000 ohms, leads to Z_0'' = 39.4 ohms and Z_c = 1.54 ohms (and Z_Q = 52.6 ohms).

A further method for obtaining a low VSWR bias T over a broad bandwidth is to use ferrite beads as shown in Fig. 3-14. A single hole miniature ferrite bead is put close to the center conductor so the increase in series inductance in the main transmission line due to the proximity of the ferrite bead is equalized by the shunt capacitance of the fine wire extending from the center conductor through the ferrite bead. At high frequencies the ferrite bead will cause the fine wire to have an RF open circuit in it inside the bead. The larger fer-

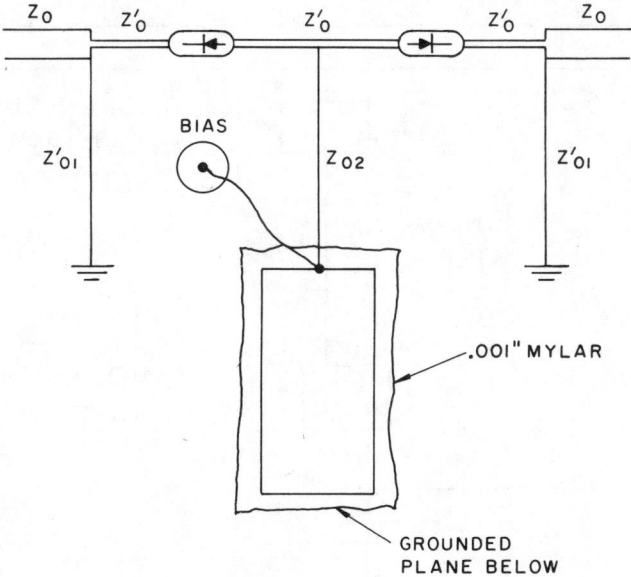

Fig. 3-12 Stripline Circuit for Octave Bandwidth Low VSWR Series Diode Switch

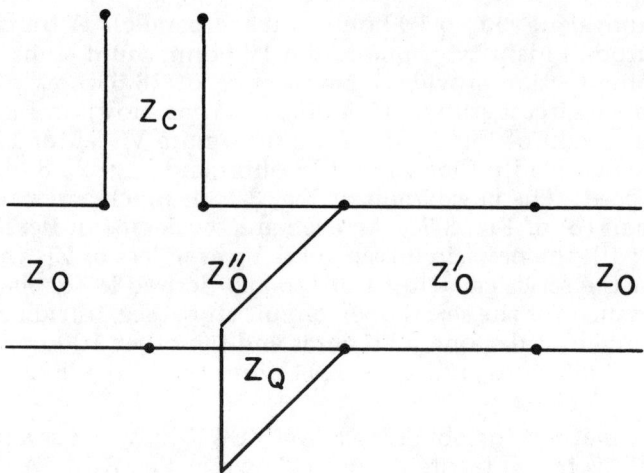

Fig. 3-13 Bias T Derived from Muehe & Young

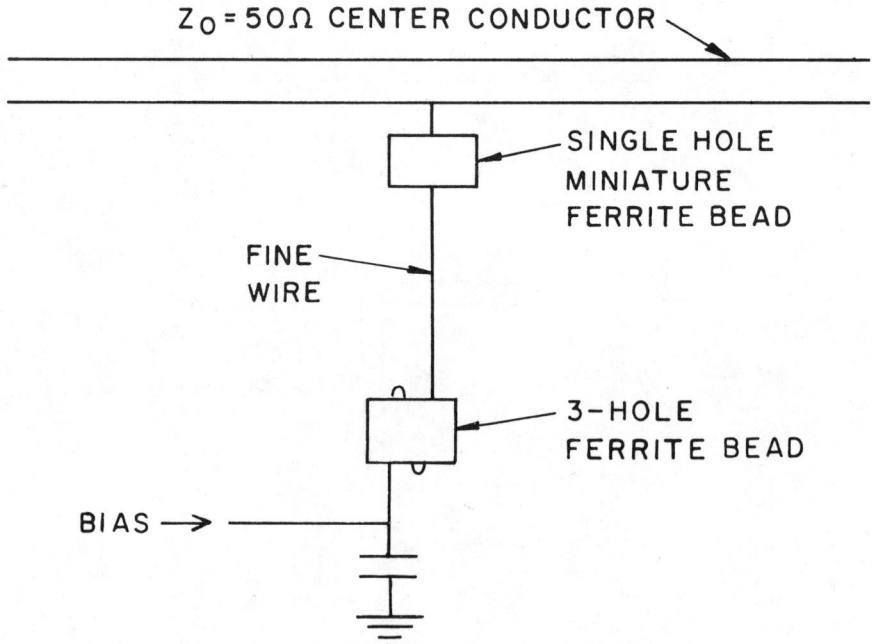

Fig. 3-14 Bias Wire Using Ferrite Beads

BIASING CIRCUITS

rite bead farther from the center conductor has several holes in it through which the fine wire is threaded. This bigger bead provides more inductance at lower frequencies permitting the bias T to work at lower frequencies without a large structure.

C. Designing for Fast Switching

The problem of limited modulation rate is that of separating modulation power and RF power which appear across the same two-terminal diode junction. When the modulation frequency is low and the RF is high, there is no problem separating the powers with simple filters. As the modulation rate increases, the modulation power has higher frequency components and the modulated RF power has lower (as well as higher) frequency components. For increasing modulation rates, modulation power spectral components and modulated RF power spectral components approach each other, requiring filters to have steeper skirts to separate the powers. When the power are not separated well enough, modulation power appears in the RF transmission line as modulation transients, and troublesome RF power can be coupled to other parts of the system via the modulation circuit. To demodulate a carrier with wide video bandwidth information on it, the demodulator must satisfy filter requirements similar to those of the modulator.

Filters and conventional transmission line methods can be used to satisfy most high speed switching requirements. For extremely high switching speeds balanced transmission lines must be used to separate modulation power and modulated power.

1. Minimum Switching Time

The general characteristics of filters for diode modulators may be seen from Fig. 3-15. Unmodulated RF power enters through port 1. Baseband modulation is supplied through port 2. The modulated RF power emerges through port 3. The voltage waveforms and the spectra of the power at each port are shown adjacent to the port numbers. The filter on port 1 needs to pass power at f_0 only; however, it should not deteriorate the waveform of the modulation power; that is, it must be a high-impedance circuit from dc up to some high frequency, as seen from port 2. The filter on port 2 must pass power at the lower frequencies and stop RF power. In addition, this filter must present a high impedance to RF power as seen from port 1 and the diode. The filter on port 3 must pass power at all frequencies above and below f_0 corresponding to the bandwidth of the modulation power. This filter must also present a high or low

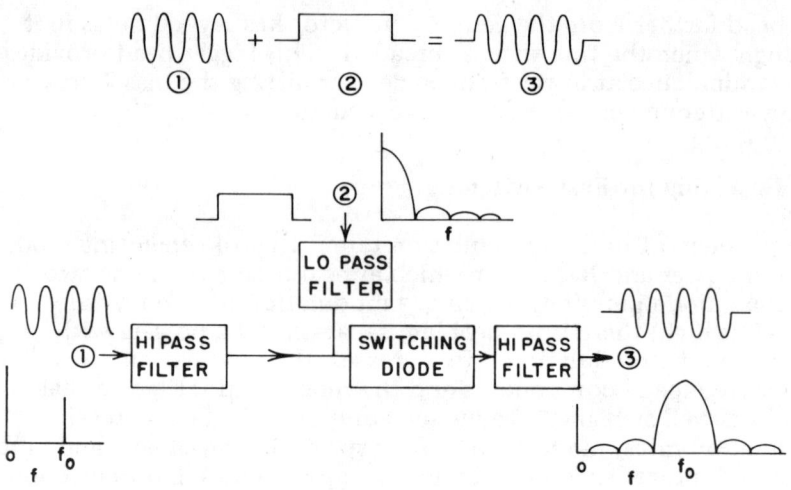

Fig. 3-15   Filters for High-Speed Diode Switches

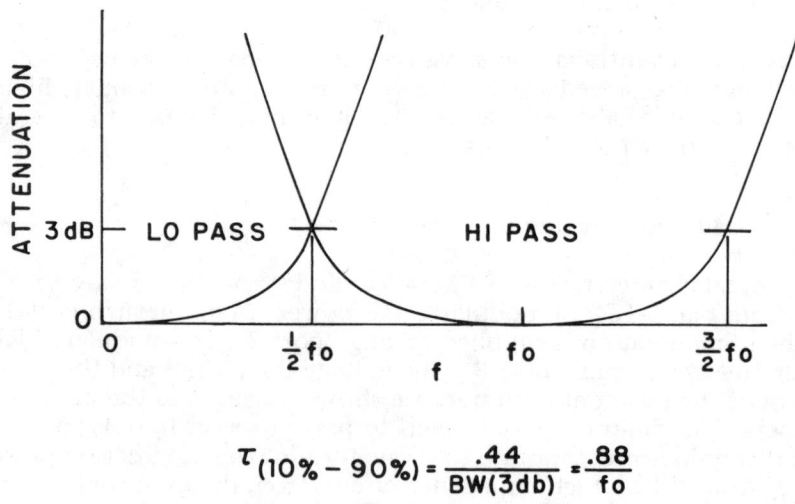

$$\tau_{(10\% - 90\%)} = \frac{.44}{BW(3db)} = \frac{.88}{f_o}$$

Fig. 3-16   Filter Characteristics for Minimum Switching Times

BIASING CIRCUITS

impedance to the diode at the modulation frequency depending on whether the diode is in shunt or in series. For pulse modulation as shown in Fig. 3-15, it is also recommended that the filters be constant-time-delay filters to prevent "ringing."

The minimum switching time may be derived with the aid of Fig. 3-16. A low-pass filter with a finite 3 dB bandwidth, BW, will cut off some of the infinite spectrum of a zero rise-time pulse giving it a 10 to 90 percent switching time of

$$\tau_F = \frac{0.44}{BW} \tag{3-31}$$

The high-pass filter must pass the spectral components of the modulated carrier or further deteriorate switching speed. Assuming that the difference between the carrier frequency and the cutoff frequency of the high-pass filter is equal to the bandwidth of the low-pass filter, then, because rise times add as the sum of their squares, the switching times of the modulated carrier are given by

$$\tau = \sqrt{2}\ \tau_F = \frac{0.44\sqrt{2}}{BW} = \frac{0.62}{BW} \tag{3-32}$$

Since the spectra have to be separated, the best arrangement of filters from a switching speed point of view is to have the 3 dB frequencies coincide at $f_0/2$ as shown in Fig. 3-16. Then the minimum switching time becomes

$$\tau_{MIN} = \frac{0.62}{f_0/2} = \frac{1.24}{f_0} \tag{3-33}$$

This equation indicates that the lowest switching time that could be generated using filters on a 1 GHz carrier is 1.24 ns, and on a 10 GHz carrier is 0.124 ns.

2. TEM Transmission-Line Modulators and Demodulators

Normally a diode modulator is designed to provide maximum RF bandwidth. The biasing lead and ground return are high-impedance quarter-wavelength lines connected to the center conductor with a lumped circuit capacitor providing an RF ground at the bias input. These bias leads allow moderate modulation speeds, but the theoretical limit may be more closely approached and the suppression of RF power out of the modulation ports may be increased by using the filter circuits disclosed by Matthei, et. al.[6] (Filter Bible pp. 991-999, "Diplexers with Continuous Pass Bands,"). The circuit shown in Fig. 3-17 comes close to the limit and is easy to remember.

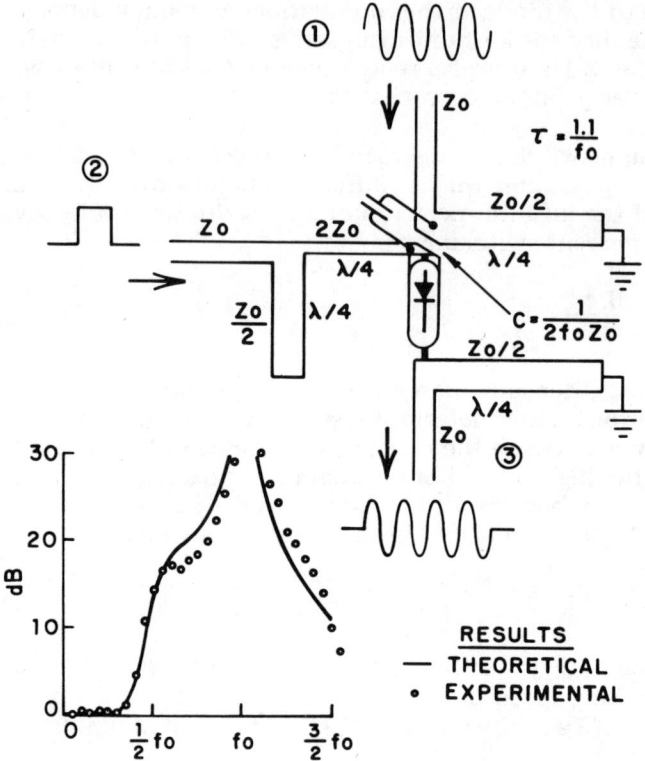

Fig. 3-17   Simplified High-Speed Bias T

The attenuation of the modulation input filter (port 2 to port 3) with the diode replaced by a Z_0 transmission-line section and the port 3, $\lambda/4$ stub removed) is shown at the bottom of Fig. 3-17. The low-frequency π equivalent circuit of the modulation input filter is a 1 dB ripple Tchebyscheff filter with a 3 dB bandwidth of $2/3\, f_0$. The same basic filter is shown in Fig. 3-18 as it would be used in a demodulator. The series inductance in the π equivalent circuit is derived from the high-impedance $2Z_0$, $\lambda_0/4$ line section ($L_{2Z_0} = 4\, Z_0/f_0$). Each shunt capacitance is derived from the low-impedance. $Z_0/2$ open-circuit stubs ($C_{Z_0/2} = 1/2f_0\, Z_0$) and half the $2\, Z_0$ line ($C_{2Z_0}/2 = \dfrac{1}{16\, f_0\, Z_0}$).

The ground returns in both figures are $Z_0/2$, $\lambda_0/4$ stubs. One of these by itself has a critically damped response to a step waveform. If the characteristic impedance of the stub were higher, the transient

BIASING CIRCUITS

to a step voltage would be a fast-rising slowly decaying pulse. If the characteristic impedance were lower, the response would be a damped oscillation at f_0. The response of one $Z_0/2$ stub to a step voltage is a pulse having half the voltage of the step and one half an RF cycle at f_0 in duration. When more than one stub is used without attenuation between them, they interact and produce transients of longer duration. The simpler analysis of multiple stubs makes use of conventional filter theory. A single $Z_0/2$ stub provides 3 dB attenuation at $f_0/2$ and has a rise time of about $0.7/f_0$. Adding rise times as before, the theoretical rise time of 2.0 ns was measured when the modulator of Fig. 3-17 was used with the demodulator of Fig. 3-18. Hot carrier diodes were used as modulator and demodulator at 1 GHz. Some difficulty was encountered in arranging filter spacing and attenuators so that transient voltages were suppressed yet pulse shape was preserved.

3. Waveguide Modulators and Demodulators

Experimental work was done in X-band waveguide at 9.3 GHz. A typical X-band diode switch is shown in Fig. 3-19. The waveguide performs the function of high-pass filters with a cutoff frequency of 6.5 GHz. Because this is 2.8 GHz below the carrier frequency, the waveguide will have a rise time of about 0.16 ns. The 10 pF capacitor performs the function of the low-pass filter. For a 50 ohm pulse generator driving a 50 ohm diode switch the filter response is as shown in Fig. 3-20. The 3 dB point is at 0.6 GHz which for a capacitor provides a rise time of 0.6 ns.

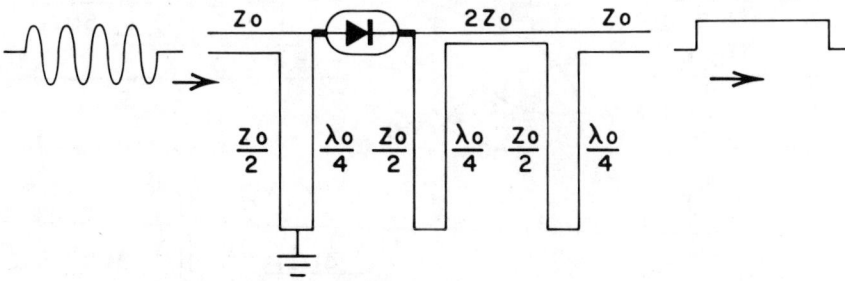

Fig. 3-18 Simplified High-Speed Bias T as Adapted to a Demodulator

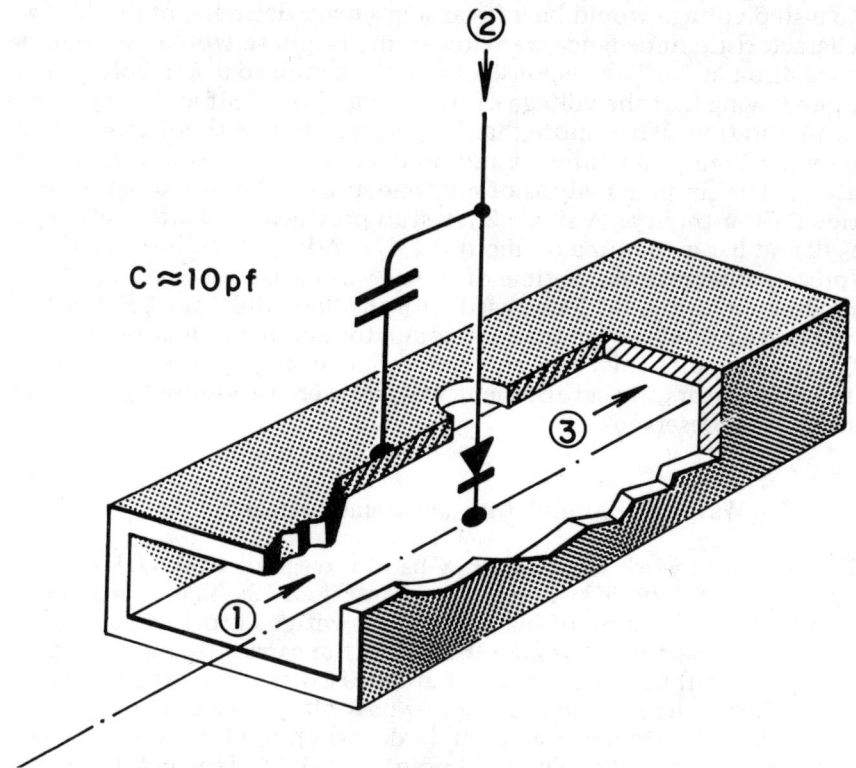

Fig. 3-19 Conventional Waveguide Diode Switch

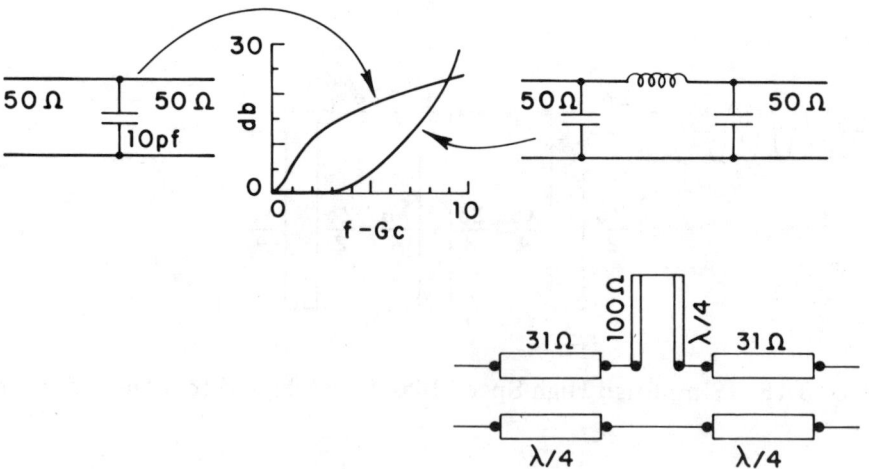

Fig. 3-20 Choke Responses for Waveguide Switches

BIASING CIRCUITS

The rise time of the low-pass filter can be decreased to 0.1 ns by building the π section filter also shown in Fig. 3-20. This filter has an additional advantage over the capacitor in that its attenuation of RF power becomes very high at the design frequency. The physical embodiment of this filter is shown in Fig. 3-21. Because of the capacitance existing between the two ends of the center conductors of the 31 ohm sections, the length of the 100 ohm section had to be made shorter than $\lambda_0/4$ to present an inductance for parallel resonance with this capacitance at the design frequency.

To evaluate this technique, a pulse modulator and a demodulator were made up of such filters in conjunction with 1N263 point-contact germanium diodes[7]. The demodulated waveforms are shown in Fig. 3-22. The observed OFF-ON time is 0.30 ± 0.05 ns and the ON-OFF is 0.18 ± 0.02 ns. A 0.25 ns rise-time pulse was used in conjunction with a 0.1 ns rise-time oscilliscope. The theoretical rise time of the combination pulse generator, modulator input filter, waveguide (as a high-pass filter), demodulator output filter, and oscilloscope is 0.34 ns.

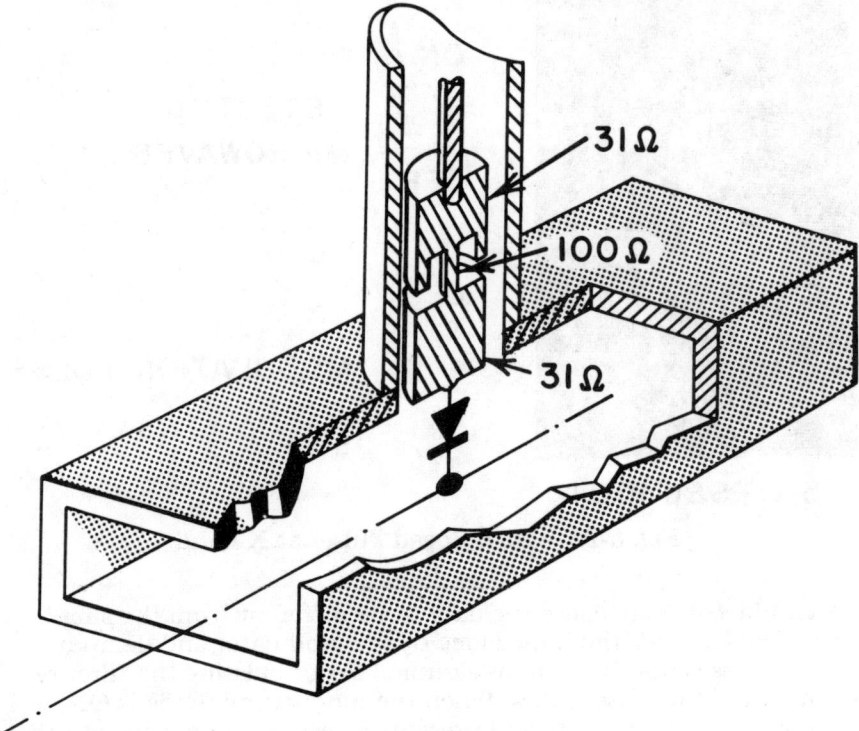

Fig. 3-21 High-Speed Bias Choke for Waveguide Diode Switches

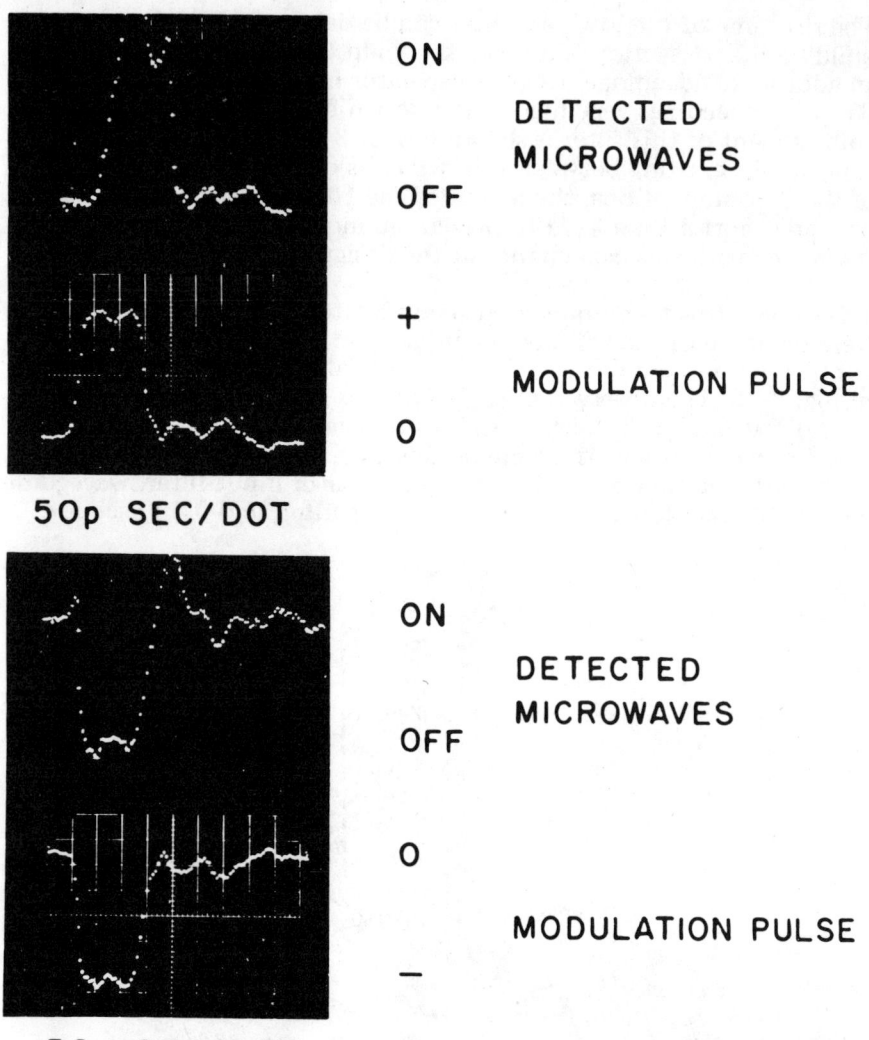

Fig. 3-22   High-Speed Pulses at X-Band

Several factors may cause the data to be different from the simplified theory. First, the impedance of the modulator and demodulator diodes is not 50 ohms as assumed for calculating the filter responses and their rise times. When the modulation diode is ON, it is in the conduction state and presents a very low impedance to the modulation source. Similarly when the demodulator diode is providing rectified current (also during ON time), it has a low imped-

BIASING CIRCUITS

ance. The high impedance of the diodes associated with OFF is also different from the characteristic impedance of the filter as portrayed in Fig. 3-20. Thus, the pulse is reflected by the diode as an open circuit and travels back to the 50 ohm pulse generator. To prevent these mismatches from deteriorating the data, the pulse generator was padded by either attenuators or a time delay (the line length between pulse generator and diode modulator).

Bias to the diode modulator, opposite in polarity to the pulse, was supplied through the other broad-waveguide wall in a circuit similar to Fig. 3-19 (turned upside-down). The longer the pulse, the larger the required capacitance. This capacitance is part of a diode mount used for measuring diodes. The waveguide walls are thicker than normal and a block is cut from the bottom of the waveguide and replaced with a thin dielectric insulator where the cut was made. The broad waveguide wall is cut $\lambda g/4$ in front of and behind the diode. The narrow waveguide wall is cut through to the inside face of the bottom broad waveguide wall. The waveguide wall thickness is a quarter-wavelength in the insulation dielectric. This capacitor provides minimum interference with the RF power and can be added to externally for longer pulses. This capacitor should not deteriorate the pulse wave-form.

A second factor that might alter the pulse wave-form is the nonlinearity between voltage into the modulator and voltage out of the demodulator. This nonlinearity can cause the experimental switching times to be less than the theoretical switching times.

A third factor possibly altering the demodulated pulse waveform is dispersion of the low-pass filter and in the waveguide. A more sophisticated system would have constant-time-delay low-pass filters and a waveguide delay equalization circuit.

Theoretically a zero rise-time pulse generator used with these filters would impart a rise time to the microwave power of 0.135 ns.

A more direct filter synthesis has been used by Adams, et. al.[8] to obtain high speed switching in waveguide. They used a 3-element filter giving 0.1 dB ripple up to 3 GHz. The stop band was centered at 10 GHz with greater than 40 dB attenuation from 8 to 12 GHz. Most of the filter was realized in stripline. Kurado's identity was used to obtain a low impedance series $\lambda_0/4$ coaxial section on the end of the filter immediately adjacent to the diode and thus provide an RF short circuit at the waveguide wall bias point. Their structure is shown in Fig. 3-23.

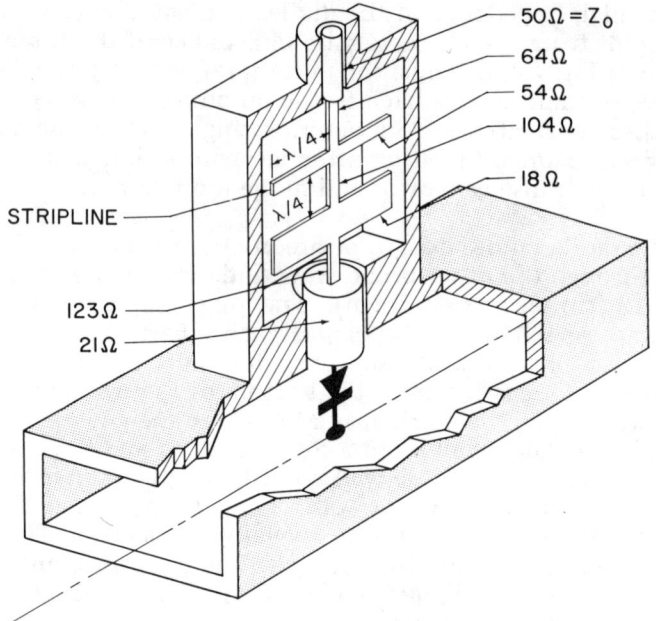

Fig. 3-23   High-Speed Bias Choke of Adams

D. Suppression of Modulator Transients

Ideally the application of step voltage modulation to port 2 of Fig. 3-15 would cause no voltage to appear at port 3. Due to the broad spectrum of the step and the finite attenuation of the combination of low-pass filter and high-pass filter, the step causes a transient to emerge from port 3 as shown in Fig. 3-24. Transmission measurements of combined low-pass and high-pass filters on a time domain reflectometer have demonstrated that the transient response is similar to that generated by the application of a step voltage to a bandpass filter. This combination of low-pass filter and high-pass filter also gives attenuation similar to that of a bandpass filter but with high attenuation inside the pass band as shown approximately in Fig. 3-25 (n is the number of elements in the prototype filter). The cut-off frequency (3 dB point) of the low-pass filter is f_1 and the cutoff frequency of the high-pass filter is f_2. The attenuation formulas shown are approximations for maximally flat filters with n elements in their lumped-element low-pass prototypes. The attenuation for constant-time-delay filters is slightly lower and for Chebychev filters is slightly higher. Note that increasing the separation between f_1 and f_2 increases the attenuation inside the pass-

BIASING CIRCUITS

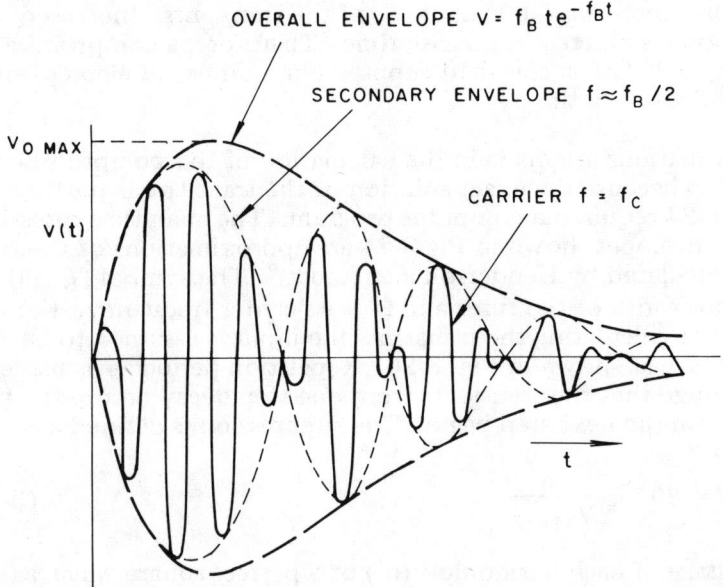

Fig. 3-24 Typical Switching Transient Induced on RF Line

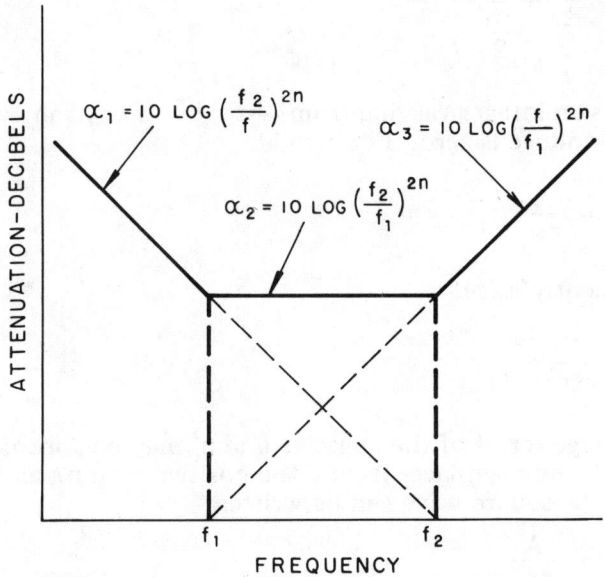

Fig. 3-25 Attenuation of Overlapping Low-Pass and High-Pass Filters

band, thus reducing the magnitude of the transients. Increased separation also increases the rise time. Therefore, a compromise has to be made between transient suppression, number of elements in the filter, and rise time.

Approximations are made in the calculation of this compromise relationship because the exact solution of the transient as portrayed in Fig. 3-24 required a computer program. The analytic expressions for the envelopes shown in Fig 3-24 are approximations of the transients calculated by Henderson and Kautz.[9] The symbol f_B is the 3 dB bandwidth of the filter and f_c is its center frequency. For the purpose of calculating the transients the input is assumed to be a square wave as shown in Fig. 3-26. Repetition period τ_p is made long enough that the transients from one step decay before the transients from the next step begin. The suppression is defined as

$$S = 20 \log \frac{A}{V_{0\,MAX}} \quad (3\text{-}34)$$

The voltage of each harmonic $V(n')$ of a perfect square wave, as shown in Fig. 3-26 is

$$V(n') = \frac{A}{2} \frac{\sin \frac{\pi n'}{2}}{\frac{\pi n'}{2}} \quad (3\text{-}35)$$

in which n' is an integer varying from $-\infty$ to $+\infty$. For n' an even number, the voltage is zero. For n' odd,

$$|V(n')| = \frac{A}{\pi n'}. \quad (3\text{-}36)$$

For n' sufficiently large,

$$\frac{A}{\pi n'} \approx \frac{A}{\pi(n'+1)} \quad (3\text{-}37)$$

Thus the voltage terms of the negative odd n' may be approximately folded into the empty places left by the positive even n', and the spectrum of the square wave can be written

$$|V(n')| \approx \frac{A}{\pi n'} \quad (3\text{-}38)$$

for n' from 0 to ∞. Thus the power at each harmonic can be written

BIASING CIRCUITS

$$P(f\tau_p) = \frac{A^2}{\pi^2 (f\tau_p)^2}. \qquad (3\text{-}39)$$

The power past the filters can be found by adding up the power at each $f\tau_p$ after it has been attenuated. To facilitate integration, the attenuation of a maximally flat filter

$$\alpha = 10 \log [1 + (f/f_0)^{\pm 2n}] \qquad (3\text{-}40)$$

is approximated by

$$\alpha = 10 \log (f/f_0)^{\pm 2n} \qquad (3\text{-}41)$$

as shown in Fig. 3-25 in the attenuation band, or $\alpha = 0$ otherwise. The approximation introduces negligible error for

$$(f/f_0)^{\pm 2n} > 10 \qquad (3\text{-}42)$$

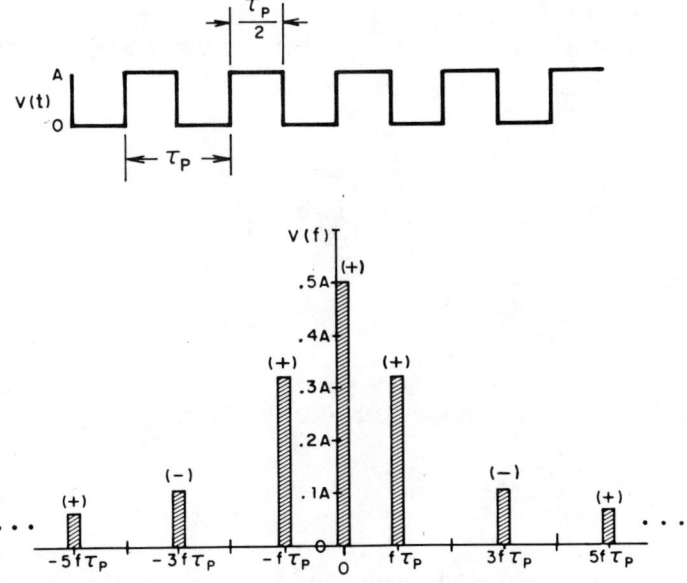

Fig. 3-26 Square Wave and Spectrum for Calculation of Transient Suppression

which corresponds to attenuation for each filter greater than 10 dB. The average power getting past the filters is obtained by integration over the sections of Fig. 3-25 as follows:

$$P_{AV} = \int_0^{f_1 \tau_p} \frac{P(f\tau_p)}{\left(\frac{f_2 \tau_p}{f\tau_p}\right)^{2n}} d(f\tau_p) + \int_{f_1 \tau_p}^{f_2 \tau_p} \frac{P(f\tau_p)}{\left(\frac{f_2 \tau_p}{f_1 \tau_p}\right)^{2n}} d(f\tau_p) + \int_{f_2 \tau_p}^{\infty} \frac{P(f\tau_p)}{\left(\frac{f\tau_p}{f_1 \tau_p}\right)^{2n}} d(f\tau_p)$$

$$= \frac{A^2}{\pi^2} \left(\frac{f_1 \tau_p}{f_2 \tau_p}\right)^{2n} \left[\frac{f_2 \tau_p}{f_1 \tau_p} \frac{1}{1 - \frac{1}{2n}} - \frac{1}{1 + \frac{1}{2n}}\right] \frac{1}{f_2 \tau_p} \quad (3\text{-}43)$$

The general output transient response is assumed to be of the form

$$V_0 = K(f_B t) \, \epsilon^{-f_B t} \cos \omega_c t \quad (3\text{-}44)$$

Dropping the secondary envelope term from the expression does not significantly alter the average power in each wave packet or the maximum voltage of the wave packet. The maximum voltage is given by

$$V_{0\,max} = \frac{K}{\epsilon} \,. \quad (3\text{-}45)$$

The energy in each wave packet is given by

$$\int_0^{\infty} V_0^2 \, dt = \frac{K^2}{8 f_B} \quad (3\text{-}46)$$

Since there are $2/\tau_p$ envelopes per second, the average transient power is

$$P_T = \frac{\epsilon^2 V_0^2{}_{max}}{4 f_B \tau_p} \quad (3\text{-}47)$$

The f_B must be determined. Referring to Fig. 3-25, the minimum attenuation of the two overlapping filters is the ratio $(f_2/f_1)^{2n}$. Referring to Fig. 3-27, the 3 dB bandwidth will be the difference between f_H and f_L as determined from the following equations:

BIASING CIRCUITS

$$10 \log \left(\frac{f_H}{f_1}\right)^{2n} = 3.0 + 10 \log \left(\frac{f_2}{f_1}\right)^{2n} \tag{3-48}$$

$$10 \log \left(\frac{f_2}{f_L}\right)^{2n} = 3.0 + 10 \log \left(\frac{f_2}{f_1}\right)^{2n} \tag{3-49}$$

Solving

$$f_B = \left(2^{1/2n} \frac{f_2}{f_1} - \frac{1}{2^{1/2n}}\right) f_1. \tag{3-50}$$

The ratio of A and $V_{0_{max}}$ which gives the attenuation may be derived by equating $P_{AV} = P_T$

$$\alpha = 10 \log \frac{A^2}{V_{0_{max}}^2} = 10 \log \left\{ \frac{\pi^2 \epsilon^2 (f_2/f_1)^{2n+1}}{4 \left(2^{1/2n} \frac{f_2}{f_1} - 2^{-1/2n}\right)\left(\frac{1}{1-\frac{1}{2n}} \frac{f_2}{f_1} - \frac{1}{1+\frac{1}{2n}}\right)} \right\}. \tag{3-51}$$

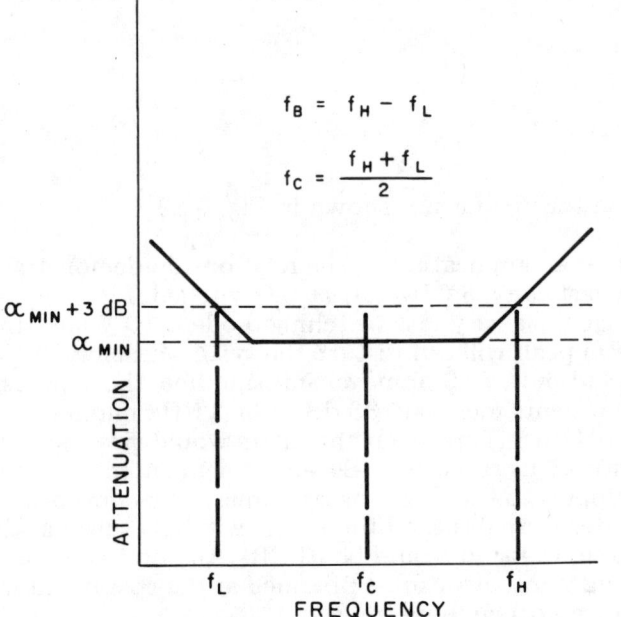

Fig. 3-27 Definition of 3-dB Bandwidth for Calculating Transient Suppression

The expression is not correct for n = 1 because the waveform is not as assumed. Using Laplace transforms, the attenuation of two single-element filters consisting of a series inductance and a shunt inductance is given exactly by

$$\alpha = 20 \log \left\{ \left(2 \frac{f_2}{f_1} + 1 - M\right) \left[\frac{2 \frac{f_2}{f_1} + 1 + M}{2 \frac{f_2}{f_1} + 1 - M} \right]^{\frac{1}{2}(2(f_2/f_1) + 1 + M)/M} \right\},$$

(3-52)

$$M = \sqrt{\left(2\frac{f_2}{f_1}\right)^2 + 1}.$$

(3-52)

The attenuation is a function of n and f_2/f_1. A more convenient relationship would give attenuation as a function of increased switching time.

From the equations shown in Fig. 3-28 the following relationship is derived

$$\frac{f_2}{f_1} = 2 \frac{\tau}{\tau_{\min}} - 1.$$

(3-53)

The attenuation for various n is shown in Fig. 3-29.

As an example of an application of the relationship demonstrated in Fig. 3-29, how fast can a 3 GHz carrier be switched if the required switching voltage for a very fast switching diode is 10 V and transients below –100 dBm peak will not disturb the system? Since 10 V correspond to +33 dBm in a 50 ohm transmission line, the suppression of switching transients must be 133 dB. For 3 GHz minimum rise time (τ_{MIN}) is 0.41 ns. Three-element filters would give the required suppression with 41 ns rise times. Seven-element filters would give the required suppression with 1.9 ns rise times. If a conventional two-element filter were designed for 10 ns switching speed at this frequency, the suppression would be 51 dB. Any desired suppression of switching transients can be obtained at the cost of filter complexity and increased switching time.

BIASING CIRCUITS

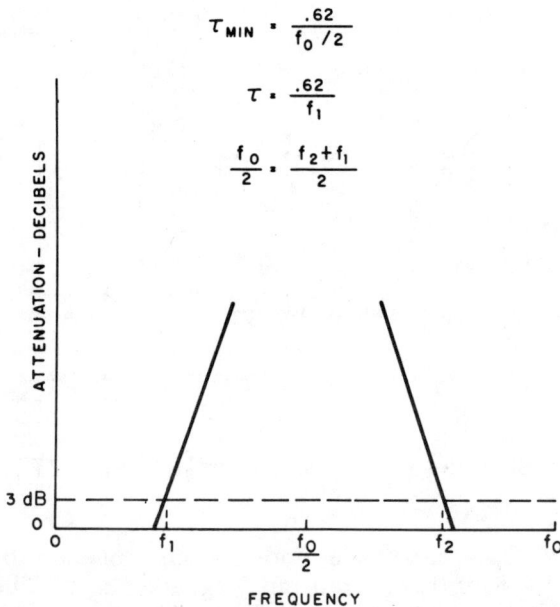

Fig. 3-28   Rise Time Dependence on Filter Cut-off Frequencies

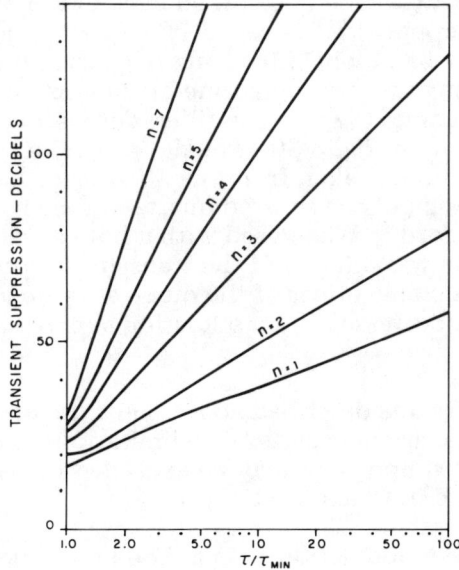

Fig. 3-29   Transient Suppression Dependence on Rise-Time and Number of Filter Elements

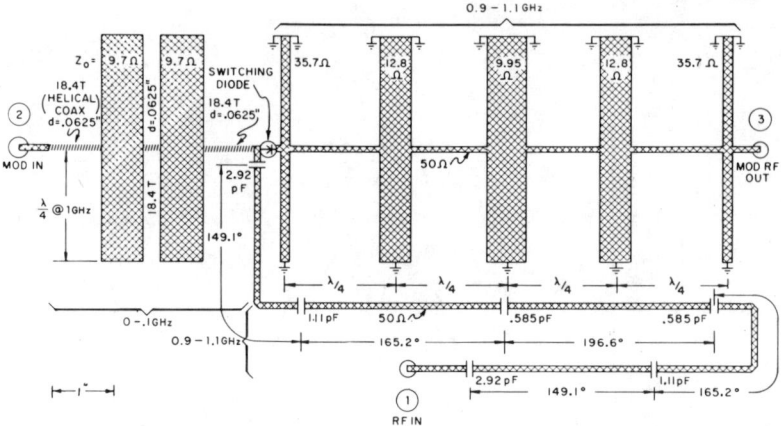

Fig. 3-30 Stripline Filters for 1 GHz High-Speed Diode Switch

As another example, suppose a diode switch is needed to protect a 1 GHz radar receiver from "transmit" signal leakage. The diode requires 10 V to cause switching and the transients must be below 200 μV to prevent receiver saturation. The switching pulse is generated by a matched 50 ohm generator so that any of the pulse reflected by filters or the diode is absorbed back in the pulse generator. Therefore, 94 dB suppression is needed. Fig. 3-29 indicates that 100 dB suppression may be obtained by using five-element filters at a sacrifice of increasing the switching time by a factor of 5. Thus the switching speed becomes 6.2 ns. The filters for a series diode switch in stripline are shown in Fig. 3-30. The design of the low-pass and series capacitor filters are taken from the *Microwave Engineers Handbook*,[10] and the design of the shorted quarter-wavelength stub filter is taken from Mumford.[11] When used with a hot-carrier diode as a switch, the rise time was 3.6 ns and the transient suppression was 50 dB. The pulse-shaping effect of the diode as evidenced by the shortened rise time decreased the modulation suppression by a factor of two.

All of the biasing circuits described above make use of conventional series or shunt diode arrangements. The limitations in switching speed and transient suppression may be exceeded by using balanced switches as discussed in Chapter XI.

To obtain high-speed modulation with a diode the following rationale is suggested:

1. To keep modulation transients low, a modulating diode should

BIASING CIRCUITS 81

be used requiring only very low voltages for modulation, and modulation voltage should be kept as low as possible.

2. The filters for further suppressing the modulation transients may be designed with the help of Fig. 3-29.

3. Modulation circuit simplification may be accomplished by using a modulation source with a rise time no faster than possible according to Fig. 3-29. The slow rise time then accomplishes the same effect as the low-pass filter.

4. Waveguide for a high-pass filter should be used if possible, since this corresponds to an $n = \infty$ type filter.

5. Balanced transmission-line techniques may be used to obtain additional suppression of modulation transients when required.

6. Diodes no faster than required should be used otherwise they will cause pulse shaping and reduction of desired transient suppression. Diodes demonstrating step-recovery characteristics should not be used when transients will be a problem. The transients of a step-recovery diode will not be reduced by a slowly rising pulse whether from the pulse generator or a low-pass filter.

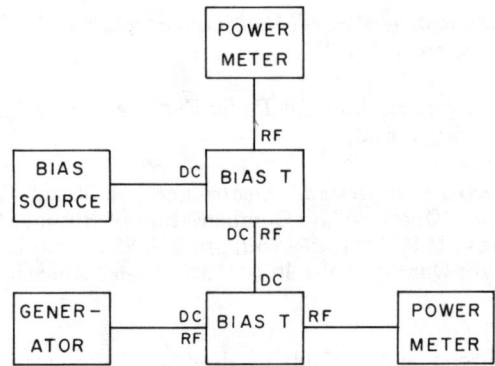

Fig. 3-31 Microwave Circuit for Measuring RF Leakage out the Bias Port

E. Suppression of RF Leakage out of the Bias Port

The design of high RF leakage suppression out of the bias port is straightforward using the filter synthesis methods from the previous sections. Several points warrant special note. The "ON" diode will have the normal microwave voltages and currents across and through it as the input to the bias port filter. The "OFF" diode will have double the voltage or current across it as an input to the bias port filter depending on whether the diode is in series or shunt. The decoupling of the bias port filter is easily measured using two bias "T's" covering the same frequency range as shown in Fig. 3-31. When no diode is in the circuit to add insertion loss the ratio of powers from the power meters gives the decoupling. When a diode is in the circuit the incident power to the diode is used instead of the power transmitted past the T to calculate decoupling.

As a final note, when high decoupling is required a lossy dielectric may be used for the dc bypass capacitor in a coaxial structure. The dielectric losses at low frequencies are practically zero. But by controlling lossy conductor particle size in the dielectric the losses can be quite high at RF. A very short (1 cm) length of coax using Eccosorb R-1 produced more than 100 dB suppression at X-band.

REFERENCES

1. R.V. Pound, "Microwave Mixers" *Rad. Lab. Series*, vol. 16, McGraw-Hill, N.Y., N.Y., 1948, pp. 122-128.

2. H.P. Westman, *Reference Data for Radio Engineers*, IT&T, New York, N.Y., 4th ed., 1950, p. 600.

3. L. Young, "Coaxial Stub Design," *Electronics*, vol. 30 July 1, 1957, p. 188. Also C.E. Muehe, "Quarter-Wave Compensation of Resonant Discontinuities," *IRE Trans.*, MTT-7, April 1959, pp. 296-297. Also L. Young, "Design Chart for Quarter-Wave Stubs," *Microwave Journal*, May 1961, p. 92.

4. R.B. Mouw, "Broadband DC Isolator-Monitors," *Microwave Journal*, vol. 7, November 1964, pp. 75-77.

5. M.M. McDermott and R. Levy, "Very Broadband Coaxial DC Returns Derived by Microwave Filter Synthesis," *Microwave Journal*, vol. 8, February 1965, pp. 33-36.

BIASING CIRCUITS

6. G.L. Matthei, L. Young and E.M.T. Jones, *Microwave Filters Impedance-Matching Networks and Coupling Structures*, McGraw-Hill, N.Y., N.Y., 1964, pp. 991-999.

7. R.V. Garver and T.H. Mak, "Filters for High-Speed Diode Modulators and Demodulators," *IEEE Trans, MTT*, vol. MTT-15, No. 7, July 1967, pp. 390-397.

8. D.K. Adams B.M. Schiffman, and R.B. Larrick, "A Sub-Nanosecond X-Band Pulse Modulator," *1967 IEEE International Microwave Symposium Digest*, pp. 177-179.

9. K.W. Henderson and W.A. Kautz, "Transient Responses of Conventional Filters," *IRE Trans. Circuit Theory*, vol. CT-5, December 1958, pp. 333-347.

10. *The Microwave Engineers Handbook*, Horizon House, Dedham, Mass. 1965, pp. 88-91.

11. W.W. Mumford, "Tables of Stub Admittances for Maximally Flat Filters Using Shorted Quarter-Wave Stubs," *IEEE Trans. Microwave Theory and Techniques*, vol. MTT-13, September 1965, pp. 695-696.

Basic Limitations

Chapter 4

Some of the properties of diode switching elements are limited by the basic properties of the diode rather than the circuit being used. The Q of the diode limits the insertion loss and sometimes the isolation that can be obtained at a given frequency. Q is defined by

$$Q \equiv \frac{\omega_c}{\omega_0} \tag{4-1}$$

in which ω_0 is the frequency of operation and the cutoff frequency ω_c for a varactor diode is defined as the frequency at which the junction reactance is equal to the series resistance:

$$\omega_c \equiv \frac{1}{R_s C_D} \tag{4-2}$$

The insertion loss and isolation bandwidths available from a single diode are limited by the diode parasitics which in turn are limited by the diode cartridge physical dimensions or ultimately by the diode junction physical dimensions. The power handling capacity of a diode switching element is limited by the maximum RF voltages and currents the diode can tolerate. RF current is limited by RF resistance and the capacity of the diode to store or eliminate heat generated in this resistance (diode thermal resistance).

Other limits exist for unique ways in which the switching elements are used. In digital phase shifters the insertion loss and power handling capacity are also a function of phase shift per diode. In a high power limiter the voltage no longer contributes to setting the upper limit on power capacity but it is set by maximum RF current, the coefficient of RF conductivity, and the characteristic impedance of the main transmission line. The switching speed of diodes is limited by the bias circuits as discussed in Chapter III and also limited in how fast carriers may be injected into and removed from the depletion region of the diode.

These limits will be derived for the more common diode switching circuits because the derivation will indicate the adjustment of circuit parameters to give the limit. Comparison will be made to the fundamental limits derived by others to substantiate the fact that the limit has been reached.

A. Limits Imposed by Diode Q

The diode impedances at forward and reverse biases are shown in Fig. 4-1. When the diode is placed in a circuit that optimizes insertion loss and isolation the diode parasitic elements are tuned out by the optimizing circuit.

1. Non-Absorbing Diode Switch

The basic forms of non-absorbing diodes switches are the series diode and shunt diode. The attenuation of a series diode is given by

$$\alpha = 10 \log\left[\left(\frac{R}{2 Z_0} + 1\right)^2 + \left(\frac{X}{2 Z_0}\right)^2\right] \qquad (4\text{-}3)$$

When the diode is tuned to the operating frequency the reactance is tuned out and the attenuation is given by

$$\alpha = 20 \log\left(\frac{R}{2 Z_0} + 1\right) \qquad (4\text{-}4)$$

This tuning is automatically present in Mode 2 operation at the resonant frequency, 5 GHz for the Mode 2 switch shown in Fig. 2-10. Tuning of Mode 1 is possible by putting a capacitor in series with the diode to tune out the diode inductance at forward bias and by putting an inductor in parallel with the diode to tune out the diode capacitance at reverse bias.

When tuned the resistive components given in Fig. 4-1 determine the insertion loss δ and isolation η as follows:

$$\eta = 20 \log\left[\frac{R_s}{2 Z_0}(1 + Q^2) + 1\right] \text{ or } \frac{R_s}{2 Z_0}(1 + Q^2) = 10^{\eta/20} - 1$$
$$(4\text{-}5)$$

$$\delta = 20 \log\left[\frac{R_s}{2 Z_0} + 1\right] \text{ or } \frac{R_s}{2 Z_0} = 10^{\delta/20} - 1 \qquad (4\text{-}6)$$

BASIC LIMITATIONS

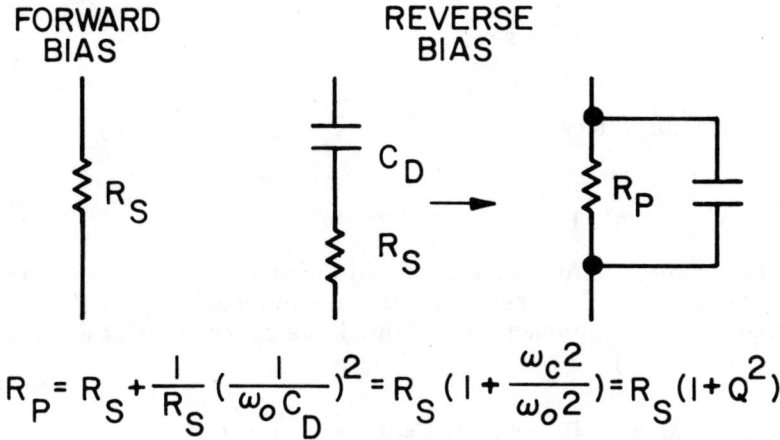

Fig. 4-1 Reverse Biased Diode Resistance

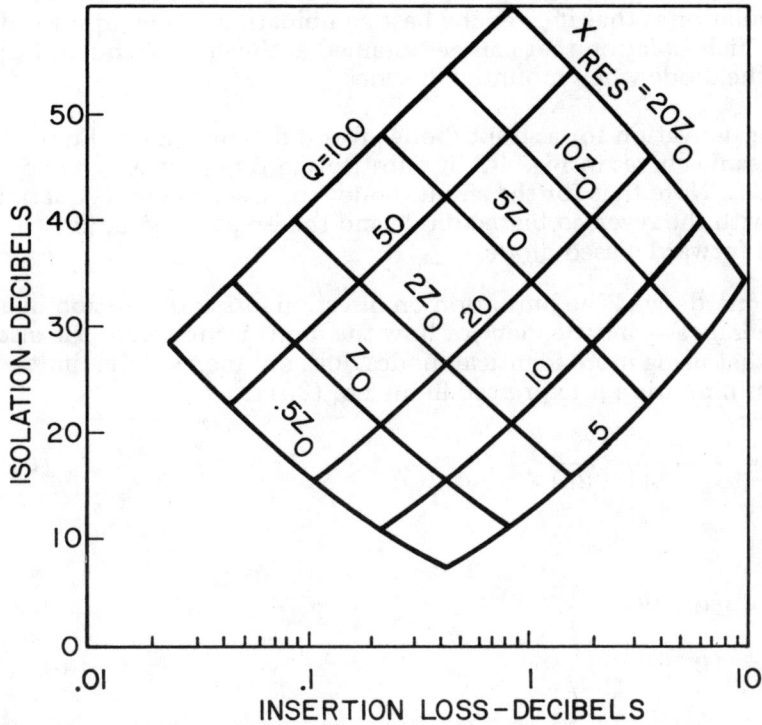

Fig. 4-2 Diode Switching Range Limit Imposed by Diode Q-Non-Absorbing Switch

Dividing Eq. (4-5) by Eq. (4-6) gives

$$1 + Q^2 = \frac{10^{\eta/20} - 1}{10^{\delta/20} - 1} \qquad (4\text{-}7)$$

Solving for Q produces

$$Q = \left[\frac{10^{\eta/20} - 1}{10^{\delta/20} - 1} - 1\right]^{1/2} \qquad (4\text{-}8)$$

The relationship is shown graphically in Fig. 4-2. For a given Q an exchange can be made between insertion loss and isolation. The level of exchange is set by the reactance of the diode capacitance at resonance:

$$X_{RES} = \frac{1}{\omega_0 C_D} \qquad (4\text{-}9)$$

For example assume $R_S = 1\Omega$, $C_D = 1$ pF, $f_0 = 3.18$ GHz, and $Z_0 = 50\Omega$. Then $X_{RES} = Z_0$ and $Q = 50$. Using Eq. (4-6) $\delta = .08$ dB; and using Eq. (4-5) $\eta = 28.3$ dB which agrees with the intersection of the curves for $Q = 50$ and $X_{RES} = Z_0$ in Fig. 4-2. The significance of this calculation is that it gives the best combination of low insertion loss and high isolation that can be obtained at the given frequency by tuning the diode when mounted in series.

A similar derivation for a shunt diode gives the same limit as Eq. (4-8) and the same curves in Fig. 4-2 by substituting B_{RES} for X_{RES} and Y_0 for Z_0. Note that for the shunt diode the insertion loss equation is used with the reversed biased diode and the isolation equation with the forward biased diode.

Kawakami[1] derived the limitation on insertion loss and isolation in a very general case, independent of how the diode is mounted. He also proved that using more identical diodes does not increase the limitation. His limitation is expressed in his Eq. (30) as

$$N = \frac{n(1 + M)}{1 - M(1 - 2n)} \qquad (4\text{-}10)$$

in which

$$\left.\begin{array}{l} N = 10^{-\delta/20} \\ n = 10^{-\eta/20} \\ M = \dfrac{Q^2}{2 + Q^2} \end{array}\right\} \qquad (4\text{-}11)$$

BASIC LIMITATIONS

Substituting Eqs. (4-11) into Eq. (4-10) and solving for Q gives

$$Q = \sqrt{\frac{N-n}{n(1-N)}} \qquad (4\text{-}12)$$

The same equation can be derived by substituting the first two equations (4-11) into Eq. (4-8). Thus the limit expressed in Eq. (4-8) and Fig. 4-2 is very basic for non-absorbing diode switches.

2. Absorbing Diode Switch[2]

To make an absorbing diode switch the diode must be arranged in a circuit that matches the diode to the transmission line in one bias state and is a good reflector in the other bias state. This could be simply accomplished by tuning out the diodes parasitics as for Fig. 4-1 and adjusting Z_0 so that either $Z_0 = R_s$ or $Z_0 = R_p$. When the diode is placed on a reflection output device such as a circulator or 3 dB coupler as in Section II C, it becomes an absorption diode switch. Assuming a perfect match the isolation will be infinite. The attenuation of this type circuit is given by

$$\alpha = 10 \log \left[\frac{\left(\frac{R}{Z_0} + 1\right)^2 + \left(\frac{X}{Z_0}\right)^2}{\left(\frac{R}{Z_0} - 1\right)^2 + \left(\frac{X}{Z_0}\right)^2} \right] \qquad (4\text{-}13)$$

Adjusting Z_0 so that $Z_0 = R_s$ and the diode cartridge parasitics are tuned out then isolation is perfect. Insertion loss is set by the diode impedance at reverse bias which is $R_s - j\frac{1}{\omega C_D}$

$$\delta = 10 \log \left[\frac{\left(\frac{R_s}{R_s} + 1\right)^2 + \left(\frac{1}{\omega C_D R_s}\right)^2}{\left(\frac{R_s}{R_s} - 1\right)^2 + \left(\frac{1}{\omega C_D R_s}\right)^2} \right] \qquad (4\text{-}14)$$

Eq. (4-14) reduces to

$$\delta = 10 \log \left[1 + \left(\frac{2}{Q}\right)^2 \right] \qquad (4\text{-}15)$$

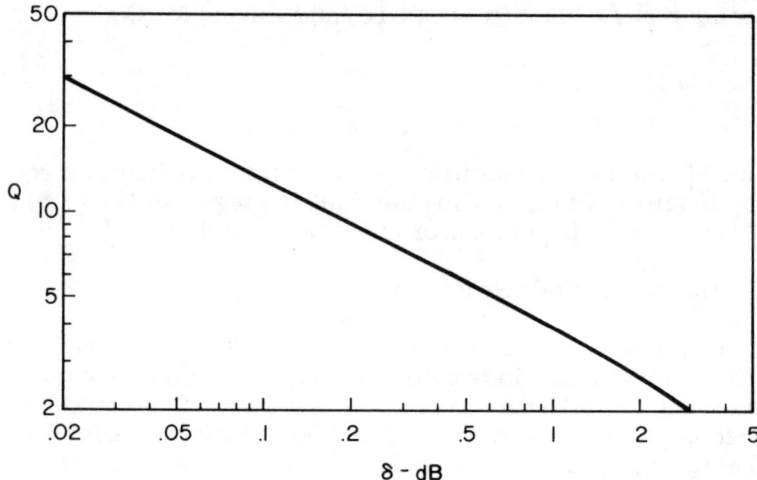

Fig. 4-3 Insertion Loss Limit Imposed by Diode Q-Absorbing Switch

Solving for Q

$$Q = \frac{2}{\sqrt{10^{\delta/10}-1}} \quad (4\text{-}16)$$

This limit is shown in Fig. 4-3. This limit is the same whether the diode was matched to the transmision line at forward bias or at reverse bias. This circuit for making switches is most useful for low Q diodes. For example, to have a switch to go between 0.1 dB and 20 dB a non-absorbing diode switch would require a diode Q of 28 while an absorbing diode switch would require a diode Q of 13.2. Eqs. (4-8) and (4-16) are approximately equivalent when

$$10^{\eta/20}-1 = 2$$
$$\eta = 9.5 \text{ dB} \quad (4\text{-}17)$$

Thus under ideal circumstances (but finite diode Q) the non-absorbing switching circuit provides better switching when $\eta < 9.5$ dB and the absorbing circuit when $\eta > 9.5$ dB. Note however that the absorbing switch circuit is seldom used for a number of reasons: it requires added components such as a circulator or 3 dB coupler, the circulator or 3 dB coupler gives added insertion loss, all incident power is absorbed by the diode in the off state, isolation greater than 20 dB is difficult to obtain over any bandwidth or other variation of para-

BASIC LIMITATIONS

meters, and most diodes do not have impedance levels that are readily matched to the characteristic impedances normally associated with circulators and 3 dB couplers. The absorbing diode switch should be most useful at mm wave frequencies where the Q is low and the waveguide impedance is high. For example matching a 400 ohm waveguide to R_p would require a Q of 20 for $R_s = 1$ ohm.

3. Binary Phase Modulators

There are a number of ways of making binary phase modulators as discussed in Chapter X. Two circuits are most commonly used: the loaded line circuit and the reflection modulator. The contribution of Q to insertion loss will be derived for the reflection modulator and then compared to the more general but less easily derived limitations.

The reflection modulator is a diode terminated circulator. The power reflected from the diode circuit is the output of the circulator. The diode is switched between $R_s + jX_{cMIN}$ at maximum reverse bias and $R_s + j0$ ohms at maximum forward bias, assuming the other reactive diode parasitics have been tuned out. Maximum phase shift is obtained when these two reactance states are equally displaced above the $0 + j0$ point on the Smith Chart as shown in Fig. 4-4. (This is

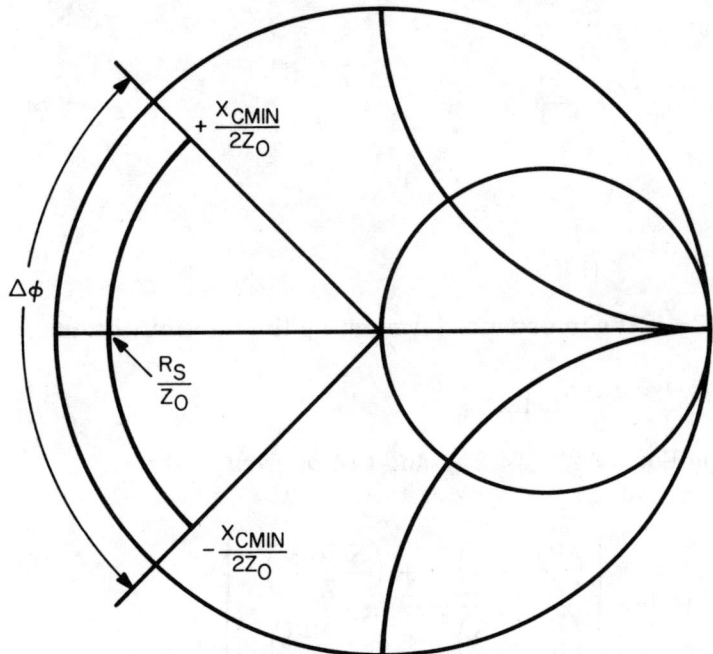

Fig. 4-4 Impedance of Reflection Phase Shifter

easily shown by setting $\frac{\partial^2 X}{\partial \theta^2} = 0$ for $X = Z_0 \tan \theta$.) The RF phase shift is the geometric angle shown in Fig. 4-4 which is twice the impedance angle, e.g., the geometric angle between $0 + j\,0$ and $0 + j\,1$ is $90° = \phi$ while the impedance angle is $45° = \theta$, therefore

$$\frac{\Delta \phi}{2} = 2\theta \qquad \theta = \frac{\Delta \phi}{4} \qquad (4\text{-}18)$$

For low loss the approximation is made that

$$\frac{X_{C\ MIN}}{2\,Z_0} = \tan \theta \qquad (4\text{-}19)$$

Combining Eqs. (4-18) and (4-19) gives

$$\frac{X_{CMIN}}{Z_0} = 2 \tan \left(\frac{\Delta \phi}{4} \right) \qquad (4\text{-}20)$$

Recall that

$$\Gamma = \frac{Z_T/Z_0 - 1}{Z_T/Z_0 + 1} \qquad (4\text{-}21)$$

and

$$\delta = 10 \log \frac{1}{|\Gamma|^2} \qquad (4\text{-}22)$$

in which Z_T is the impedance terminating the transmission line.

$$Z_T = R_s + jX_{CMIN} \qquad (4\text{-}23)$$

Combining Eqs. (4-22), (4-21), and (4-23) gives

$$\delta = 10 \log \left[\frac{\left(\frac{R_s}{Z_0} + 1 \right)^2 + \left(\frac{X_{CMIN}}{Z_0} \right)^2}{\left(\frac{R_s}{Z_0} - 1 \right)^2 + \left(\frac{X_{CMIN}}{Z_0} \right)^2} \right] \qquad (4\text{-}24)$$

BASIC LIMITATIONS

Using

$$Q = \frac{X_{CMIN}}{R_s} = \frac{\omega_c}{\omega_0} \tag{4-25}$$

and Eq. (4-20) provides

$$\delta = 10 \log \left[\frac{\left(1 + \frac{Q}{2} \cot \frac{\Delta\phi}{4}\right)^2 + Q^2}{\left(1 - \frac{Q}{2} \cot \frac{\Delta\phi}{4}\right)^2 + Q^2} \right] \tag{4-26}$$

The limitation is given by the dashed curves in Fig. 4-5. This limitation was derived by Korokawa and Schlosser[3] in a more general circuit without the low-loss approximation and is given by

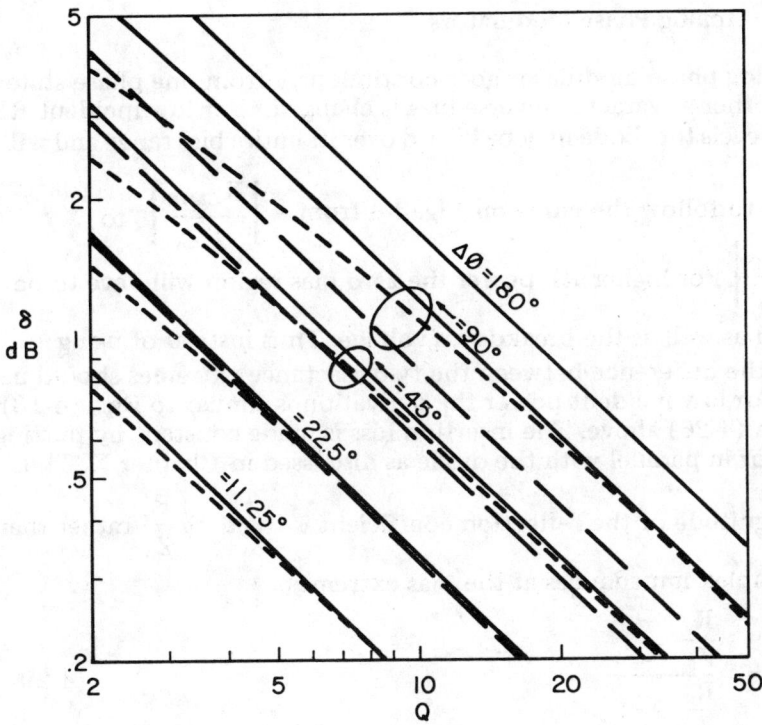

Fig. 4-5 Insertion Loss Limit Imposed by Diode Q-Digital Phase Shifter

$$\delta = 10 \log \left[\frac{1}{1 - \left(\frac{2}{Q} \sin \frac{\Delta\phi}{2}\right)^2 \left(\sqrt{\frac{Q^2}{\sin^2 \frac{\Delta\phi}{2}} + 4} - 2\right)} \right] \quad (4\text{-}27)$$

which is given by the solid curves in Fig. 4-5. Hines[4] derived an expression for this insertion loss limitation which assumes low losses and is given by

$$\delta = 10 \log \left[1 + \frac{4}{Q} \sin \frac{\Delta\phi}{2} \right] \quad (4\text{-}28)$$

which is shown by the dotted curves in Fig. 4-5.

All of the above limitations on binary phase modulators are based on having equal loss in both phase states.

4. Analog Phase Modulators

An analog phase modulator goes continuously from one phase state to the other as varactor reverse bias is changed. For low incident RF power levels the diode may be biased over its entire bias range and will be able to follow the curve on Fig. 4-4 from $-\left|\frac{X_{CMIN}}{2Z_0}\right|$ to $+\left|\frac{X_{CMIN}}{2Z_0}\right|$. For higher RF power the zero bias region will have to be avoided as well as the breakdown voltage. Thus instead of using X_{CMIN} the difference between the two reactance extremes should be used. For low incident power the derivation is similar to Eqs. (4-18) through (4-26) above. The insertion loss is made constant by putting a resistor in parallel with the diode as discussed in Chapter X. Thus the magnitude of the reflection coefficient is fixed by $\frac{R_s}{Z_0}$ rather than the complex impedances at the bias extremes.

$$\Gamma = \frac{\frac{R_s}{Z_0} - 1}{\frac{R_s}{Z_0} + 1} \quad (4\text{-}29)$$

Using Eqs. (4-22), (4-25), (4-29) and (4-20), the insertion loss is given by

BASIC LIMITATIONS

$$\delta = 20 \log \left[\frac{1 + \frac{Q}{2} \cot \frac{\Delta\phi}{4}}{1 - \frac{Q}{2} \cot \frac{\Delta\phi}{4}} \right] \quad (4\text{-}30)$$

which is shown in Fig. 4-6. Note in Chapter X that when the diode circuit is adjusted to linearize the voltage-phase relationship the insertion loss is higher than given by Eq. (4-30).

It has been shown that diode Q determines phase modulator insertion loss and switching insertion loss and isolation. Thus the cutoff frequency f_c of diodes used in control devices is an important parameter. The cutoff frequency of varactor diodes is simply calculated using Eq. (4-2). The calculation is not so simple with PIN diodes because their R_s at forward bias is usually different than at reverse bias.

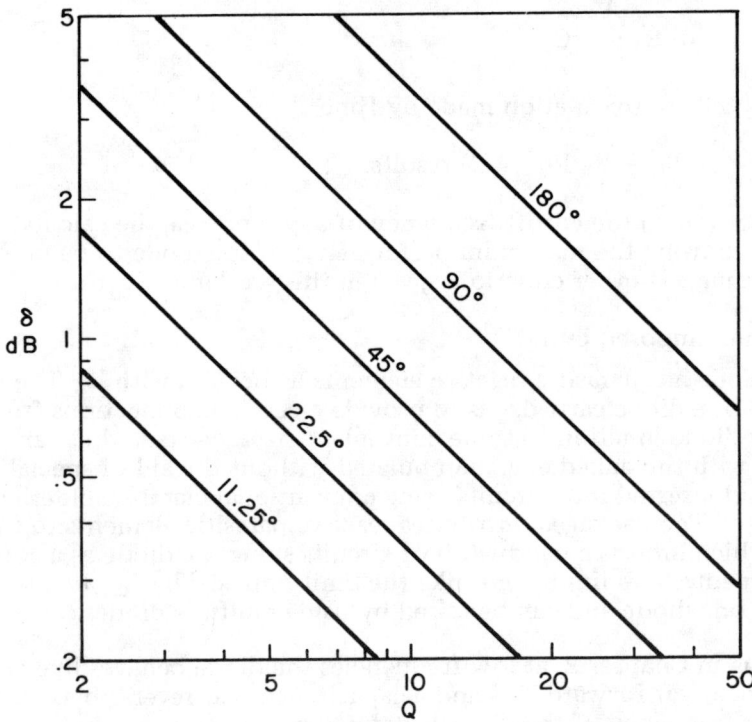

Fig. 4-6 Insertion Loss Limit Imposed by Diode Q-Analog Phase Shifter

5. Calculation of Diode Q

Kurokawa and Schlosser[3] have derived an expression that gives diode Q from its measured impedances in its two switching states independent of the circuit in which it is embedded, so long as the imbedding circuit is lossless.

$$Q = \frac{\omega_c}{\omega_0} = \frac{\sqrt{(R_1 - R_2)^2 + (X_1 - X_2)^2}}{\sqrt{R_1 R_2}} \qquad (4\text{-}31)$$

in which $R_1 + jX_1$ and $R_2 + jX_2$ are the measured impedances in the two switching states. In the special case where the imbedding circuit and diode cartridge parasitics are stripped away and where the subscript 1 corresponds to forward bias and 2 to reverse bias then $X_1 = 0$, $X_2 = \frac{1}{\omega_0 C_D}$, $(R_1 - R_2)^2 \ll (X_2)^2$ and

$$\omega_c = \frac{1}{\sqrt{R_1 R_2}\ C_D} \qquad (4\text{-}32)$$

which is the approximation made by Hines.[4]

When $R_1 = R_2 = R_s$ Eq. (4-2) results.

Using Eq. (4-31) the cutoff frequency of any diode can be calculated exactly allowing the maximum performance of the diode to be predicted using the other equations given in this section.

B. Limits Imposed by f_R

Every diode has parasitic reactive elements associated with it. The purpose of a diode cartridge is to provide electrical connections from the tiny diode junction to some convenient sized package that can be seen with the naked eye, manipulated without the aid of special tools, and inserted into circuits using conventional clamps, soldering irons, etc. The package contributes reactive parasitic elements to the diode which limit the bandwidth of circuits suing the diode as a control element. This limit is not like the limit imposed by f_c. It is the limit of one diode and can be raised by using multiple diodes.

As shown in Chapter 2, at low frequencies the diode behaves like an inductance L at forward bias and a capacitance C at reverse bias. The resonant frequency of the diode is defined by

$$\omega^2_R = \frac{1}{LC} \qquad (4\text{-}33)$$

BASIC LIMITATIONS

1. Bandwidth Limit

Referring to Fig. 4-7, the forward biased diode insertion loss can be tuned to any design frequency by a series capacitor C_T. The reactance is given by

$$X = \omega L - \frac{1}{\omega C_T} \tag{4-34}$$

In order for the insertion loss of a series diode to be less than some maximum across some band of frequencies, the forward biased reactance must not exceed some maximum X_{MAX} across the band. At the low frequency ω_1 the reactance is capacitive and thus negative.

$$\omega_1 L - \frac{1}{\omega_1 C_T} = -X_{MAX} \tag{4-35}$$

At high frequency ω_2 the reactance is positive

$$\omega_2 L - \frac{1}{\omega_2 C_T} = X_{MAX} \tag{4-36}$$

Solving both equations for ω gives

$$\omega_2 = \frac{X_{MAX}}{2L} + \sqrt{\left(\frac{X_{MAX}}{2L}\right)^2 + \frac{1}{LC_T}} \tag{4-37}$$

$$\omega_1 = -\frac{X_{MAX}}{2L} + \sqrt{\left(\frac{X_{MAX}}{2L}\right)^2 + \frac{1}{LC_T}} \tag{4-38}$$

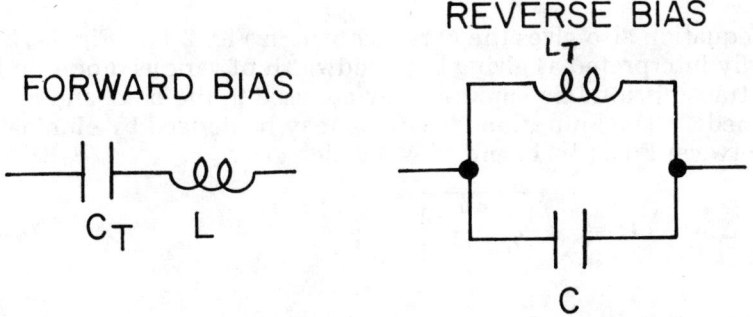

Fig. 4-7   Tuning the Diode for Low Insertion Loss and High Isolation

The difference in frequency $\Delta\omega$ is given by

$$\Delta\omega = \omega_2 - \omega_1 = \frac{X_{MAX}}{L} \qquad (4\text{-}39)$$

Thus

$$X_{MAX} = \Delta\omega L \qquad (4\text{-}40)$$

Substituting this into the series attenuation equation gives

$$\delta = 10 \log \left[1 + \left(\frac{\Delta\omega L}{2 Z_0} \right)^2 \right] \qquad (4\text{-}41)$$

assuming we are slightly off resonance so that the reactive component of impedance of the diode determines the insertion loss rather than the resistive component.

This equation is the same as that used to generate the insertion loss curves in Fig. 2-12. The baseband mode of operation can be visualized as the tuned mode where the reactance (susceptance) of the series (shunt) tuning element has been allowed to go to zero.

A similar derivation for the reverse biased diode beginning with

$$X = \frac{1}{\omega C - \frac{1}{\omega L_T}} \qquad (4\text{-}42)$$

and X_{MIN} gives

$$\eta = 10 \log \left[1 + \left(\frac{1}{\Delta\omega C 2 Z_0} \right)^2 \right] \qquad (4\text{-}43)$$

This equation also gives the curves shown in Fig. 2-12. Fig. 2-12 is broadly interpreted as giving the bandwidth of various diodes in 50 ohm transmission line whether they are used in the baseband mode or tuned. The limitation free of Z_0 may be derived by eliminating Z_0 between Eqs. (4-41) and (4-43) which gives

$$(\Delta\omega)^2 LC = \sqrt{\frac{10^{\delta/10} - 1}{10^{\eta/10} - 1}} \qquad (4\text{-}44)$$

Using the definition of ω_R given in Eq. (4-33) then Eq. (4-44) becomes

BASIC LIMITATIONS

$$\frac{\Delta \omega}{\omega_R} = \left[\frac{10^{\delta/10} - 1}{10^{\eta/10} - 1} \right]^{1/4} \tag{4-45}$$

A similar derivation for a shunt diode gives the same results. This result is shown in Fig. 4-8. Improved switching bandwidth for a single diode can be obtained by increasing ω_R. For example, if a switch were required giving 1 dB maximum insertion loss and 20 dB minimum isolation, then the resonant frequency of the diode would have to be about 5 times larger than the desired bandwidth.

The limit on the absorbing diode switch used with a circulator is different than Eq. (4-45) above. The insertion loss tends to be reduced slightly away from the design frequency because the match to the diode becomes even worse. The highest insertion loss is given by

$$\delta = 20 \log \left(1 + \frac{2}{Q^2} \right) \tag{4-46}$$

which is obtained by having R_p from Fig. 4-1 terminate a line of characteristic impedance R_s. For $Q \geqslant 3$ Eq. (4-46) is approximately

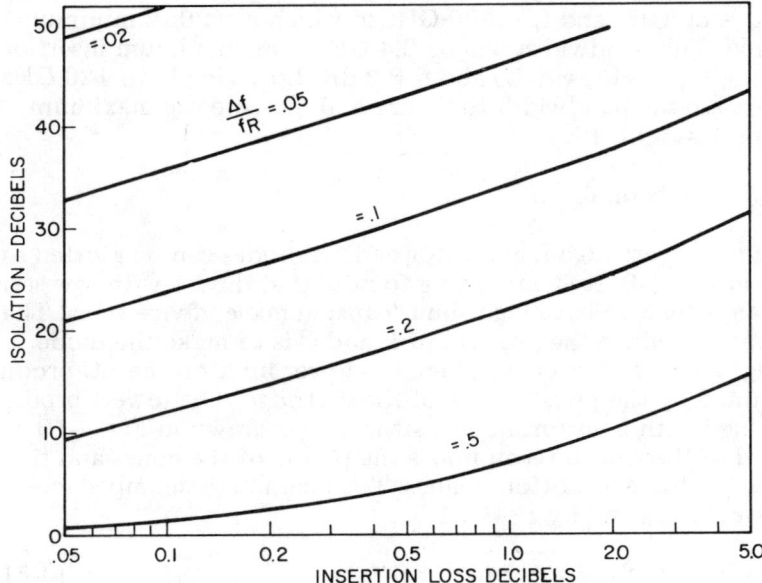

Fig. 4-8 Bandwidth Limit Imposed by Diode Resonance

equivalent to Eq. (4-15). The isolation bandwidth however is limited. The isolation is reduced by reactance in series with R_s. Thus $X_{MAX} = \Delta\omega L$ will dominate the bandwidth. Using $Z_0 = R_s$, and $X = \Delta\omega L$ in Eq. (4-13) the following isolation results

$$\eta = 10 \log \left[1 + \left(\frac{2 R_s}{\Delta \omega L}\right)^2\right] \tag{4-47}$$

From equation (4-33) Eq. (4-48)

$$\frac{1}{L} = \omega^2_R C \tag{4-48}$$

can be derived. Assuming $C = C_D$, then

$$\eta = 10 \log \left[1 + \left(\frac{2\omega^2_R}{\Delta\omega \, \omega_c}\right)^2\right] \tag{4-49}$$

which leads to

$$\Delta\omega = \frac{2\omega^2_R}{\omega_c \sqrt{10^{\eta/10} - 1}} \tag{4-50}$$

Thus ω_c influences the bandwidth so that higher ω_c gives smaller bandwidth. As an example consider a switch operating at 30 GHz with $f_R = 60$ GHz and $f_c = 300$ GHz in which $\eta = 20$dB minimum is desired. The bandwidth will be 2.4 GHz. The maximum insertion loss using Eq. (4-46) will be about 0.2 dB. Lowering f_c to 120 GHz will increase the bandwidth to 6 GHz and increase the maximum insertion loss to 1 dB.

2. Limits on f_R

The parasitic cartridge L and C for various diodes can be plotted as shown in Fig. 4-9. It is interesting to note that diodes with low series inductance tend to have high shunt capacitance and vice versa. The only way to reduce the product of L and C is to make the diode cartridge smaller physically. There is a lower limit on the LC product determined by the physical size of the cartridge. The lowest product is obtained with a uniform conic structure as shown in Fig. 4-10. One end of the transmission line is the points of the cones and the other is the top and bottom planes. The characteristic impedance of cartridge Z_0 is given by

$$Z_0 = 120 \ln [\sqrt{1 + (h/d)^2} + (h/d)] \tag{4-51}$$

which is shown in Fig. 4-11.

BASIC LIMITATIONS

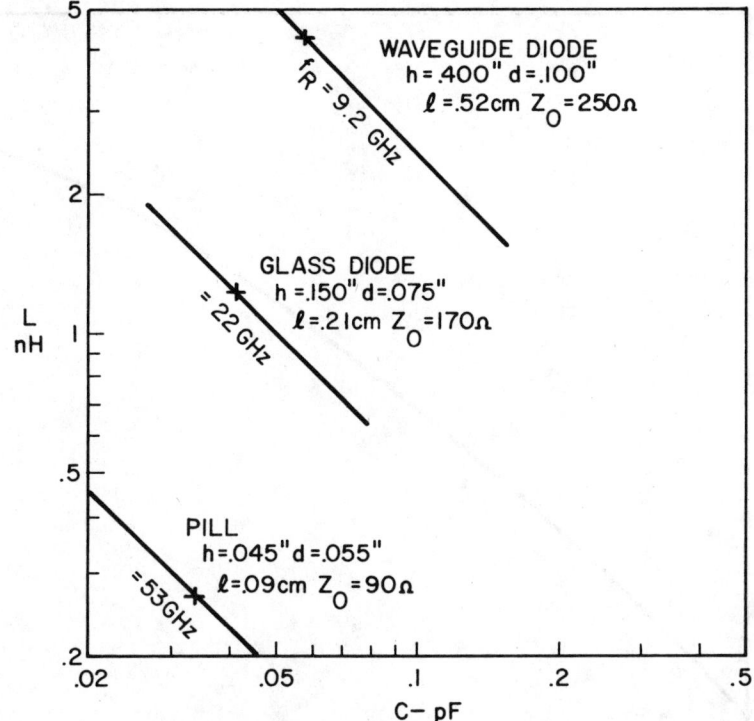

Fig. 4-9   Physical Limits on Diode Resonance

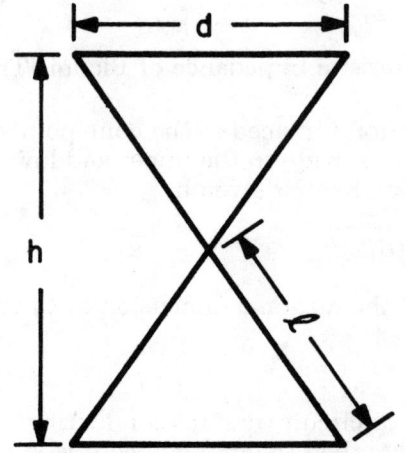

Fig. 4-10   Ideal Diode Cartridge

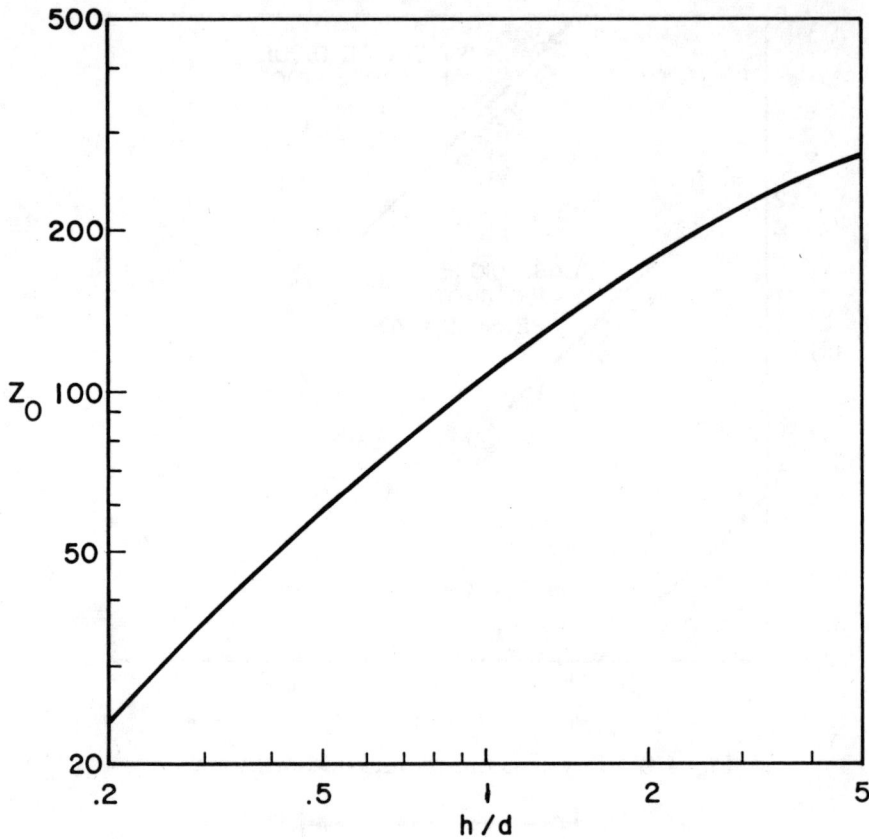

Fig. 4-11 Characteristic Impedance of Biconic Transmission Lines

The tiny diode junction is placed at the cone points and convenient electrical connection is made to the upper and lower planes. The length of transmission line ℓ is given by

$$\ell = \sqrt{(h/2)^2 + (d/2)^2} \qquad (4\text{-}53)$$

which is just half of the diagonal dimension of the package.

Assuming that a short circuit (perfect conducting diode) is placed across the cone points then the inductance L would be that of a short length of short circuit terminated transmission line given by

$$L = Z_0 \, \ell/c \qquad (4\text{-}55)$$

BASIC LIMITATIONS

in which $c = 3 \times 10^{10}$ cm/sec, the speed of light in free space (ℓ is in cm.) Assuming that an open circuit (perfect non-conducting diode) is placed across the cone points, then the capacitance C would be that of a short length of open circuit terminated transmission line given by

$$C = \ell/(Z_0 c). \tag{4-56}$$

Normally a diode cartridge has glass or ceramic holding the cones together which gives a velocity of light slower than in free space.

Reducing this velocity increases both L and C. Most diodes also use either a spring wire or some wire stitch-bonding to the junction which gives more inductance than a cone structure. Thus the cone structure in free space is the limit that can only be approached by real cartridges.

Using Eqs. (4-55) and (4-56) the limitation on ω_R is given by

$$\omega_R = \frac{1}{LC} = \frac{c}{\ell} \tag{4-57}$$

This limit was used to give the lines shown in Fig. 4-9. The limit is therefore stated

$$\omega_R = \frac{2c}{\sqrt{h^2 + d^2}} \tag{4-58}$$

This limit was used to construct Fig. 4-12. As the cartridge becomes very small it becomes possible to make very wide-band diode control elements.

The ultimate in size reduction is the physical structure of the diode junction. The capacitance of a Silicon ($\epsilon_R = 13$) junction is given by

$$C_j = .225\,(13)\,\frac{\pi d^2}{4h}\,\text{pF} \tag{4-59}$$

in which d is junction diameter in inches and h is height. To simplify inductance calculations the conducting junction is considered to be the center conductor of a coaxial transmission line having a 1 inch diameter outer conductor. Then

$$Z_0 = 60 \ln \frac{1}{d} \tag{4-60}$$

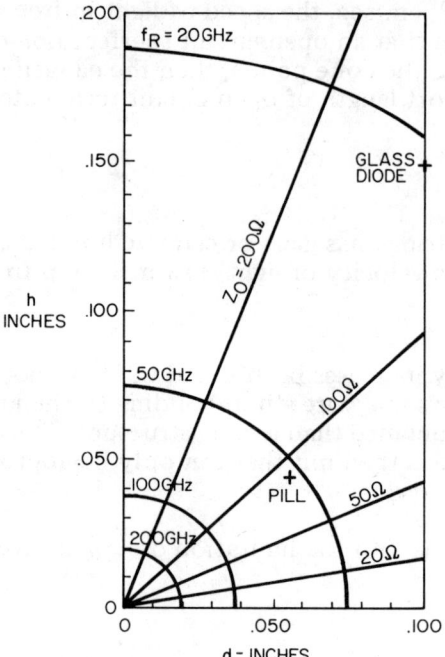

Fig. 4-12 Resonance and Impedance Limiting Properties of Diode Cartridges

and

$$L = .8h \ln \frac{1}{d} \text{ nH} \qquad (4\text{-}61)$$

These limits are shown in Fig. 4-13. A typical PIN junction will be about .0015 inches high and .005 inches in diameter giving an f_R of about 300 GHz. Fig. 4-13 gives an idea of the high capability of diode control elements when used in integrated circuits where no cartridge is needed, but where special micromanipulators and welding equipment are required.

For fully integrated circuits no limitation exists. The diode can be made as one long junction with the junction capacitance comprising the distributed capacitance of the transmission line. When the diode is biased into conduction the center conductor is shorted out over a significant length. The structure would have no reactive elements limiting its bandwidth. But as a practical matter losses of microstrip transmission lines on silicon substrates are prohibitively high.

BASIC LIMITATIONS

All of the limits imposed by f_R have been based on the assumption that one diode is used and that all tuning has been done in the same electrical plane as the diode and accomplished by adding elements. These limitations imposed by f_R can be exceeded by using multiple diodes or other elements not in the same electrical plane as will be discussed in Chapter 6. The limit holds significance not as a limit but as a reference which is used for comparing the improvement possible with more complex structures. The limit can also be exceeded by a single diode by removing some of the distributed capacitance from the transmission line in the area of a shunt diode, or by increasing the characteristic impedance of the transmission line in the region of a series diode as discussed in Chapter 3.

C. Power Limitations

Different factors limit the incident peak power $\hat{P}_i$ and incident average power $\overline{P}_i$ that a diode can control. The $\hat{P}_i$ is determined by the diode breakdown voltage E_B, the characteristic impedance of the transmission line in which it is mounted Z_o, and circuit used

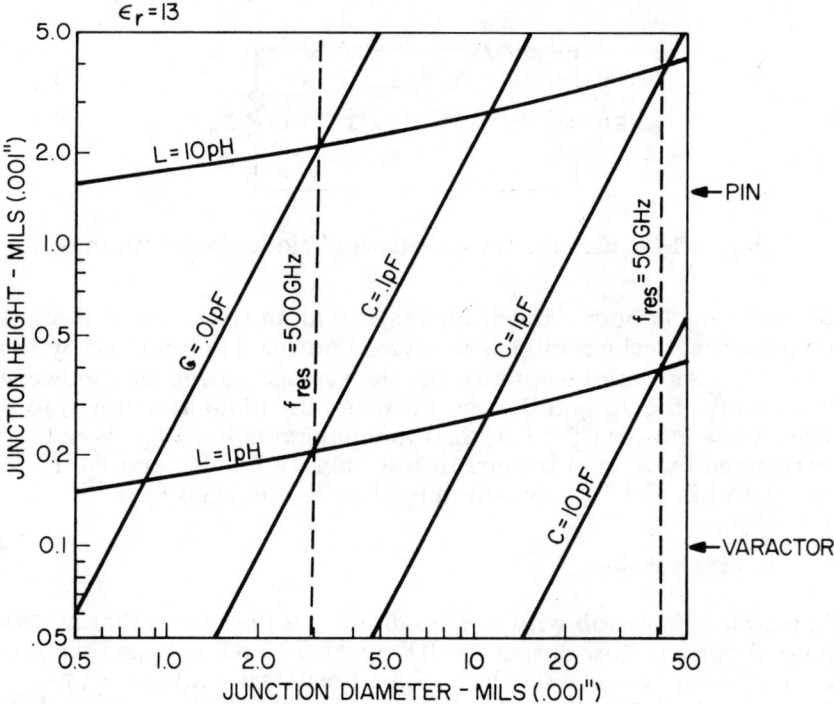

Fig. 4-13 Resonant Frequency Limits of Diode Junctions

MAXIMUM PEAK POWER

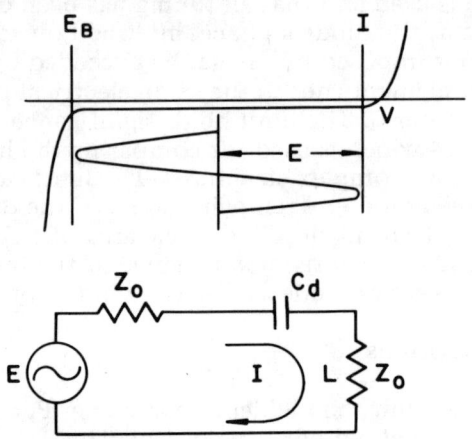

MAXIMUM AVERAGE POWER

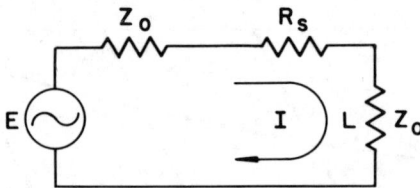

Fig. 4-14 Circuits for Calculating Diode Power Ratings

for switching (whether the diode is series mounted or shunt mounted for most practical circuits). The average power is determined by the forward biased series resistance R_s, the average power the diode can dissipate $\overline{P_D}$, the Z_o and the circuit. When the diode is required to control c.w. power ($\hat{P_i} = \overline{P_i}$), then the maximum limit P_{cw} is not determined by Z_o and the circuit but only by E_B, R_s, and $\overline{P_D}$. (The circuit and Z_o for P_{cw} are determined by E_B, R_s, and $\overline{P_D}$.)

1. Peak Power

To maintain isolation with a series diode it is important that no conduction current flow during the RF cycle. The RF voltage thus must be positioned between 0 volts and the breakdown voltage E_B as shown in Fig. 4-14. When the diode is providing high isolation all of the generator voltage appears across the high impedance of C_D,

BASIC LIMITATIONS

which is the reverse biased diode. When the diode is biased to $E_B/2$, as shown in Fig. 4-14, the magnitude of the generator voltage may be as large as $E_B/2$.

Using Eq. (2-7) the incident power is given by

$$P_i = \frac{E^2}{8 Z_0} \qquad (4\text{-}62)$$

which gives a maximum incident peak power P_i for a series diode of

$$\hat{P_i} = \frac{E_B^2}{32 Z_0} \qquad (4\text{-}63)$$

A shunt diode in the low loss state will have half of the generator voltage dropped across it; thus the maximum incident peak power $\hat{P_i}'$ for a shunt diode is given by

$$\hat{P_i}' = 4\hat{P_i} = \frac{E_B^2}{8 Z_0} \qquad (4\text{-}64)$$

Eqs. (4-63) and (4-64) were used to generate the peak power curves given as an example in Fig. 4-15. Note that lowering Z_0 increases the peak power handling capacity of the diode in both circuits.

Experience has shown that this limitation is more severe than normally required for PIN diodes. Normally the high voltage $E_B/2$ is not available in systems and when back biasing to $E_B/2$ is required it results in a significant increase in circuit complexity and system power drain. Because of the slow response of PIN diode junctions to RF currents, it is common practice to back bias PIN diodes only enough to hold off rectified current (10V-100V back bias). In this manner the RF voltage may go as high as E_B with no significant DC currents being induced and the peak power limit is higher than Eqs. (4-63) and (4-64) by a factor of 4 as shown on Fig. 4-15 "expanded peak power limit."

When bias is at less than $E_B/2$ then the PIN diode exhibits some nonlinearities contributing to harmonic generation[6,7], cross-modulation products[8], and slightly greater RF losses. When these effects will not deteriorate system performance then operation at less then $E_B/2$ is desirable.

The limitation on $\hat{P}$ is not a hard limit. If it is exceeded the diode is not destroyed unless the losses in the diode cause too much pulse

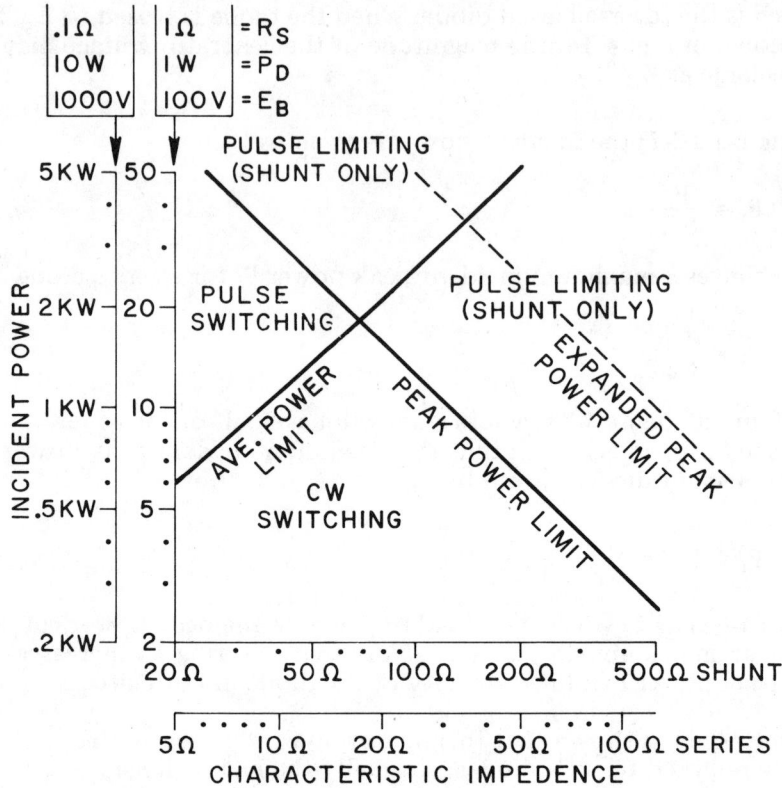

Fig. 4-15 Power Ratings of Typical Diodes

energy to be absorbed. Instead the isolation begins to deteriorate. Experience in X-band waveguide with the 1N263 ($\hat{P}$ = 2mW) used as a switch showed isolation reduced to 10dB (from 25dB) for $\hat{P}$ = 20 mW. A second diode placed after the first gave 25 dB isolation (only 2 mW incident on it) for a total isolation of 35-40dB. Even though $\hat{P}$ is exceeded, significant isolation can be obtained using multiple diodes.

2. Average Power

The current shown in the bottom of Fig. 4-14 is given by

$$I = E/(R_s + 2 Z_0) \qquad (4\text{-}65)$$

The power dissipated in the diode $\overline{P}_D$ is given by

BASIC LIMITATIONS

$$\overline{P}_D = \tfrac{1}{2} I^2 \, R_s = \frac{E^2 \, R_s}{2 \, (R_s + 2 \, Z_0)^2} \tag{4-66}$$

(Refer to the end of this section and to Appendix A for more detail on $\overline{P}_D$.)

Solving for E^2

$$E^2 = 2 \, (R_s + 2 \, Z_0)^2 \, \overline{P}_D / R_s \tag{4-67}$$

Recalling from Eq. (1-7) the incident power is given by

$$P_i = \frac{E^2}{8 \, Z_0} \tag{4-68}$$

Combining Eqs. (4-67) and (4-68) gives the average incident power $\overline{P}_i$ the series diode can control:

$$\overline{P}_i = \frac{\overline{P}_D \, Z_0}{R_s} \left(1 + \frac{R_s}{2 \, Z_0} \right)^2 \tag{4-69}$$

Substituting G_s for R_s and Y_0 for Z_0 gives the average incident power $\overline{P}_i'$ a shunt diode can control:

$$\overline{P}_i' = \frac{\overline{P}_D \, Y_0}{G_s} \left(1 + \frac{G_s}{2 \, Y_0} \right)^2 = \frac{\overline{P}_D \, Z_0}{4 \, R_s} \left(1 + \frac{2 \, R_s}{Z_0} \right)^2 \tag{4-70}$$

Eqs. (4-69) and (4-70) were used to generate the average power curves given as an example in Fig. 4-15. Note that increasing Z_0 increases the average power handling capacity of the diode in both circuits. It is assumed in Fig. 4-15 that the diode is lossless in the reverse bias state. The losses were normally much lower than in the forward bias state; however, when $Z_0 > Q R_s$ reverse bias loss begins to dominate and Eqs. (4-69) and (4-70) must be modified, substituting R_p (Fig. 4-1) for R_s.

Various regions are defined by these curves in Fig. 4-15. Below both curves the diode can control both the average power and peak power and therefore can control C.W. power. The shunt circuit allows Z_0 to be higher by a factor of 4 for the same power ratings. The maximum C.W. power the diode can control is the same whether the diode is in series or shunt. This point is discussed in the next section at length. To the left of both curves the diode can control pulsed RF. The specification of peak and average power permits the most satisfactory Z_0 to be calculated noting that the C.W. maximum rating must be able to satisfy the equation

$$\overline{P}_{CW} = \sqrt{\hat{P}_i \overline{P}_i} \qquad (4\text{-}71)$$

The upper right hand region of Fig. 4-15 is useless for a series diode in most conventional applications because isolation deteriorates significantly. The only applications where a diode could be used in that region are as expanders such as in performing the ATR function in a radar. The upper right hand region for a shunt diode is quite useful when the diode is used to perform the limiter function. The switch is in the isolation state every time it has the peak power incident to it and only has to satisfy the average power limitation.

Some PIN diodes have an extremely low R_s, in which case the tendency is to bias them with a very high current to obtain a very low total resistance and thus control very large average power. But the bias current causes bias power to be dissipated in the diode and use up some of the power dissipating capacity of the diode. How much bias current should be used to obtain maximum incident RF average power capacity? What is the power limitation of the diode under these circumstances?

When R_s is very small Eq. (4-70) reduces to

$$\overline{P}_i' = \overline{P}_D \left(\frac{Z_0}{4 R_s} \right) \qquad (4\text{-}72)$$

Power due to bias current $\overline{P}_B$ is subtracted from $\overline{P}_D$ the difference in the limit to be caused by RF power. Thus Eq. (4-72) becomes

$$\overline{P}_i' = (\overline{P}_D - \overline{P}_B) \left(\frac{Z_0}{4 R_s} \right) \qquad (4\text{-}73)$$

When the normal R_s is zero the total resistance is due to conductivity modulation in the I region giving junction resistance R_j

$$R_j = \frac{1}{\alpha_{RF} I_B} \qquad (4\text{-}74)$$

in which I_B is bias current in amps and α_{RF} is the coefficient of conductivity modulation from Chapter 2. Eq. (4-73) becomes

BASIC LIMITATIONS

$$\overline{P}_i' = (\overline{P}_D - \overline{P}_B) \left(\frac{Z_0 \alpha_{RF} I_B}{4} \right) \tag{4-75}$$

The bias power may be simply related to the bias current because the voltage across a PIN diode does not change rapidly with current. In Chapter 2 the DC current is given by

$$I_{DC} = I_0 \, (\epsilon^{\alpha_{DC} V_j} - 1) \tag{4-76}$$

in which $\alpha_{DC} = 20$ for PIN diodes. A decade change in current causes .115 volts change in junction voltage. The junction voltage will be about 1 volt for large bias current, thus

$$\overline{P}_B = V_j I_B = I_B \tag{4-77}$$

Eq. (4-75) becomes

$$\overline{P}_i = (\overline{P}_D - I_B) \left(\frac{Z_0 \alpha_{RF} I_B}{4} \right) \tag{4-78}$$

To find the bias current that gives maximum $\overline{P}_i'$ set

$$\frac{\partial \overline{P}_i'}{\partial I_B} = 0 = (\overline{P}_D - I_B) \frac{\alpha_{RF} Z_0}{4} - \frac{\alpha_{RF} Z_0 I_B}{4} \tag{4-79}$$

which gives

$$I_B = \overline{P}_D / 2 \tag{4-80}$$

In other words, half of the $\overline{P}_D$ of the diode should be used for bias and half for RF power. It is interesting to note that bias power is dissipated in the PI and IN junctions and RF power is dissipated in the I region.

Substituting Eq. (4-80) into Eq. (4-78) produces the limit

$$\overline{P}_i' = \frac{1}{16} \alpha_{RF} Z_0 \overline{P}_D^2 \tag{4-81}$$

A fused glass pill diode having $\alpha_{RF} = 16$, $P_D = 10W$ working in $Z_0 = 50$ ohm transmission line will be able to control an average incident power of 5000 W when operating as a limiter. The diode will require 5 watts or about 5 Amps bias. The incident RF peak power may be any value just so the energy of short pulses does not cause too much heating before the power can be dissipated. Diode limiters have pulse power limitations similar to that shown in Fig.

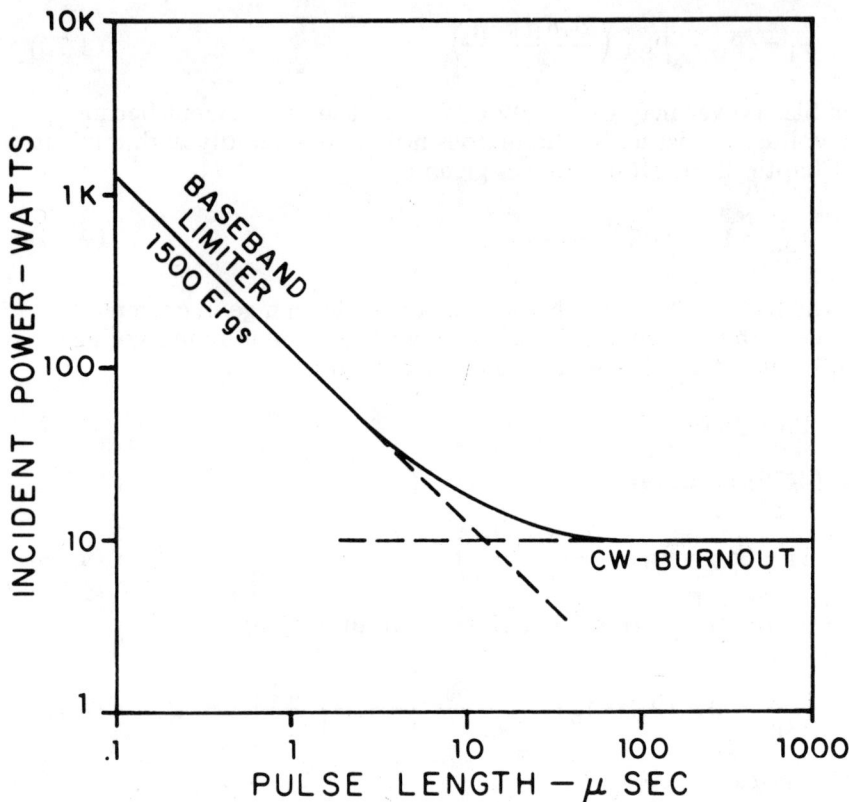

Fig. 4-16 Typical Pulse Power Absorption Limitation of Diode Switch or Limiter

4-16. For short length pulses the heat energy does not have time to leave the I region and the amount of heating is determined by the heat capacity of the I region and the maximum temperature that will not permanently change the I region properties. For short enough pulses and an .001 duty ratio the single fused glass diode should be able to limit 5 MW. Diodes can be placed in parallel to increase power capacity. Three diodes will have three times the dissipation capacity of one diode and using Eq. (4-81) will be able to control 9 times the RF power of one diode. Thus three fused glass PIN diodes should be able to control about 5 MW at an .01 duty ratio.

The pulse energy that the I region can absorb can be estimated. The specific heat of Silicon is 0.2 calories per gram-degree centi-

BASIC LIMITATIONS

grade. The density, 2.6 grams per cm^3. One Joule is 1 watt-second or 10^7 Ergs. One calorie is 4.186 Joules. For 100° temperature rise in the I region the junction can absorb 220 watt-seconds of energy per cm^3 or 3.5 W-μsec per mill^3. A typical fused glass junction is .030" in diameter and .004" high which can absorb 10,000 W-μsec for 100° temperature rise. Since half of the absorbed power capacity will be used by the bias, this is also halved to 5,000 W-μsec. As calculated above one of these diodes controlled 5,000 watts average power while dissipating 5 watts. Thus, using the same ratio of controlled to dissipated power (power control ratio) the diode could withstand RF pulses having 5,000,000 W-μsec energy. A figure like 4-16 could be made giving cw burnout at 5 KW, a 1 μsec pulse length power of 5 MW and a time constant of 1 msec (intersection of the two straight lines). As another example a typical packaged PIN diode has a junction diameter of .004" and a height of .0015". Using the same arguments as above the junction can withstand 35 W-μsec. Estimating the power control ratio to be 100 (which is typical of these smaller junction diodes) 3500 W-μsec pulse energy can be controlled by the one diode.

The average power a diode can dissipate, $\overline{P}_D$, is determined by the thermal resistance from the junction to a heat sink, θ, and the temperature above the heat sink that will deteriorate the diode, ΔT. A stout fuzed glass pill has $\theta = 2°$C/watt and can easily tolerate 200°C above a heat sink of 25°C. Thus a $\overline{P}_D$ is obtained of $\Delta T/\theta = 100$ watts. The heat sink must be removing this power and staying at 25°C. The smaller fused glass pill has $\theta = 18°$C/watt and will be calculated at $\Delta T = 100°$C which will not melt solder and allow the heat sink some temperature rise. Then $\overline{P}_D$ is 5.5 watts. A glass whisker type diode has $\theta = 200°$C/watt giving ½ watt for $\Delta T = 100°$C. Normally the thermal resistance from diode to heat sink is added to θ to determine the $\overline{P}_D$ of the diode in the mount.

3. C.W. Power

As was seen in Fig. 4-15 the maximum power the diode could control as a switch was independent of whether the diode was mounted in series or in shunt. For the series diode, Eq. (4-69) is modified by dropping the $R_s/(2Z_0)$ term as being negligible; thus,

$$\overline{P}_i = \frac{\overline{P}_D Z_0}{R_s} \qquad (4\text{-}82)$$

Using Eq. (4-63) setting

$$\widehat{P}_i = \overline{P}_i = \overline{P}_{CW} \qquad (4\text{-}83)$$

and solving for Z_0 gives

$$Z_0 = E_B \sqrt{\frac{R_s}{32\overline{P}_D}} \tag{4-84}$$

Substituting this into Eq. (4-63) or Eq. (4-82) gives $\overline{P}_{CW}$

$$\overline{P}_{CW} = E_B \sqrt{\frac{P_D}{32 R_s}} \tag{4-85}$$

The same result is obtained by using Eqs. (4-64) and (4-70) for a shunt diode.

Hines[4] has derived a very general relationship for cw switching that produces Eq. (4-85), as well as another relating to digital phase shifting.

$$\overline{P}_{CW} = \frac{\tfrac{1}{2} I_{sc} V_{so}}{|\Gamma_{so} - \Gamma_{sc}|} \tag{4-86}$$

Γ_{so} is the reflection coefficient of the diode and load when the diode is non-conducting. Γ_{sc} is the reflection coefficient with the diode conducting. I_{sc} is the maximum rms current through the diode in the conduction state; and V_{so} is the maximum rms voltage the diode can hold off in the non-conduction state. In the case of a single-pole, single-through switch $\Gamma = 1$ in one switching state and $\Gamma = 0$ in the other. Thus

$$\overline{P}_{cw} = \tfrac{1}{2} I_{sc} V_{so} \tag{4-87}$$

Using $I_{sc} = I_{RMS} = \sqrt{P_D/R_s}$ and $V_{so} = V_{RMS} = E_B/(2\sqrt{2})$ Eq. (4-87) reduces to Eq. (4-85).

Hines' equation is very useful for digital phase modulators: $|\Gamma| = 1$ for both phase states and

$$|\Gamma_{so} - \Gamma_{sc}| = 2 \sin \frac{\Delta\phi}{2} \tag{4-88}$$

Thus

$$\overline{P}_{CW\phi} = \frac{I_{sc} V_{so}}{4 \sin \frac{\Delta\phi}{2}} = \frac{\overline{P}_{CW}}{2 \sin \frac{\Delta\phi}{2}} \tag{4-89}$$

in which $\Delta\phi$ is the phase shift.

BASIC LIMITATIONS

This relationship is shown in Fig. 4-17. When pulse power is controlled Eq. (4-71) is used giving

$$\overline{P}_{CW\phi} = \sqrt{\widehat{P}_{i\phi}\overline{P}_{i\phi}} \qquad (4\text{-}90)$$

Very high powers can be controlled by having low phase shift per diode phase controlling element.

D. Other Limitations

1. Switching Speed

In Chapter 3 the limitation of switching speed without suppression of transients was set at $\tau = \dfrac{1.24}{f_0}$. By using balanced structures, this limitation can be exceeded. Another limitation on switching speed is caused by the time required to change the conductivity state of

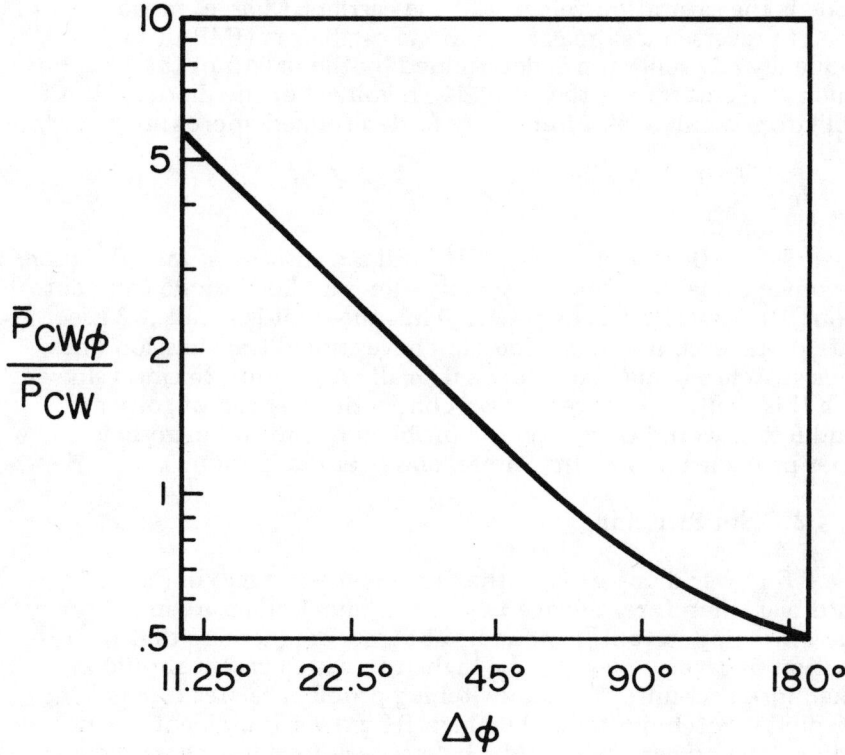

Fig. 4-17 Average Power Rating of Phase Shifters

the I region in the PIN diode. The thicker the I region, the longer it takes to get carriers into and out of it. Thus the higher the breakdown voltage of the diode the more switching time that will be required. And the peak power the diode can control is dependent on the breakdown voltage. Thus the switching time is directly proportional to the peak power the diode can control. The units that have been made using filter techniques to overcome the bandwidth limit have obtained ½ nsec/watt/diode up to 1 KW (½ μsec for 1KW) An improvement by about a factor of 10 should be possible by partially differentiating the switching pulse before it is supplied to the diode and making the pulse generator impedance low. Another improvement by a factor of 10 should be possible by changing the transmission line characteristic impedance so that E_B is smaller by a factor of 10 for the same incident RF power. By using 10 diodes, wave shaping, and Z_0 changing, it may be possible to obtain ½ nsec/KW.

The limit of the speed of removing charge from the I region of a diode is the saturation velocity of the carriers. Charges move at this speed in reverse bias breakdown diode oscillators (IMPATT). The frequency of oscillation is determined by the length of the I region which also determines the breakdown voltage of the diode. IMPATT oscillators oscillate at a frequency f_0 determined approximately by

$$f_0 \approx \frac{500}{E_B} \text{ GHz} \tag{4-91}$$

When E_B is 100 V then f_0 is 5 GHz and a transit must take 0.2 nsec. In conventional 50 ohm transmission lines a 100 V diode can control about 10 W incident peak power. Thus the limit is about .02 nsec/watt/diode = 20 nsec/kW/diode in conventional transmission lines. Thus switches being made conventionally are about 25 times slower than this limit. Using breakdown conduction instead of forward conduction would overcome the problem but would introduce other problems of stability, noise, and heat dissipation.

2. Hot Switching

The RF power dissipation in the diode is usually maximum at forward bias, therefore, the average power calculations are made for that switching state. It is possible at higher frequencies that the dissipation of reverse bias may be higher, in which case it should be taken into account. The hot switching problem relates to changing the diode switching state while high RF power is incident on the switch. The power absorbed P_a by a series diode having impedance R is

BASIC LIMITATIONS

$$P_a = \frac{\frac{1}{2} E^2 R}{(R + 2 Z_0)^2} \quad (4\text{-}92)$$

in which E and Z_0 are the same as in Eq. (2-4). Using Eq. (2-7) for P_i, the fraction of incident power observed by the diode is given by

$$\frac{P_a}{P_i} = \frac{R/Z_0}{(1 + R/2 Z_0)^2} \quad (4\text{-}93)$$

Solving Eq. (2-23) for R/Z_0 and substituting it into Eq. (4-93) gives

$$\frac{P_a}{P_i} = \frac{2(10^{\alpha/20} - 1)}{10^{\alpha/10}} \quad (4\text{-}94)$$

which is shown in Fig. 4-18. Note that for this purely resistive switch the dissipation at .5 dB is the same as at 25 dB. Also note that the maximum dissipation of incident power is 50% at 6 dB attenuation. Thus when the switch is moved slowly between states it will be absorbing 50% of the incident power at midrange, and the switch can handle only twice the power it can dissipate at the midrange point. This power absorption at midrange can be reduced by switching fast for two reasons. First, by dwelling on the midrange for a brief moment only a brief moment's worth of energy will be absorbed. If switching is too slow the diode will heat up in midrange, increasing the leakage current, and thus increasing the power absorption at reverse bias. It is possible that switching too slowly could trap the diode in a lossy state and cause burnout. This could happen only in going into the reverse bias state, since the increased temperature lowers the resistance in the conduction state (and thus lowers the fraction of incident power absorbed). Another mechanism[5] also reduces power absorption when the diode is switched quickly. When the PIN diode bias current is changed slowly the impedance of the diode junction goes from a low resistance to high resistance resistively. The diode acts like a variable resistor. When the diode is pulsed from the reverse bias state to the forward bias state, the voltage drop across the I region is low. This results in a low electric field which causes the carriers to drift slowly into the I region filling it up as shown in Fig. 4-19. The charge distribution is the same for the DC bias states and gives a resistance changing with time. Thus, the forward biased transient of a PIN diode is resistive, the RF impedance going across the middle of the Smith Chart as shown by curve A in Fig. 4-20.

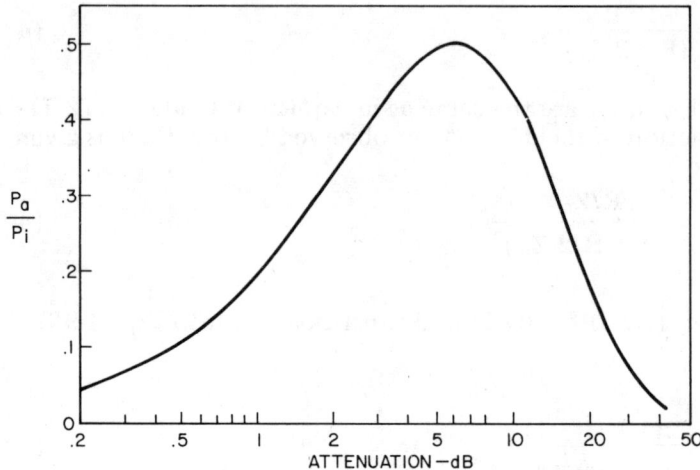

Fig. 4-18 Power Absorption by a Resistive Variable Attenuator Diode

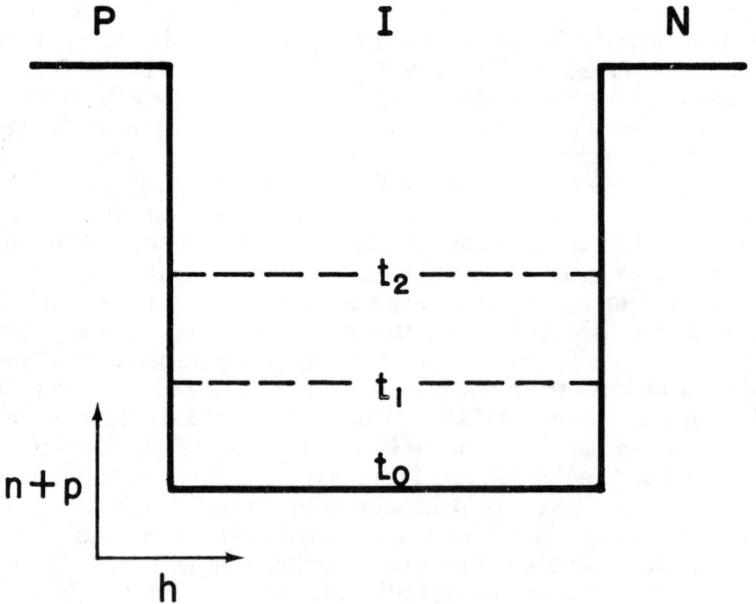

Fig. 4-19 Charge Carrier Build-up in I-Region for Turn-On

BASIC LIMITATIONS

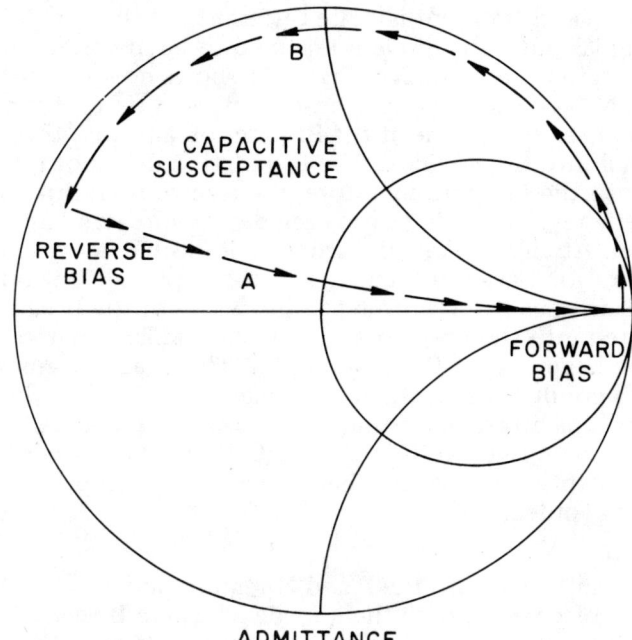

Fig. 4-20 Diode Admittance for Turn-On (a) and Turn-Off (b) Transients

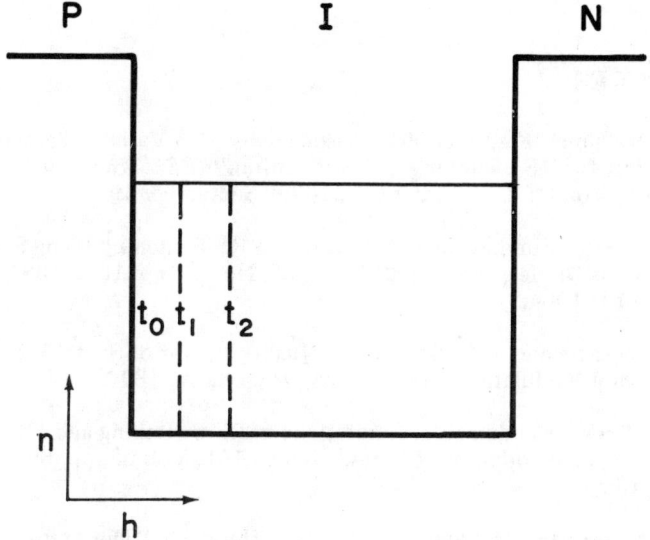

Fig. 4-21 Charge Carrier Exit for Turn-Off (only n-type shown)

The reverse biased transient is quite different. At the beginning of
the reverse bias pulse the diode is assumed to be in conduction;
thus, the I region is conductive. Fig. 4-21 shows how the distribution of the N-type carriers (electrons) changes in time. It is assumed
that all electrons recombine at the PI junction and that there is no
diffusion tail into the P region. The PI junction does not inject
electrons into the I region; therefore, the electrons all drift back
toward the N region, leaving a growing depleted region adjacent to
the P region. At any instant the structure looks like a capacitor
with a separation between the plates of the depleted I region. In time,
the separation goes from nothing to the width of the I region. The
holes injected into the I region experience a similar progressive
distribution beginning at the IN junction. Thus, rapidly going from
the forward conduction to the reverse bias state, the PIN diode behaves like a capacitor going from $C = \infty$ to $C = C_{MIN}$ and gives the
impedance shown as curve B in Fig. 4-20. Since the impedance is not
resistive, absorption is low and the power absorption problem of hot
switching is avoided.

If diffusion tails exist into the P and N regions and good sharp step
recovery is not present, then the impedance curve B veers slightly
away from the edge of the Smith Chart, especially near the end
of the transient. Thus, good step recovery implies good hot switching.

REFERENCES

1. S. Kawakami, "Figure of Merit Associated with a Variable-Parameter One-Port for RF Switching and Modulation," *IEEE Trans. on Circuit Theory*, Vol. CT-12, No. 3, pp. 321-328, September 1965.

2. M.E. Hines, "Fundamental Limitations in RF Switching Using Semiconductor Diodes," *Proc. IEEE*, Vol. 52, No. 11, pp. 1382-1384, November 1964.

3. K. Kurokawa and W.O. Schlosser, "Quality Factor of Switching Diodes for Digital Modulation," *Proc. IEEE*, Vol. 58, pp. 180-181, January 1970.

4. M.E. Hines, "Fundamental Limitations in RF Switching and Phase Shifting Using Semiconductor Diodes," *Proc. IEEE*, Vol. 52, pp. 697-708, June 1964.

5. D.F. Ciccoletta, R.L. Johnston, and B.C. DeLoach, "The Transient Microwave Impedance of PIN Diodes," *Microwave Diode Research Report No. 18*; Contract DA 36-038 SC-89205, 20 December 1964.

6. J.K. Hunton and A.G. Ryals, "Microwave Varriable Attenuators and Modulators Using PIN Diodes," *IRE Trans. on Microwave Theory and Techniques*, Vol. MTT-10, pp. 262-273, July 1962.

7. H. Mott and D.M. McQuiddy, "The Harmonics Produced by a PIN Diode in a Microwave Switching Application," *IEEE Trans. on Microwave Theory and Techniques*, Vol. MTT-15, pp. 180-181, March 1967.

8. R.L. Sicotte and R.N. Assaly, "Intermodulation Products Generated by a PIN Diode Switch," *Proc. IEEE*, Vol. 56, pp. 74-75, January 1968, also "Intermodulation Product and Switching Noise Amplitudes of a PIN Diode Switch in the UHF Band," *IEEE Trans. on Microwave Theory and Techniques*, Vol. MTT-18, pp. 48-50, January 1970.

Control Element Design

Chapter 5

Many switching requirements can be met simply by using the concepts developed in Chapters II and III. Packaged diodes work quite well in mode 1, without any tuning up to about 4 GHz. Unpackaged diodes work well in miniaturized microstrip up to about 20 GHz. Many practical designs have been created which greatly expand the range over which diodes can be used. As a matter of fact practically any diode can be made to switch at any frequency as long as its Q is high enough at that frequency.

A number of problems are not solved by the circuits given in Chapters II and III. What are the practical circuits for using packaged diodes in stripline at frequencies above 4 GHz? The same question can be asked concerning the application of unpackaged diodes in miniaturized microstrip for use above 10 GHz. How should switches be designed for use in waveguide? High power switches have large junctions, large capacitances, and necessarily have narrow bandwidths. What are the most practical circuits for making them switch well at high frequencies? Most systems have bandwidths in the 10-percent range. A switch optimized over a narrow bandwidth can have performance superior to a broad-bandwidth switch.

There is a wide variety of ways of combining diodes with transmission lines to obtain good switching. Every diode switch must include a diode and the bias circuit. These were considered separately in Chapters II and III, except for the two examples in the design of low VSWR ON switches. When the circuit requirements for the diode and bias circuit are considered separately, their interactions may be quite unfavorable when combined. When both bias circuit and diode are considered together, it frequently becomes possible to use the elements of the bias circuit to optimize the diode insertion loss and/or isolation at the design frequency. In fact when diode and bias circuit are considered together, the wide selection of possible circuits reduces to just a few which are superior to the others in all ways.

In the following pages, most discussion will be centered on TEM transmission lines because of their increasing utilization in systems. Waveguide switches are included because waveguide continues to be the most used transmission line at the higher (mm wave) frequencies.

A. TEM Transmission Line — Stripline

Stripline is usually two Teflon-Fiberglass dielectric boards 1/16" thick sandwiched together. The outer surfaces of the boards are metalized with a thin layer of copper comprising the ground planes which are the outer conductors. The center conductor is made up of a strip usually about .090" wide for 50 ohms characteristic impedance and about .001" thick. Screws or eyelets run on each side of each center conductor to prevent the excitation of waveguide modes in the dielectric between the ground planes, which cause undesired coupling between lines. The eyelets are less than $\lambda/8$ away from the center strip and have less than $\lambda/8$ between them. For quick experimental work the center strips can be cut on a precision sheer from .001" shim brass and held to one stripline board with scotch tape. For production assembly the strips can be etched on a board using conventional photographic etching methods. The relative dielectric constant of Teflon-Fiberglass varies from 2.55 to 2.65 for fiberglass weave type boards and is 2.22 for random fiber boards. Charts for calculating characteristic impedance are given in most microwave handbooks.

Most systems operate in 50 ohm characteristic impedance. A typical requirement on a diode switch is to provide less than 1 dB insertion loss and more than 20 dB isolation over the system bandwidth of 10 percent.

1. Switches

The selected optimum configurations for mounting diodes in stripline to make switches are shown in Fig. 5-1. In most cases the bias circuit elements are chosen to work in conjunction with the diode reactances to provide optimum insertion loss or isolation at the design frequency. Wide strips indicate characteristic impedance $Z_0 \leq 50$ ohms, while thin strips indicate $Z_0 \geq 50$ ohms. Z_0 and Y_0 in the equations of Fig. 5-1 represent the source-and-load impedance and admittance as "seen" by the diode. Capacitors are realized by placing thin Mylar between a printed strip and a short strip fabricated from thin shim stock or by using miniature ceramic capacitors.

CONTROL ELEMENT DESIGN   125

a. Series Diode

When a series diode has a low impedance, most of the incident RF current passes through it into the terminating load behind it. It then provides a low insertion loss, δ. When it has a high impedance, practically no RF current passes through it, and the device demonstrates high attenuation termed isolation, η. Since a capacitor must be put in series with the series diode to isolate the bias voltage from other parts of the system, the same capacitor can be made to series resonate with the inductance of the forward-biased diode to provide a low impedance (and thus low insertion loss) at the design frequency. Putting an inductor in parallel with a diode to parallel resonate with the reverse bias capacitance to obtain high impedance (and thus maximum isolation) is practical only when the required inductance is small, so that it is comprised of a short length of high-impedance transmission line connecting the quarter-wavelength sections on both sides of the diode. Normally such an inductor will not be used, dictating that the series configuration be used only when the diode reverse bias impedance is high enough to provide the desired isolation by itself.

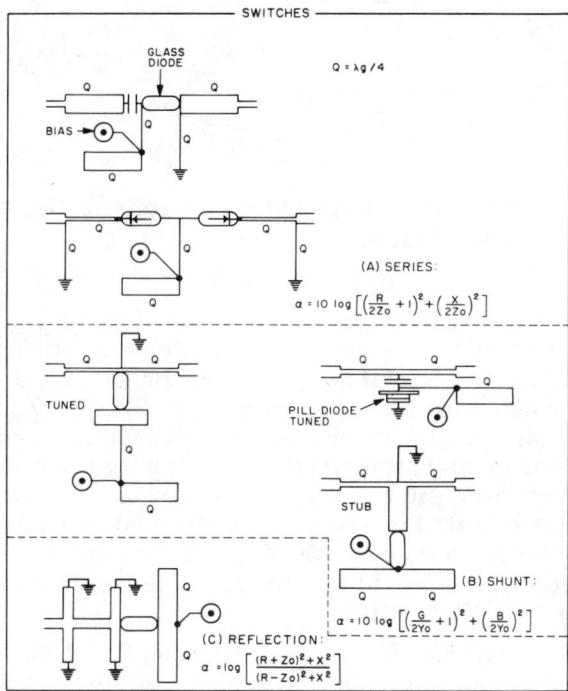

Fig. 5-1   Circuits for stripline switches (only center conductor strips are shown).

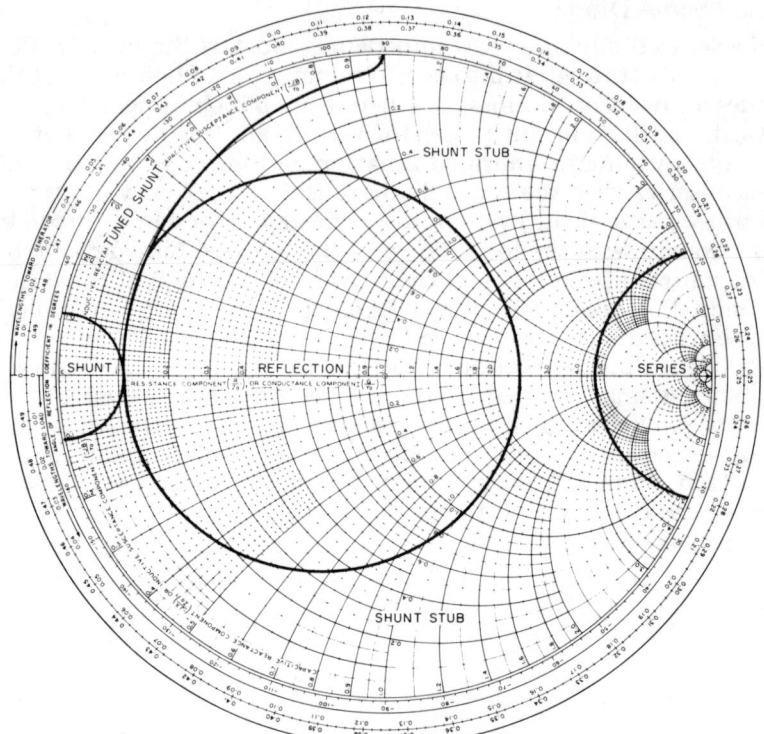

Fig. 5-2 Normalized impedance plot for selecting the most suitable circuit.

Since minimum insertion loss is easily obtained by tuning the series capacitance, the isolation is the only switching state difficult to satisfy. The isolation is dictated by the impedance of the diode at reverse bias, which must be high compared to Z_0. The Z_0 in which the diode is mounted can be lowered by putting quarter-wavelength transformers before and after the diode. Allowing the transformer stripline characteristic impedance to be as low as 25 ohms permits the diode to be mounted in a region transformed to 12.5 ohms. This gives the series diode curve on Fig. 5-2 defining the impedance region of a diode measured in 50-ohm transmission line that will provide at least 20-dB isolation.

Using the formula of Fig. 5-1(A) and assuming $R = 0$, $Z_0 = 12.5$, the reactance X corresponding to an isolation of 20 dB is ± 250 ohms, or ± 5 normalized to a 50 ohm line.

CONTROL ELEMENT DESIGN

The remainder of the circuit shown in the upper portion of Fig. 5-1(A) is for biasing. The high Z_0 quarter-wavelength line to ground provides a d-c ground return for bias with very little contribution to insertion loss over a wide bandwidth. The high-Z_0 quarter-wavelength line section from the bias point similarly provides wide bandwidth. The open-circuit low-Z_0 quarter-wavelength line from the bias point provides an RF ground at the point and prevents RF from being coupled to the external biasing circuit. This biasing arrangement provides very fast switching speeds close to the theoretical minimum of $1.24/f_0$. (Ref. Eq. (3-33).)

Another circuit is shown in Fig. 5-1(A) that uses two diodes and has the advantage that no series tuning capacitor is required. The reflections of two identical discontinuities cancel when their electrical separation ℓ/λ_g satisfies the equation.

$$\frac{\ell}{\lambda_g} = \frac{1}{2\pi} \tan^{-1}\left(\frac{2}{X/Z_0}\right) \quad \ell = \text{the distance between diode centers.} \quad (5\text{-}1)$$

A spacing of $\ell/\lambda_g = 0.2$ is approximately correct for tuning out the inductive reactance of most forward-biased diodes. This same spacing also increases the isolation. For example, if one diode in a given Z_0 provides only 10 dB, then two of them properly spaced in series and in the same Z_0 will provide 26 dB (ref. Fig. 5-18 and Eq. (6-15)). Although this lower circuit in Fig. 5-1(A) could permit the region for series diode isolation on Fig. 5-2 to be expanded, it is not expanded making Fig. 5-2 a conservative guideline. In this circuit the transformers in series with the diodes may be needed to reduce the insertion loss contributed by the series resistance of the forward biased diodes by transforming the series resistance to lower effective values.

b. Shunt Diode

Lead-type or pill-packaged diodes may be used for making shunt diode switches as shown in Fig. 5-1B. A shunt diode must have low impedance to provide high isolation and high impedance for low insertion loss. The insertion loss of the lead-type shunt diode switch may be reduced by making the d-c ground return shorter than $\lambda_g/4$.

The isolation of the lead-type diode is maximized by adjusting the lengths of the shunt capacitative stubs on the bias end of the diode. Since the forward-bias impedance of a diode is usually inductive, these open-circuit stubs are made shorter than $\lambda_g/4$ to provide the series capacitance for resonance.

The maximum isolation of the pill diode is obtained by varying the capacitor between the diode and the center strip of the stripline. These two circuits perform well as diode switches when their impedance normalized to 50 ohms falls within the regions labeled "shunt" and "tuned shunt" in Fig. 5-2.

A large portion of the outer edge of the Smith chart lies outside the regions so far defined in Fig. 5-2 by the circuits. This unclaimed region is easily within range of the shunt stub configuration.

When a diode at forward bias has inductance L and at reverse bias capacitance C, it has the same impedance as a shorted or opened length of transmission line of characteristic impedance

$$Z_0' = \sqrt{L/C} \tag{5-2}$$

and length

$$\ell' = \frac{\lambda_g}{2\pi} \tan^{-1} \omega\sqrt{LC} \tag{5-3}$$

By adding more length in front of the diode of characteristic impedance Z_0', the line can be made $\lambda_g/4$ long and perform as a switch having optimized isolation and insertion loss.

The requirement may be stated more generally in terms of diode reactance. For forward conduction, the diode reactance is X_s, and for reverse bias the reactance is X_0.

Then

$$Z_0' = \sqrt{-X_s X_0} \tag{5-4}$$

and

$$\ell' = (\lambda_g/2\pi) \tan^{-1} \sqrt{-X_s/X_0} \tag{5-5}$$

Note that X_s and X_0 must have opposite signs to provide real values. In the event Z_0' is too high or too low for practical construction, or if both reactances have the same sign, the d-c ground return may be adjusted to a length other than $\lambda_g/4$ to provide the desired tuning for optimum insertion loss, the stub having been adjusted for high isolation.

CONTROL ELEMENT DESIGN

When the Z_0' for any diode is out of the 12.5-to 100-ohm range, the impedance level of the diode is such that 10-percent bandwidth becomes very difficult to obtain in any switching circuit. It is important that the diode impedance not change rapidly with frequency, as when the cartridge goes through resonances, since this will also diminish bandwidth.

c. Reflection Diode

When the impedance of the diode is close to 50 ohms in one bias state, none of the circuits discussed thus far will provide satisfactory switching. Good switching may be obtained by connecting the diode to the second port of a circulator. Then power in the first port is absorbed by the diode in the 50-ohm impedance state and does not come out port 3. In the other bias state the diode is presumed to be

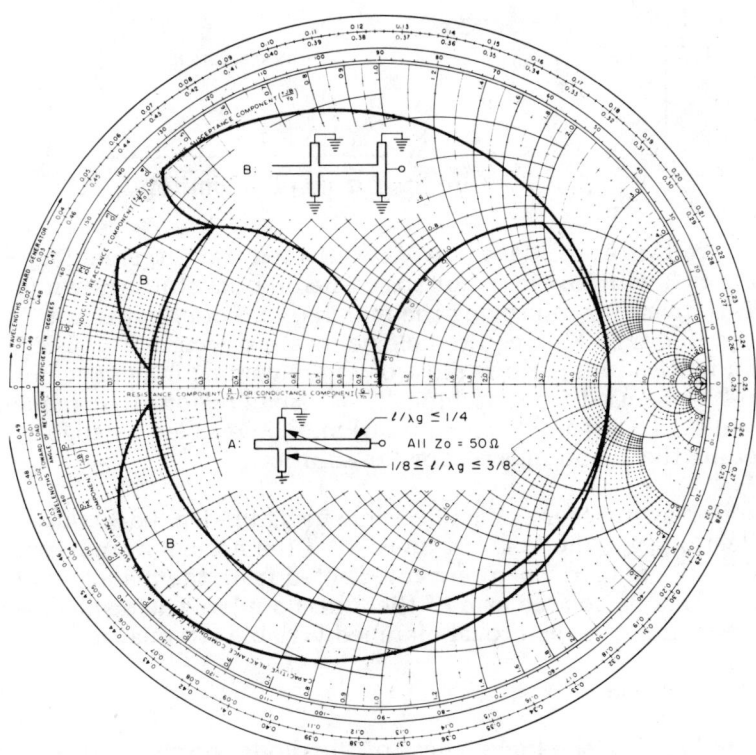

Fig. 5-3 Normalized impedance plot for selecting a matching circuit.

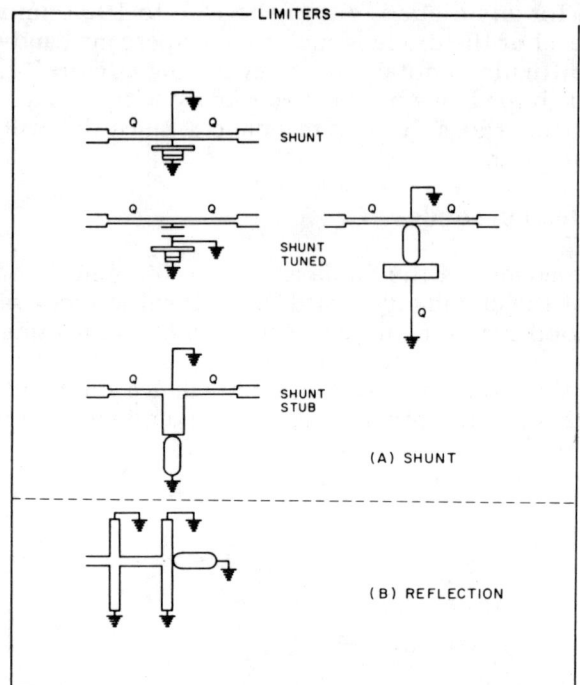

Fig. 5-4   Circuits for stripline limiters.

a better reflector; incident power is reflected from it, emerging from port 3 with low insertion loss. The 50-ohm stubs shown in Fig. 5-1(C) are for improving the match of a diode that is not exactly 50 ohms. The required elements are shown in Fig. 5-3. This matching structure permits most of the remaining impedance on Fig. 5-2 to be used for switching. Those small regions that remain (around $0.14 + j\, 0$ and $0.15 + j\, 0.36$) may be reached by slightly exceeding the Z_0 bounds set for the solutions.

Reflection switches may also be made using 90-deg 3-dB couplers. A reflection-type diode is put on each of the normal output arms of the 3-dB coupler. Then power reflected from them adds in phase at the normally isolated arm of the 3-dB coupler. The device has the characteristics of a reciprocal matched variable attenuator.

d. General

In using Fig. 5-2, the impedances of the diodes normalized to 50 ohms are plotted on the Smith chart for forward and reverse bias. The areas, as described in Fig. 5-2 in which the impedance points lie, determine

CONTROL ELEMENT DESIGN

the most suitable circuit for making a switch. For shunt or series diodes, that bias state providing the highest VSWR generally provides the highest isolation. For reflection devices, that bias state providing the lowest VSWR should be used for the isolation state.

2. Limiters

Any switch that delivers high isolation while in the conduction state will function passively as a limiter. The switching terminals are short-circuited and the rectified current from high incident power simply causes the switch to turn itself off. Because the bias leads are not needed, the circuits may be simplified as shown in Fig. 5-4. The use of elements for tuning is the same as for switching except with the shunt stub configuration. The total electrical stub length must now be $\lambda_g/2$ to provide the proper switching polarity for limiting.

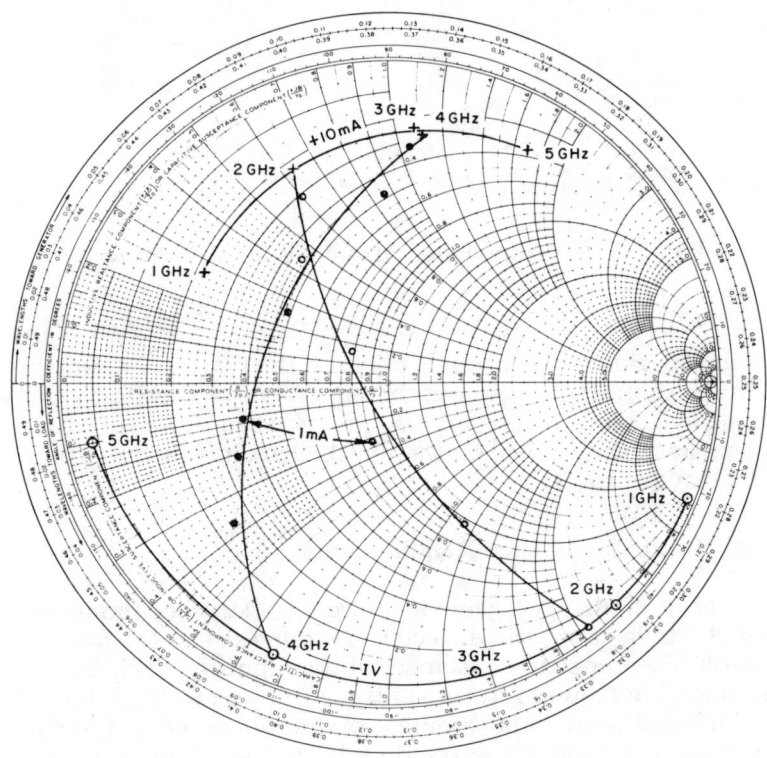

Fig. 5-5   Impedance of the MA 4850 hot-carrier diode.

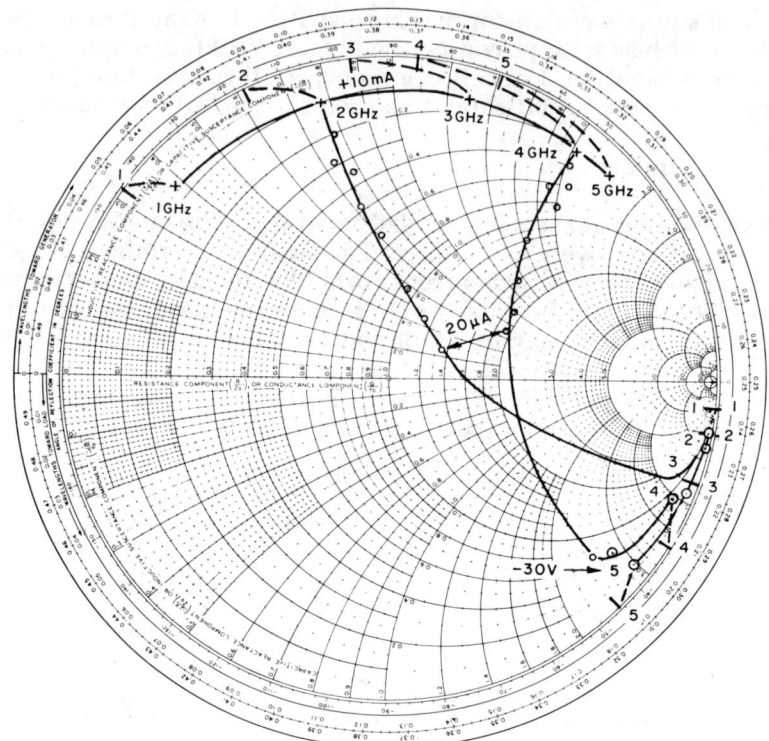

Fig. 5-6   Impedance of the HP 3001 PIN diode.

Generally speaking, exception is taken to the $\ell/\lambda_g \leq \frac{1}{4}$ requirement only in the shunt stub region in the upper half of the Smith chart. Allowing excursions in Z_0 and ℓ/λ_g greater than those specified above expands the regions on the Smith chart in which the various circuits may be used.

3. Diode Impedance Requirements

A number of diode types were measured in a stripline mount consisting of a hole cut in the dielectric the same size as the diode; the 50-ohm center strip continued on the input to touch the diode glass; the ground return was three times as wide as the diode hole and went to both outer conductors at the end of the diode; the diode hole was covered on both sides by ground planes. The impedance of each diode was measured in reference to an input plane corresponding to the beginning of the glass.

CONTROL ELEMENT DESIGN 133

The impedance of an inexpensive Schottky barrier diode is shown in Fig. 5-5. The effective f_c for switching is about 30 Ghz when measured up to 4 GHz; however, it drops to 6.5 GHz when measured at 5 GHz. The diode will thus be useful up to about 3.5 GHz. It can be seen using Fig. 5-2 that as a switch or limiter the diode can be used in either polarity in the shunt stub configuration over most of the useful range. The higher isolation could be achieved by having the switch in the high isolation state when the diode is reverse biased.

Similar calculations can be made on the 3001 PIN diode, T7G gold-bonded diode, D5151 switching diode, the impedances of which are shown in Figs. 5-6, 5-7, and 5-8.

In comparing Fig. 5-5 through 5-8 with Fig. 5-2, it can be seen that the shunt stub configuration is the most often recommended circuit for making switches and limiters. This configuration is the one most used for digital diode phase modulator bits as illustrated in Fig. 5-9 and has been used by White[1] for designing double-throw switches as shown in Fig. 5-9. In all of these designs using stubs,

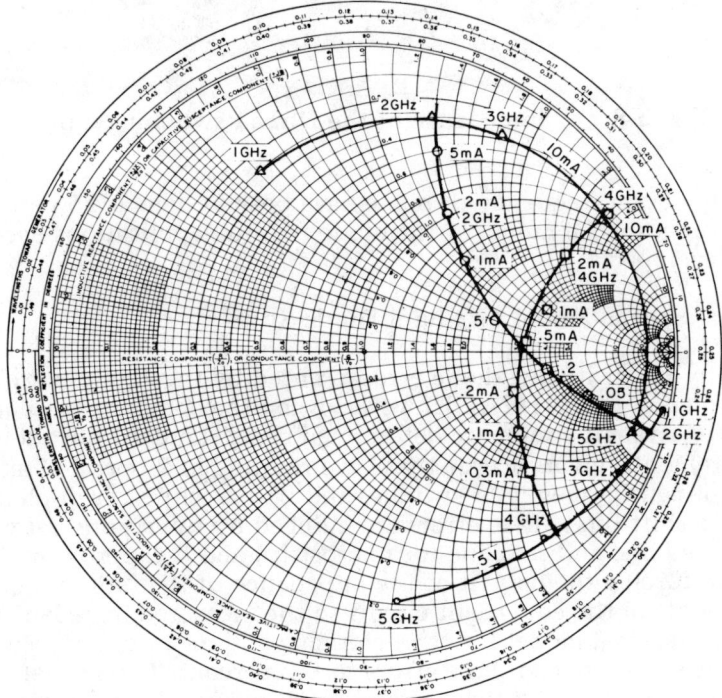

Fig. 5-7   Impedance of the T7G gold-bonded diode.

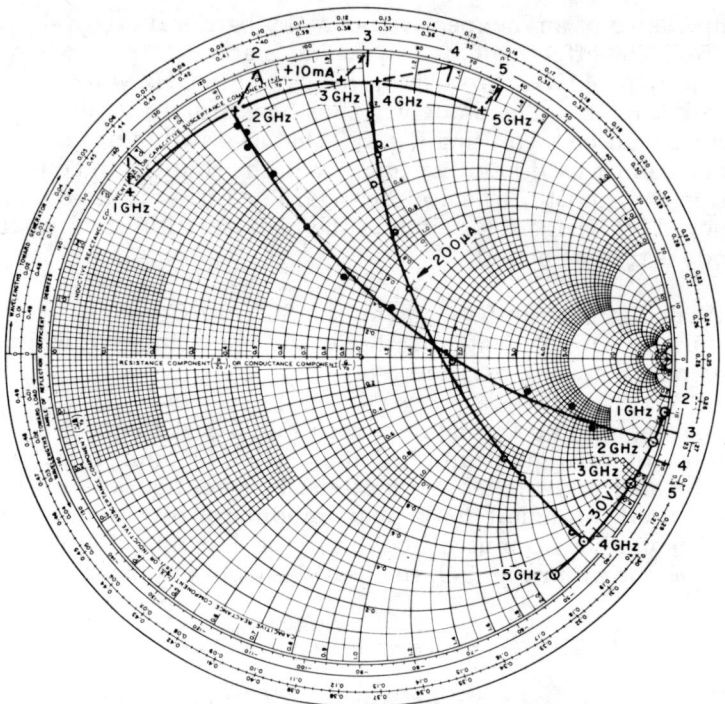

Fig. 5-8   Impedance of the D5151 switching diode.

errors in design frequency or phase have been as high as 25 percent. Therefore a special model of the diode for stub applications has been developed.

4. Special Model for Shunt Diode Stubs

Normally the switching diode is represented by the equivalent circuit shown in Fig. 5-10(A). However, when the component values of this equivalent circuit are determined using standard slotted-line measurements, errors accrue that have resulted in 25-percent error in the design frequency of switches and 22-percent error in phase of phase shift bits. These errors are induced by transitions between the slotted line and stripline (or microstrip) and by approximations made in forcing the best fit of the impedance of the finite number of elements in the equivalent circuit to the measured impedance points. A special model for characterizing and measuring the switching diodes has reduced these errors to 0.2 percent. This model eliminates the need for the slotted line in making the measurements. Accuracy is obtained using a cavity-type circuit in which frequency

CONTROL ELEMENT DESIGN

measurements determine the reactance.

a. The Model

The new model for the switching diode is a length of two-wire transmission line as shown in Fig. 5-10(B). The characteristic impedance of the transmission line representing the switching diode can be defined to be equal to the square root of the product of the input impedances when the diode junction is conducting and when it is nonconducting (as in Eq. (5-4)). The effective length of the line is defined by the reactance at either bias state and is then the same for both bias states. Alternatively, the diode can be represented with equal accuracy by a line of any characteristic impedance when the effective lengths are properly chosen for conduction and nonconduction. In general, these lengths vary with frequency. It is more convenient to fix the characteristic impedance and allow only the effective lengths to vary. This approach is taken here because it facilitates measurement and design. When the diode is forward biased, it is considered to be a length of short-circuit-terminated transmission line. When the diode is reverse biased, it is considered to be different length of open-circuit-terminated transmission line. The effective lengths of the diode are obviously a function of the

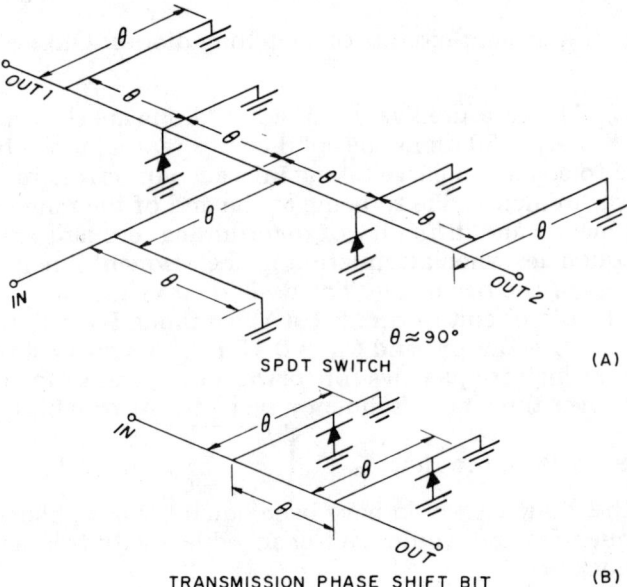

Fig. 5-9   Stub switching elements as used in double-throw switches (A) and phase shift bits (B).

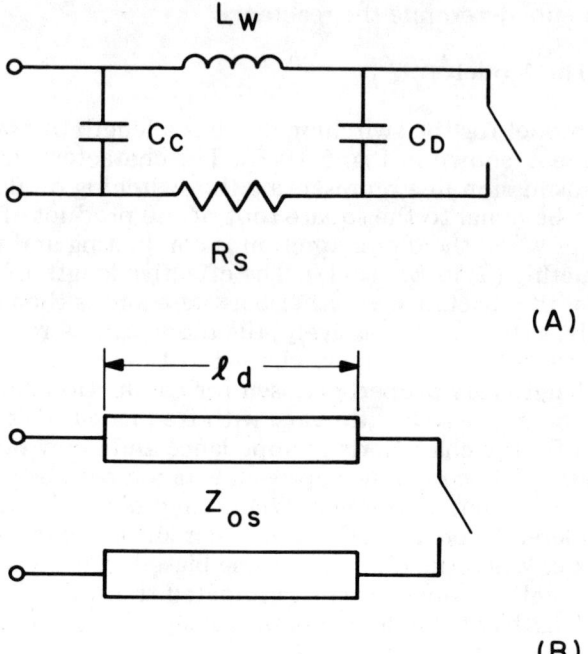

Fig. 5-10 Equivalent circuits of switching diodes: Old (A), New (B).

assumed Z_0. The measured values of a PIN diode are shown in Fig. 5-11. For $Z_0 = 50$ ohms the solid curves represent the best fit of the data to a smooth curve taking into account errors in frequency measurements, due to being at the end of the range of the frequency meters and difficulty of determining resonant frequency at high frequencies when attenuation in the resonant line has reduced the sharpness of the resonance. The dashed curve indicates the response of the old circuit of Fig. 5-10(A) assuming $L_w = 2.48$ nH, $R_s = 2$ ohms, $C_c = 0.0$ pF, and $C_D = 0.15$ pF. The greatest deviations occur at high frequencies for forward bias, and serious deviation occurs over the whole frequency range for reverse bias.

Had Z_0 been selected so that $Z_0 = \sqrt{\dfrac{L_w}{C_c + C_D}}$, the effective (5-6) length of the diode ℓ_d would have been equal for both bias states at low frequencies and would have changed less with frequency at higher frequencies.

CONTROL ELEMENT DESIGN

The diode is measured by placing it on the end of a calibrated shunt stub (Z_0 = 50 ohms for Fig. 5-11). The stub is calibrated by measuring the frequencies at which it provides maximum attenuation to the through line and its attenuation at each frequency. The response of a 30 cm. (equivalent length in air) stub in Teflon-fiberglass stripline is shown in Fig. 5-12. The attenuation at resonance falls off at higher frequencies owing to increased attenuation in the stub. Isolation peaks occur at all multiples of $\lambda_g/2$ for the short-circuit-terminated line and at odd multiples of $\lambda_g/4$ for the open-circuit-terminated line. The through attenuation at the resonant peak may be calculated as follows:

A transmission line having a perfect reflector (open or short) terminating it will have its input VSWR determined by the attenuation α_s of the line between the input and termination. An input wave is attenuated by α_s from input to termination and again by α_s from termination to input. The total attenuation of the reflected wave at the input port is thus $2\alpha_s$. The input voltage reflection coefficient magnitude will be $\Gamma = 10^{-2\alpha_s/20}$. The input VSWR will be

$$\rho = \frac{1 + 10^{-\alpha_s/10}}{1 - 10^{-\alpha_s/10}} \tag{5-7}$$

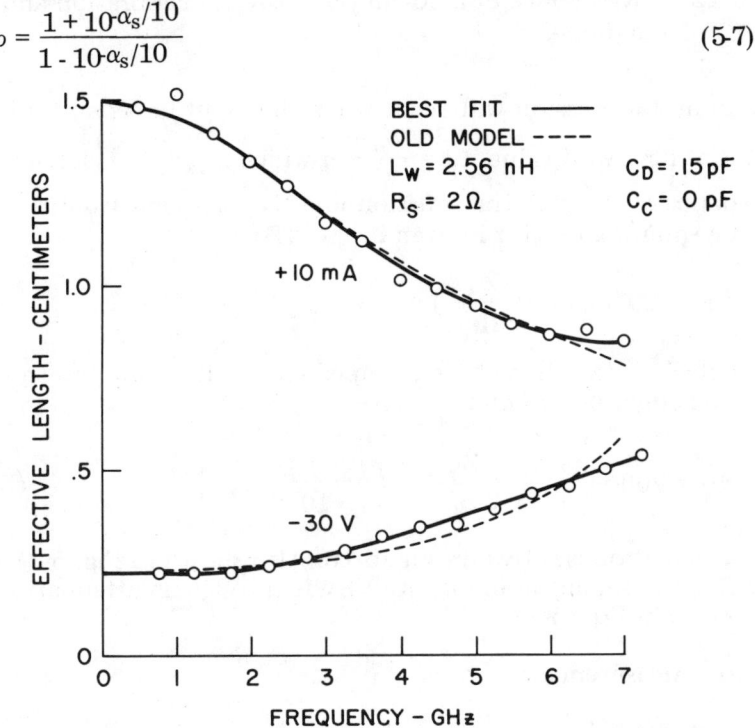

Fig. 5-11 Effective length of HP 3001 in Z_0 = 50Ω transmission line.

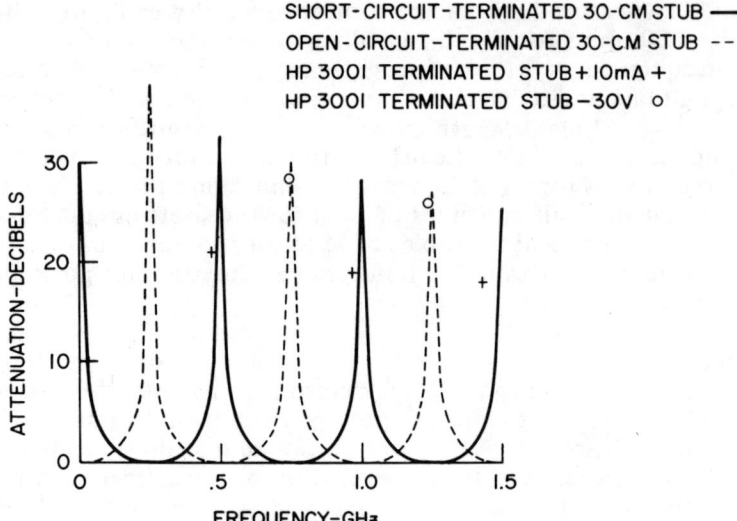

Fig. 5-12   Resonance of a 30-cm (air) stub for calibration and with a diode.

Minimum transmission will occur when the input impedance of the stub is at its lowest value $R_p = \dfrac{Z_{os}}{\rho}$ in which Z_{os} is the characteristic impedance of the stub transmission line. The through attenuation α_T of a shunting resistor is given by Eq. (2-9)

$$\alpha_T = 20 \log \left(1 + \frac{Z_0}{2R_p}\right) \tag{5-8}$$

in which Z_0 is the characteristic impedance of the main transmission line. The equations combine to give

$$\alpha_T = 20 \log \left[1 + \frac{Z_0}{2 Z_{os}} \left(\frac{1 + 10^{-\alpha_s/10}}{1 - 10^{-\alpha_s/10}}\right)\right] \tag{5-9}$$

This attenuation is shown in Fig. 5-13. Also shown in Fig. 5-13 is the VSWR seen looking at an infinite VSWR through an attenuation of α_s as given in Eq. (5-7).

b. Measurement

When the diode is put on the end of the stub in place of the open or short circuit, the resonances are shifted down in frequency and cause

CONTROL ELEMENT DESIGN

less through attenuation at the resonant peaks, as illustrated in Fig. 5-12. The effective length of the diode is obtained by calculating the length of a 50-ohm stub that would resonate at the new frequency and subtracting from it the calculated length of the reference stub. The VSWR of the diode is obtained by subtracting α_s (reference) from α_s (with diode) and using the dashed curve in Fig. 5-13. The diode impedance can be calculated exactly using these lengths and VSWR in the transmission line equation (or Smith chart) or the reactances of the diode can be obtained approximately from

$$X_F = Z_{0s} \tan 2\pi \frac{\ell_d}{\lambda_g} \quad \text{(forward bias)} \tag{5-10}$$

and

$$X_R = -Z_{0s} \cot 2\pi \frac{\ell_d}{\lambda_g} \quad \text{(reverse bias)} \tag{5-11}$$

but it is not necessary to make these calculations for designing stub switching elements.

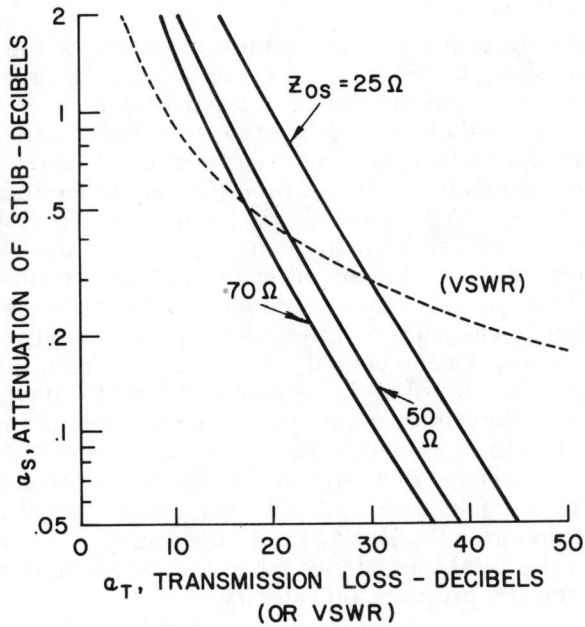

Fig. 5-13 Stub attenuation as a function of resonant stub transmission loss for a 50Ω main transmission line.

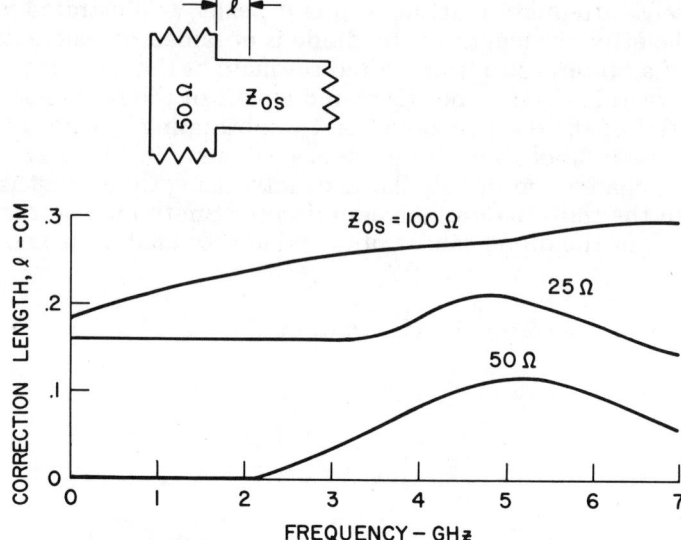

Fig. 5-14 T-junction corrections for 1/16 in. Teflon-fiberglass stripline.

It is sometimes necessary to make some correction for the T-junction.[2] The question is, "Where does the shunt-stub begin electrically?" It does not begin in the center of the main line or on the edge of it. It begins a short distance down the stub. The distance is determined by using the following measurements. The resonant frequencies of one stub cut to two precisely measured different lengths are observed. After the electrical lengths of the stubs are computed from their resonant frequencies, the ratio of the mechanical to electrical length increment gives the propagation constant of the stub. Then all of the computed electrical lengths can be compared with an electrical length obtained from the physical length and the propagation velocity. This distance from the edge of the main line is shown in Fig. 5-14. Because the experiments used 1/16-in. thick Teflon-fiberglass stripline boards, there is an uncertainty in this length of $\pm 0.0012\, Z_{0s}$ cm due to microscopic inhomogeneities. When the edge of a strip lies over a glass fiber in the substrate, the edge capacitance is different from that when it lies over Teflon. For thin strips (higher Z_{0s}) this edge capacitance is a greater percentage of the total capacitance (than for thick strips) which dictates Z_{0s} and the propagation velocity.

CONTROL ELEMENT DESIGN

The resonant method of measuring diodes is precise, because the reactive component of the diode is based on frequency measurements; and there are no intervening mismatches to grossly alter the measured reactances. The technique is especially attractive for stripline and microstrip transmission lines in integrated circuits because it obviates the difficulty in making low mismatch transitions to these transmission lines. It is also attractive because the only precise instrument needed for the measurements is a frequency meter. Conventional resonant-cavity frequency meters have proved to be sufficiently accurate because the uncertainty in tuning the oscillator to the resonance of the stub is the limiting factor.

Although any Z_{os} may be used with the diodes once they have been measured by using the reactance equations, the most accurate results are obtained by using the same Z_{os} for measurement that will be used in the final design. The structure of the diode mount also must be the same because it will influence the measured impedance. For the data of Fig. 5-11 the diode was mounted in a hole 4.4 mm long with the shim brass covering both outside planes of the hole. The deviation observed from measuring the same diode, removing it, and replacing it in the measurement mount are shown as A in Fig. 5-15. Measurements of several of the same type of diode are also shown. The ± 1-mm variation in effective length of randomly selected

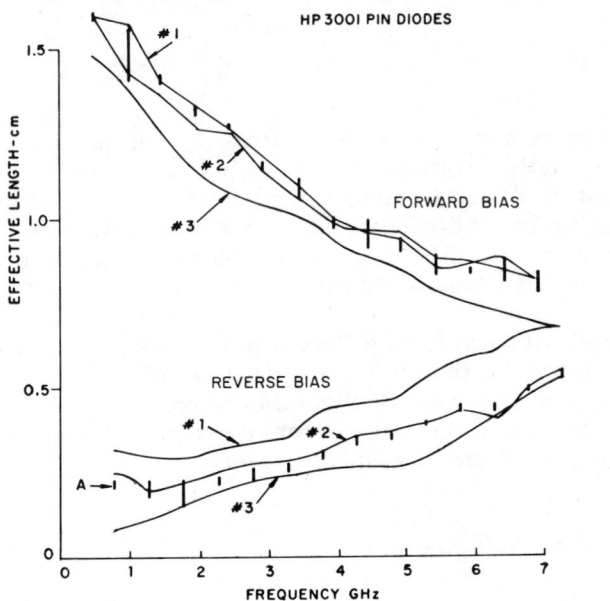

Fig. 5-15   Effective length of several HP 3001 PIN diodes ($Z_{os} = 50\Omega$).

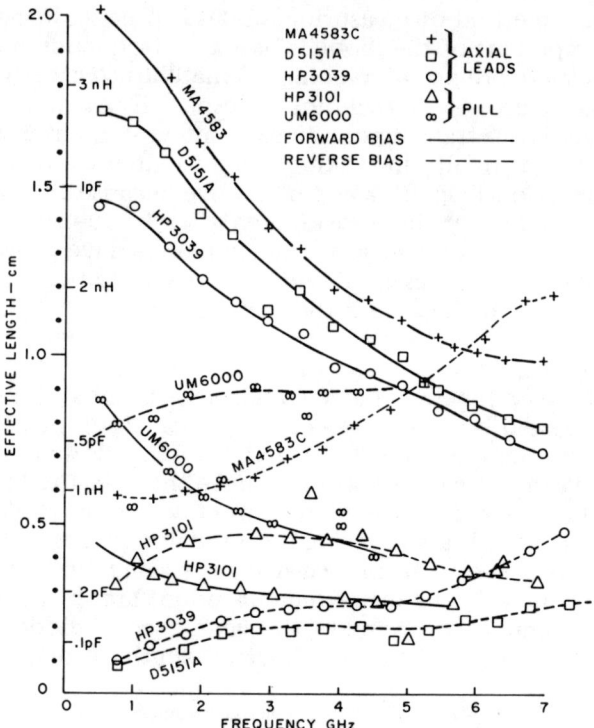

Fig. 5-16   Effective length of sample diodes ($Z_{os} = 50\Omega$).

diodes will make a contribution to errors depending on how much of the stub length is contributed by the ±1 mm. The relative dielectric constant of Teflon-fiberglass stripline with fiberglass weave varies from 2.55 to 2.65. This results in a propagation constant varying from .625 to .614. For precise design, a measured diode should be used in a previously calibrated board.

Measurements of a group of different diodes are shown in Fig. 5-16. The same mount for the diodes was used as in the data for Figs. 5-5 through 5-8. The data for these diodes are also plotted on Figs. 5-6 and 5-8 as ticks around the outer edge of the Smith chart for comparison with slotted line measurements.

5. Design Examples

Two examples are given to demonstrate the use of the design procedure. Both use HP 3001 PIN diodes, the measurements of which

CONTROL ELEMENT DESIGN

are shown in Figs. 5-6, 5-11, and 5-15. Normally, it is difficult to make switches with inexpensive diodes at high frequencies. These designs were therefore made at 4 GHz to demonstrate the power of this design procedure.

a. Series Diodes

Fig. 5-15 indicates that the effective length of the reversed biased diode at 4 GHz varies from 0.45 to 0.26 cm. These correspond to 0.060 and 0.035 wavelengths of open-circuited transmission line, which in turn give $X/Z_0 = -2.5$ to $X/Z_0 = -4.4$. Both of these points are out of the range of "series" on Fig. 5-2, but the double diode switch provides more isolation than does the single diode switch. The circuit for the switch is shown in Fig. 5-17 and the performance is shown in Fig. 5-18. (Refer to Fig. 3-12 for the bias circuit parameters.) The irregularity of the data points at 5GHz is attributed to a resonance caused by having too few suppressor screws holding the boards together for the higher frequencies.

b. Stub Switch

Because of the uncertainty and small size of the correction for the T-junction for 50-ohm transmission line, this correction was made instead by adjusting the propagation constant of the stub. Thus a reverse-biased diode having an effective length at 4 GHz of 0.385 cm was used to produce the switch shown in Fig. 5-19.

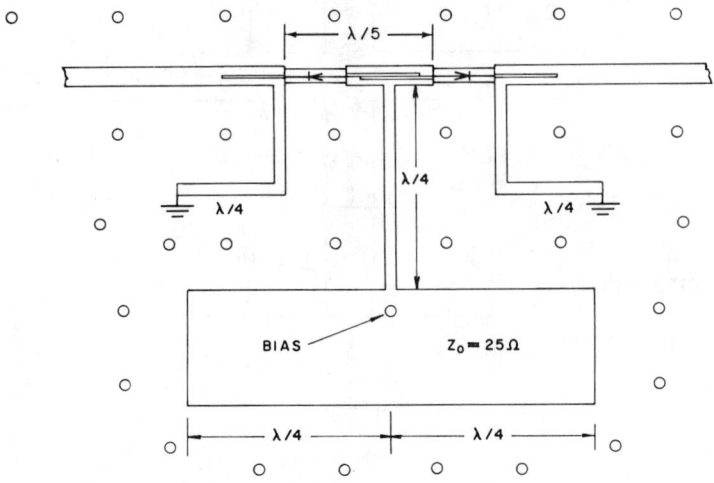

Fig. 5-17   The 4-GHz switch using two Series HP 3001 PIN diodes.

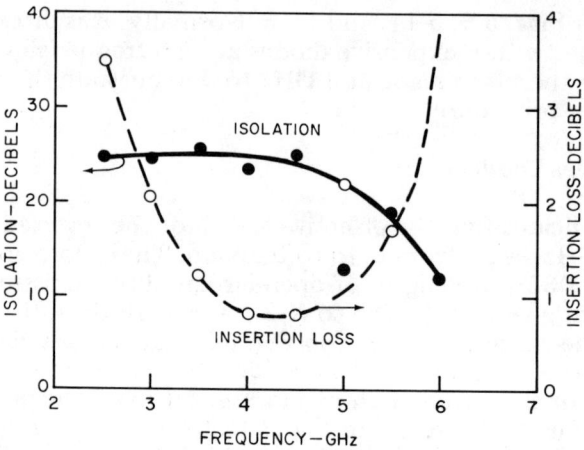

Fig. 5-18   Performance of the 4-GHz series PIN diode switch.

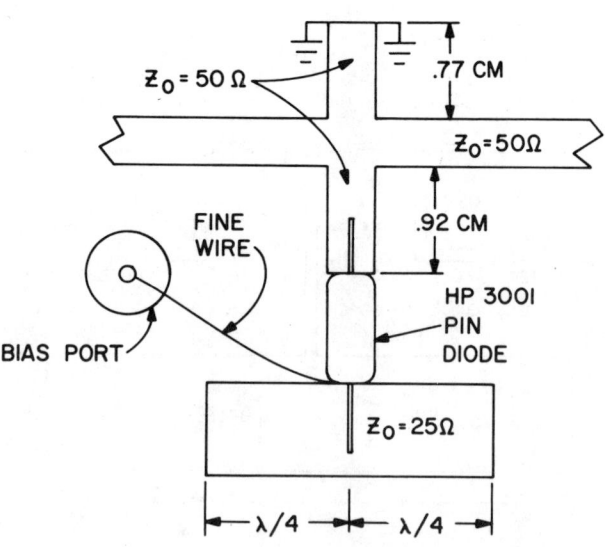

Fig. 5-19   The 4-GHz stub switch.

CONTROL ELEMENT DESIGN

When the diode is reverse-biased, the length of the stub between the main line and the diode is calculated so that it and the diode length add up to $\lambda_g/4$ at 4 GHz. An open circuit at the diode junction is thus transformed to a short circuit across the main transmission line, which results in high isolation. When the diode is forward biased, the RF current goes to ground via the two parallel 25 ohm, $\lambda_g/4$ open-circuit stubs so that the diode stub appears as a short-circuit-terminated stub slightly greater than $\lambda_g/4$ long. Thus it appears as a capacitor across the main transmission line. This capacitance is tuned out by the inductance of the upper stub, which is less than $\lambda_g/4$, and terminated in a short circuit. The stub provides d-c return for the bias current.

The performance of this switch is shown in Fig. 5-20. Peak isolation occurred at 4.014 GHz which is 0.35 percent higher than the design frequency of 4.00 GHz. Some of the error in design frequency was caused by finite fabrication tolerances in the length of the stub to the diode.

B. TEM Transmission Line — Microstrip

The difference between microstrip and stripline is that microstrip has the ground plane and dielectric only on one side of the center conductor strip. Normally alumina or sapphire are used for a dielectric instead of Teflon-fiberglass. When heat is to be dissipated, beryllia is used for a substrate; for monolithic integrated circuits, silicon is used. The comparison of the various media is shown in Table 5-1.

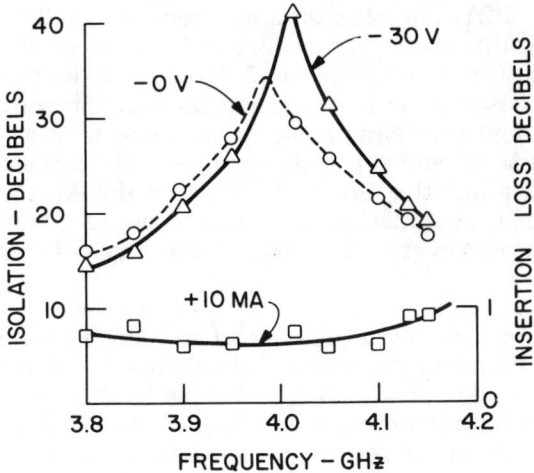

Fig. 5-20   Performance of the 4-GHz stub switch.

TABLE 5-1

Transmission Line	Dielectric	ϵ_R	Propagation Velocity	Miniaturization Ratio
Coax	air	1	3×10^{10} cm/sec	1.0
Stripline	Teflon-fiberglass	2.65 -2.22	1.85×10^{10}	1.62
Microstrip	alumina	8.5	1.03×10^{10}	2.92
Microstrip	sapphire	8.6-10.6	$.97 \times 10^{10}$	3.09
Microstrip	beryllia	6.7	1.18×10^{10}	2.54
Microstrip	silicon	11.7	$.88 \times 10^{10}$	3.41

Comparison of TEM Transmission Line Media

The propagation velocity in microstrip is slightly dependent on frequency. For substrates with a high dielectric constant most of the electric field is concentrated in the dielectric, thus causing the dielectric constant to dominate the propagation velocity. Note that components in microstrip are about half the size of their counterparts in stripline and one third the size of their counterparts in air filled coax.

Because of the high dielectric constant in microstrip and consequent reduction of the need for mode suppression screws, it is practical to "snake" the transmission lines for a further reduction in size as illustrated in Fig. 5-21. The $.25\lambda$ section is reduced to $.05\lambda$ in length which gives an additional reduction in size by a factor of 5. The total miniaturization ratio becomes 15. How close the conductors can be placed to each other is dependent on the strip width, which is in turn dependent on characteristic impedance, relative dielectric constant of the substrate, and substrate thickness (all in relation to wavelength). Although the "snaked" strip uses the same area as a straight strip, the size reduction is significant because "unsnaked" stripline and microstrip circuits usually make very poor use of substrate area.

Not only is size reduced by using microstrip transmission line but performance of control elements is also enhanced. The reason for the improvement can be appreciated by referring to Figs. 4-12 and 4-13. The smaller the diode structure is, the higher the self-resonant frequency (ω_R) is and hence the greater the bandwidth that is possible (Fig. 4-8). The cost of using microstrip is that of requiring special tools to handle and make electrical connection to the diodes.

CONTROL ELEMENT DESIGN

Instead of using fingers, tweezers, and a soldering iron, the diodes must be handled using vacuum pick-up micro-manipulators under microscopes and connected using thermo–compression bonders, ball bonders, ultrasonic welders, or two-point spot welders. In some cases the microstrip substrate must also be placed in a hermetically sealed container to prevent water vapor or oxygen from introducing troublesome surface states around the semiconductor junction.

The two most common types of diode used in microstrip are shown in Fig. 5-22. Diodes intended for shunt use have larger capacitance ($\sim$ .5 pF) than those intended for series use (.05 pF). The beam lead diode has extremely low capacitance and series resistance of 1-2 ohms. It is very good for series mounting up to 20 GHz. Power dissipation is not very high because the heat must be carried away from the junction via the very thin beam leads. Typical applications are limited to the 1 watt average power range. The mesa chip diode, on the other hand, has a very good thermal path to ground (through the thin semiconductor chip) and lower series resistance. It is, therefore, better for higher power applications and performs best in shunt.

Fig. 5-23 shows a microstrip limiter using two chip diodes in a small metal channel. No substrate is used. The connectors are bonded directly to the diodes, and the strap (center strip) between the diodes is suspended in air being attached at the ends to the diodes. The limiter has low insertion loss through X-band in the manner of the description for Fig. 6-5 and Fig. 9-11.

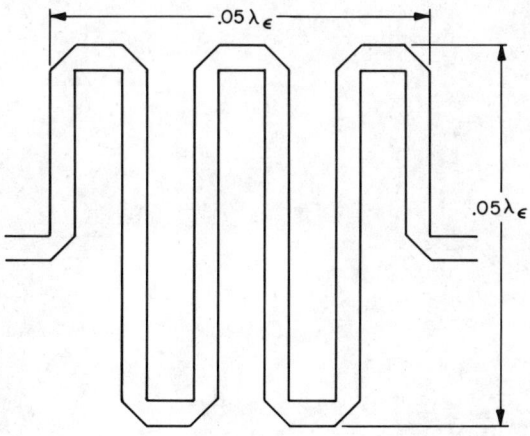

Figure 5-21
$\lambda/4$ Stub Reduced in Size by "Snaking."

148 MICROWAVE DIODE CONTROL DEVICES

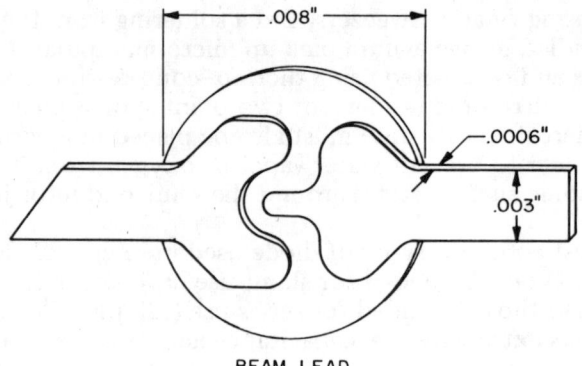

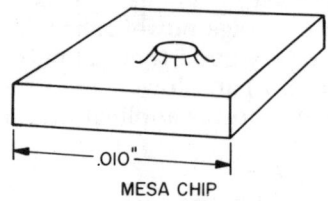

Figure 5-22
Integrated Circuit Diodes.

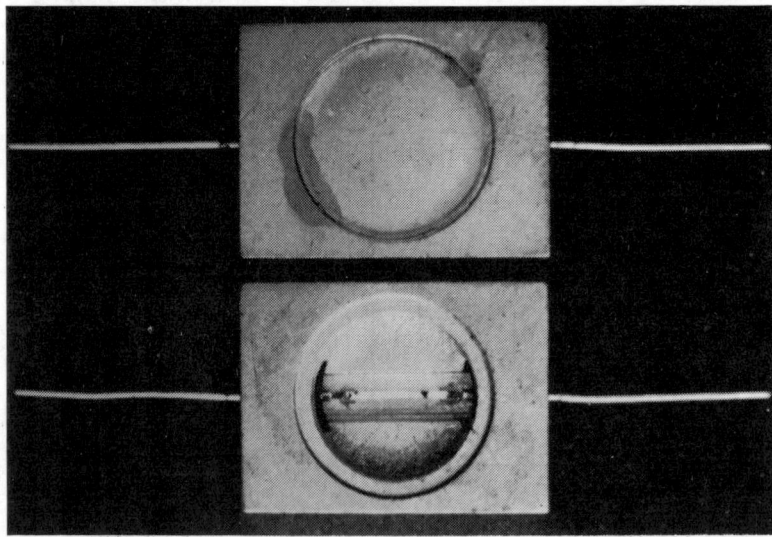

Figure 5-23
Microstrip Limiter (courtesy Hewlett-Packard).

CONTROL ELEMENT DESIGN

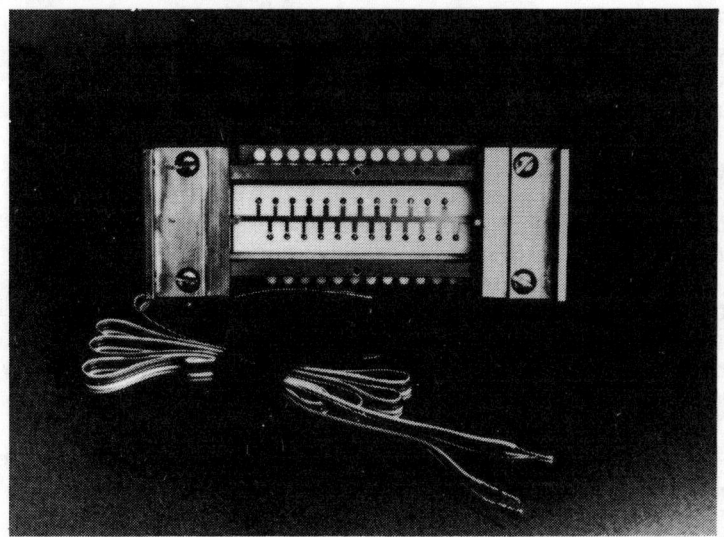

Figure 5-24
Loaded Line Phase Shifter (courtesy Microwave Associates).

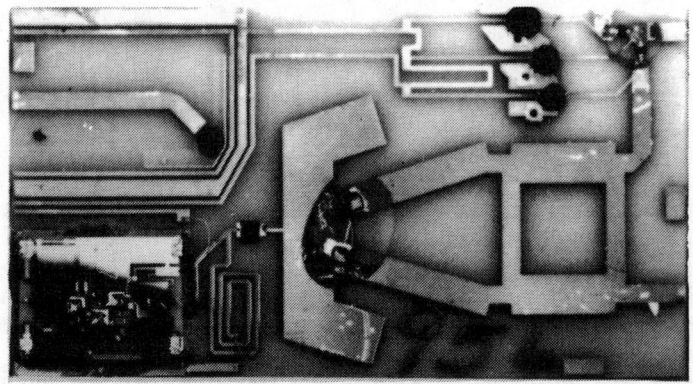

Figure 5-25
MERA Receiver Chip (courtesy Texas Instruments).

Fig. 5-24 shows a loaded line phase shifter using mesa chip diodes. Holes are drilled through the alumina substrate so the diodes can be mounted to the metal studs (capacitively isolated for bias) below the substrate. Input and output are X-band waveguide with transitions to microstrip, making the center strip at DC ground. The phase shifter controls 3 kW peak, 30 W average, 4 bits, 360°, with an average insertion loss of 2 dB and average VSWR of 1.3. Note that no stub pairs are used for 180° and 90°. Instead several 45° and 22.5° pairs are combined to give these bits. This technique avoids the problem of narrow bandwidth associated with large phase shift bits in loaded line phase shifters and also allows the diodes to control higher average incident power in accordance with Eq. (10-127) (by keeping shunt stub characteristic admittance low).

Fig. 5-25 shows a receiver hybrid IC chip from a Miniature Experimental Radar (MERA). The chip is dominated by the branch line mixer in the center and upper left areas. A tiny SPDT switch is shown in the upper right hand corner. The transmitter power enters from the right, and the antenna is located above. The switch connects the antenna element alternately to the transmitter and receiver. The diode bias lines lie along the right of the chip, and the IF amplifier is in the lower left corner. Beam lead diodes are used in the switch. Series capacitors can be seen adjacent to them. The switch works at X-band giving 1.1 dB insertion loss, 1.2 VSWR and greater than 24 dB isolation.

A switched line phase shifter is shown in Fig. 5-26 using beam lead diodes. The chip contains a 5-bit switched line phase shifter (on the left), a TR switch (single diode in the middle), and a balanced detector for phase comparison (right side).

Microstrip circuits are quite easy to make at frequencies up to 5 GHz and are useful up to 20 GHz. At frequencies above 5 GHz greater care has to be exercised that the inclosure in which the microstrip substrate is mounted does not permit propagation of waveguide modes. Even at the lower frequencies the theoretical isolation is frequently not obtained because the input side of an OFF switch excites a waveguide mode which is coupled to the output side of the inclosure. Even when the inclosure is a waveguide below cutoff, it has finite attenuation permitting some coupling. High isolation can be obtained only by making the inclosure very small to operate far below waveguide cutoff as in Fig. 5-23. Another solution to obtaining high isolation is to use multiple inclosures with a small coaxial transmission line connecting them. Most applications, however, require only modest isolation ($\sim$ 20 dB).

CONTROL ELEMENT DESIGN

Figure 5-26
Substrate with Switched Line Phase Shifter,
TR Switch, and Balanced Detector (courtesy Westinghouse).

C. Waveguide

The basic waveguide diode switch and performance are shown in Fig. 5-27. A cross-sectional view of it was shown in Fig. 2-13. The basic diode switch has proven practical for a number of reasons. It can be made inexpensively using automatic machines designed to make conventional waveguide mixer diodes. The bandwidth for one diode for 1 dB to 20 dB control range is greater than 20-percent which is sufficient for most systems. (In Chapter VI it is shown how multiple diodes provide even greater bandwidths.) The operating frequency and impedance level are determined by the diode whisker inductance and junction capacitance; thus, no sensitive tuning of the mount is required; one mount can work for a wide range of frequencies simply by using any of the commercially available diodes tuned to work over 20-percent bands in the waveguide bandwidth (3 GHz - 24 GHz). The diodes can be easily replaced in the field. The diode can be optimized for high power or high speed and still be used in the same mount. These diodes have given switching speeds as low as 0.1 nsec or switched power as high as 10 Watts.

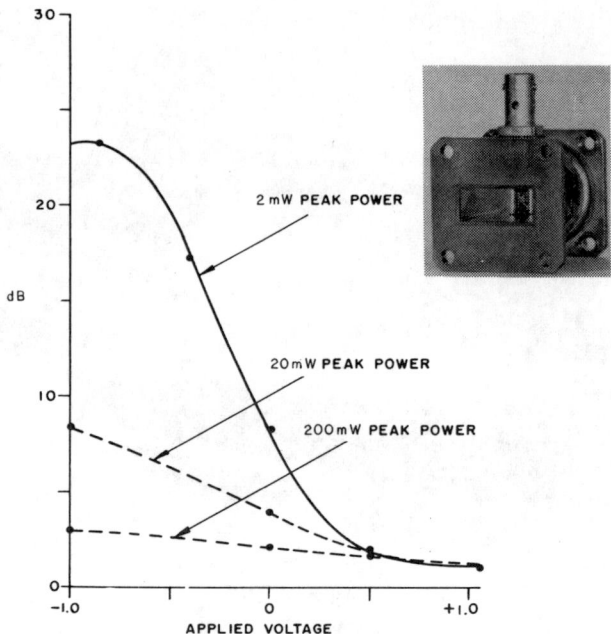

Fig. 5-27   Switching characteristic of the IN263 Germanium diode.

A circuit that permits a switching diode, which is being used below its self-resonant frequency[3] to be tuned to self-resonance, is shown in Fig. 5-28. The tuning screws adjust the inductance in series with the diode which adjusts the frequency at which maximum isolation occurs (resonance between the reverse biased junction capacitance and total inductance). The capacitative iris resonates with the inductance of the forward-biased diode to cause minimum insertion loss to occur at the desired frequency. The height of the waveguide is selected to give the desired impedance level. This circuit has advantages over the first in that one diode can be made to work over a wider range of frequencies. Its disadvantages are that the diode-to-diode variations make it necessary for the tuning screw to be readjusted whenever the diode is replaced, and it is not always convenient to work with the lower waveguide impedance.

The switching polarity of the basic switch of Fig. 5-27 is wrong for making a limiter. For the correct polarity, the switch should provide high isolation with diode conduction and low insertion loss with non-conduction. Then, rectification from high power forces the switch into the high loss state. The switch of Fig. 5-27 reflects power which it does not transmit, thus, the reflected power has the proper

CONTROL ELEMENT DESIGN

polarity for making a limiter. When one or two of these are used with a circulator or 3 dB coupler as in Fig. 2-17, then the unit will function passively as a limiter. More information on limiters is given in Chapter 10.

Another method for obtaining more freedom in the design of waveguide switches is to use series or shunt waveguide T junctions[4-6] as shown in Fig. 5-29. Two degrees of freedom are considered: the distance ℓ_1 from diode center to short circuit behind it and the distance ℓ_2 from main waveguide wall to diode center. These two degrees of freedom allow the insertion loss and isolation to both be optimized at the design frequency. There is no degree of freedom for the impedance level thus either one of the insertion loss or isolation tends to be broad band at the expense of the other. Greater circuit optimization might be possible by considering a third degree of freedom the distanced from the diode to the sidewall. A dual T switch[7] is shown in Fig. 5-30 giving only about ½ percent bandwidth for 20 dB isolation. It is noteworthy that by adjusting the four dimensions either polarity of switching could be achieved and that even silicon mixer diodes could be made into good switches using this structure.

The waveguide stub switches are also useful for very high RF power levels. By putting an iris at the waveguide input to the stub, the amount of coupling to the stub can be controlled. A large number

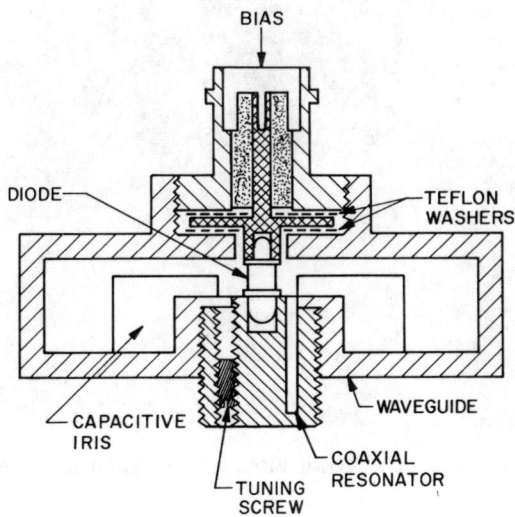

Figure 5-28   Tunable Waveguide Diode Mount.

154 MICROWAVE DIODE CONTROL DEVICES

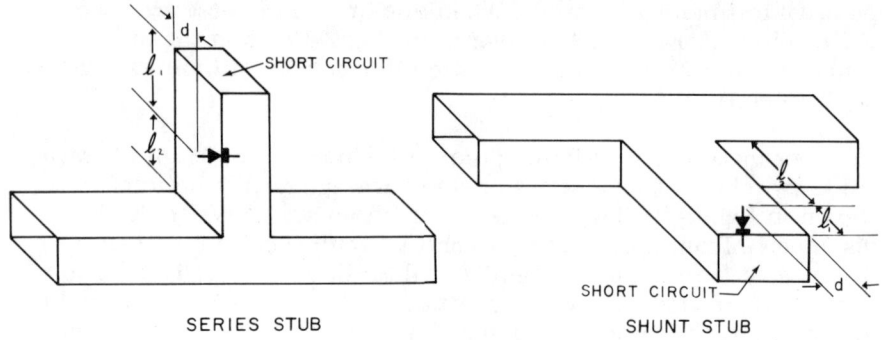

Figure 5-29
Waveguide Stub Switches.

Fig. 5-30   External view of hybrid-tee switch.

CONTROL ELEMENT DESIGN 155

of stubs loosely coupled to the main transmission line can produce high isolation with low insertion loss and reduce the power each diode has to control.

A similar method uses loops[8] to couple to the control diodes as shown in Fig. 5-31. The coupling is easily adjusted by changing the angle of the loops. Diodes with large junctions can be used with the loops to control very high power levels over narrow bandwidths.

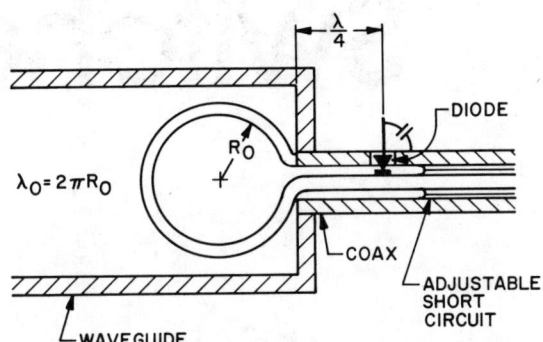

Figure 5-31
Loop Coupled Waveguide Diode Switch.

REFERENCES

1. J.F. White, "High Power, PIN Diode Controlled, Microwave Transmission Phase Shifters," *IEEE Trans. on Microwave Theory and Techniques*, Vol. MTT-13, March 1965, pp. 233-242.

2. A.G. Franco and A.A. Oliver, "Symmetric Strip Transmission Line Tee Junction," *IRE Trans. on Microwave Theory and Techniques*, Vol. MTT-10, March 1962, pp. 118-124.

3. R.N. Assaly, "Some Designs of X-Band Diode Switches," *IEEE Trans. on Microwave Theory and Techniques*, Vol. MTT-14, November 1966, pp. 553-563.

4. D.L. Rebsch, "A Low-Loss, Semiconductor Microwave Switch," *IRE Trans. on Microwave Theory and Techniques*, Vol. MTT-9, March 1961, pp. 644-645.

5. V.J. Higgins, "X-band Semiconductor Switching and Limiting Using Waveguide Series Trees," *The Microwave Journal*, November 1963, pp. 77-82.

6. H.J. Peppiatt, A.V. McDaniel, Jr., and J.B. Linker, Jr., "A 7-Gc/s Narrow-Band Waveguide Switching Using PIN Junction Diodes," *IEEE Trans. on Microwave Theory and Techniques*, Vol. MTT-13, January 1965, pp. 44-47.

7. R.V. Garver, E.G. Spencer, and M.A. Harper, "Microwave Semiconductor Switching Techniques," *IRE Trans. on Microwave Theory and Techniques*, Vol. MTT-6, October 1958, pp. 378-383.

8. W.C. Jakes, Jr., "Analysis of Coupling Loops in Waveguide and Application to the Design of a Diode Switch," *IEEE Trans. on Microwave Theory and Techniques*, Vol. MTT-14, April 1966, pp. 189-200.

Multiple Diode Switches

Chapter 6

The main reason for using multiple diodes in switches is the very substantial increase in bandwidth possible over that which can be obtained with a single diode. When one or more identical diodes are mounted in the same electrical plane, all in series or all in shunt, their bandwidths together will be the same as the bandwidth of one of them, according to the relationships derived in Chapter IV, Section B. But when they are properly mounted in different electrical planes, their insertion loss reactances interact in such a manner as to provide an even lower insertion loss then each diode would give separately and the isolation reactances interact in such a manner as to provide more isolation than each diode would provide separately. This enhancement of insertion loss and isolation may be understood by considering the filter structures shown in Fig. 6-1.

A. Filter Theory (Lumped Circuits)

   1. Insertion Loss

In the insertion loss state the diode is a series inductance (or shunt capacitance) and thus becomes an element in a low pass filter. The maximum insertion loss and cutoff frequency are fixed by the inductance (or capacitance) and characteristic impedance of generator and load. The capacitance (or inductance) in the filter must be selected to complement the diode inductance (capacitance) to give the desired maximum insertion loss and cutoff frequency with the given characteristic impedance. The elements may be selected using Eqs. C-1 and C-2. A single diode has an insertion loss given by

$$\delta = 10 \log [1 + (\tfrac{1}{2} B_N)^2] \qquad (6\text{-}1)$$

where $B_N = \omega C/Y_0$ for a shunt diode and which may be replaced by

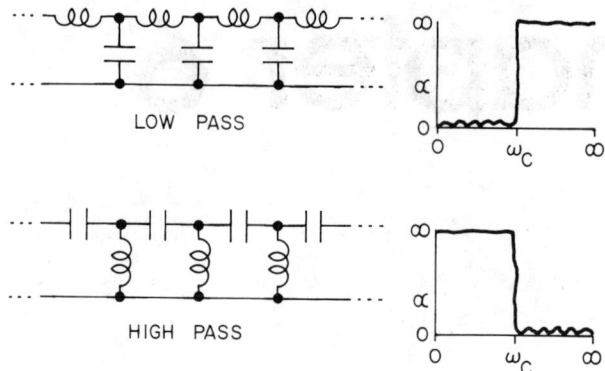

Fig. 6-1   Low Pass and High Pass Filters

$X_N = \omega L/Z_0$ for a series diode. This insertion loss is shown for one diode in Fig. 6-2. By tuning the insertion loss at some higher frequency the bandwidth remains the same: ω is replaced by bandwidth $\Delta\omega$. The curve for the Bode limit (refer to appendix C section 3) shows how much more bandwidth can be obtained using filter design methods. The Bode limit is

$$\ln\left|\frac{1}{\Gamma}\right| = \frac{\pi}{(\Delta\omega)CR} \qquad (6\text{-}2)$$

in which the reflection coefficient is constant at $|\Gamma|$ over a frequency range of $\Delta\omega$ and then goes to unity outside the frequency range. R is the load resistance, and C is the capacitance being matched into. In terms already defined the Bode limit can be written

$$|\Gamma| = \epsilon^{-\pi/B_N} \qquad (6\text{-}3)$$

The insertion loss becomes

$$\delta = 10\log\left[\frac{1}{1-|\Gamma|^2}\right] = -10\log\left[1 - \epsilon^{-2\pi/B_N}\right] \qquad (6\text{-}4)$$

which is shown as the Bode limit in Fig. 6-2. The improvement in bandwidth ranges from 10 to 1 for the lowest insertion loss shown to 5 to 1 for higher insertion loss.

The Bode limit is rapidly approached with a small number of elements in the filter. Using a 3-element filter with 2 diodes for the 2 shunt capacitors, the "ends of 3-element filter" curve in Fig. 6-2 can be drawn. From Eq. C-1

MULTIPLE DIODE SWITCHES

$$C_K = \frac{Y_0}{\omega_1'} g_K$$

$$C_1 = \frac{Y_0}{\omega_1'} g_1 \qquad (6\text{-}5)$$

$$B_{CN} = g_1$$

where $g_1 = g_3 =$ the coefficients for the prototype filter using here the coefficients for the Tchebyscheff (equal ripple) filter.

The insertion loss bandwidth of multiple diode rapidly approaches the Bode limit. After about 3 diodes the bandwidth is not significantly increased but neither is it decreased. Continuously increasing the number of diodes will not change the insertion loss bandwidth but will increase the isolation (at a slight increase of in-band insertion loss due to the finite losses of the ON diodes).

2. Isolation

In the isolation state the diode is a series capacitance (or shunt inductance) and thus becomes an element in a high pass filter as shown in Fig. 6-1. Selecting elements to give a Tchebyscheff response

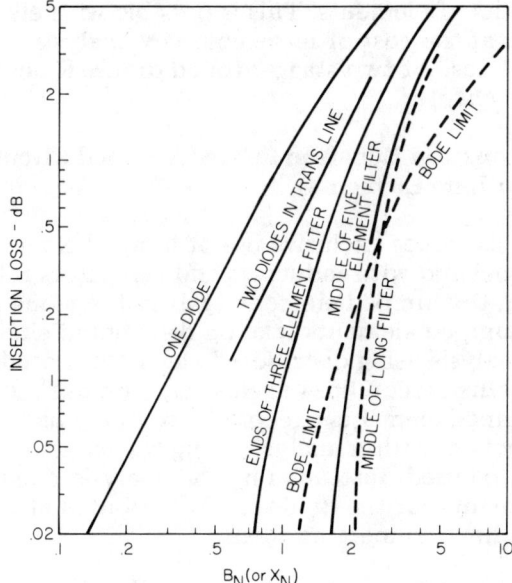

Fig. 6-2 Insertion Loss Bandwidth Limits

would only guarantee that the insertion loss ripple in the high pass is well behaved, which is of no significance in a diode switch that is supposed to work up to ω_C. The isolation slope of a Tchebyscheff filter is steeper than that of a maximally flat filter; however, the characteristics of the maximally flat filter are simpler and, therefore, will be used to illustrate how isolation is improved by adding more diodes. The attenuation of a maximally flat filter is given by

$$\alpha = 10 \log \left[1 + \left(\frac{\omega_C}{\omega}\right)^{2n}\right] \approx 20 \, n \log \left(\frac{\omega_C}{\omega}\right) \qquad (6\text{-}6)$$

in which n is the number of elements in the prototype and ω_C is the 3 dB frequency. The isolation of a single shunt diode is given by

$$\eta = 10 \log \left[1 + (½ B_N)^2\right] = 10 \log \left[1 + \left(\frac{Z_0}{2\omega L}\right)^2\right] \qquad (6\text{-}7)$$

which has the form of a maximally flat filter in which n = 1 and $\omega_C = \frac{Z_0}{2L}$. Eq. 6-6 thus indicates that each element added to the high pass filter structure adds more isolation equal to that of a single element. Greater isolation bandwidth is obtained by having many elements and the highest possible ω_C. Greater isolation bandwidth is possible using fewer elements by replacing elements of the filter with resonant pairs of elements. This is possible with elliptic function filters but is at the cost of more complex analysis and structures. The simple case of two stagger tuned diodes is developed in Section 4 of this chapter.

B. Hybrid Attenuation Equation (Mixed Lumped Circuit and Transmission Line Elements)

The above analysis makes exclusive use of lumped elements. As long as the length associated with inductors and capacitors is less than 1/10 wavelength, the lumped element analysis is adequate. Above 5-10 GHz even lumped elements take on distributed circuit characteristics; thus, an analysis using distributed elements is needed. It is not possible to use transmission lines exclusively because the diode parasitics are lumped elements except at extremely high frequencies. Therefore conventional filter analysis using transmission line elements cannot be used. Instead, a hybrid analysis is made which uses lumped elements for the diodes and distributed elements for the transmission lines connecting them.

MULTIPLE DIODE SWITCHES

1. Derivation

The circuit composed of two lumped elements jX_1 and jX_2, separated by a distributed element transmission line θ and Z_0 is shown in Fig. 6-3 ($\theta = 2\pi\ell/\lambda$). The ABCD matrix for the circuit is given by

$$\begin{bmatrix} A & B \\ C & D \end{bmatrix} = \begin{bmatrix} 1 & jX_1 \\ 0 & 1 \end{bmatrix} \begin{bmatrix} \cos\theta & jZ_0\sin\theta \\ jY_0\sin\theta & \cos\theta \end{bmatrix} \begin{bmatrix} 1 & jX_2 \\ 0 & 1 \end{bmatrix} \quad (6\text{-}8)$$

$$= \begin{bmatrix} \cos\theta - X_1 Y_0 \sin\theta & j(X_2\cos\theta + Z_0\sin\theta - X_1 X_2 Y_0 \sin\theta + X_1\cos\theta) \\ jY_0 \sin\theta & -X_2 Y_0 \sin\theta + \cos\theta \end{bmatrix}$$

The attenuation is given by Eq. (2-12)

$$\alpha = 10 \log \tfrac{1}{4} |A + BY_0 + CZ_0 + D|^2$$

$$= 10 \log \tfrac{1}{4} \left\{ 4 + (X_{1N} - X_{2N})^2 \sin^2\theta + \left[(X_{1N} + X_{2N})\cos\theta - X_{1N} X_{2N}\sin\theta\right]^2 \right\} \quad (6\text{-}9)$$

in which $X_{1N} = X_1/Z_0$, etc. The equation may be used for shunt elements by substituting $B_{1N} = B_1/Y_0$ for X_{1N}, etc.

For identical diodes $X_{1N} = X_{2N} = X_N$ and Eq. (6-9) reduces to

$$\alpha = 10 \log \frac{1}{4} \left[4 + (2X_N \cos\theta - X_N^2 \sin\theta)^2 \right] \quad (6\text{-}10)$$

The isolation for two 20-dB isolation switches is shown in Fig. 6-4 as their spacing, $\theta = 2\pi\ell/\lambda$ is varied. Both curves have a broad maximum. The reactive diodes isolation (eq. 6-10) has a very narrow low isolation region while the resistive diodes isolation (eq. 8-34) has minima of 26 dB. These curves indicate that the spacing for optimum isolation is not critical while $\lambda/2$ spacing should be avoided.

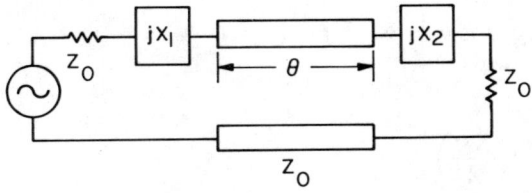

Fig. 6-3 Circuit for calculation of isolation of two lumped elements

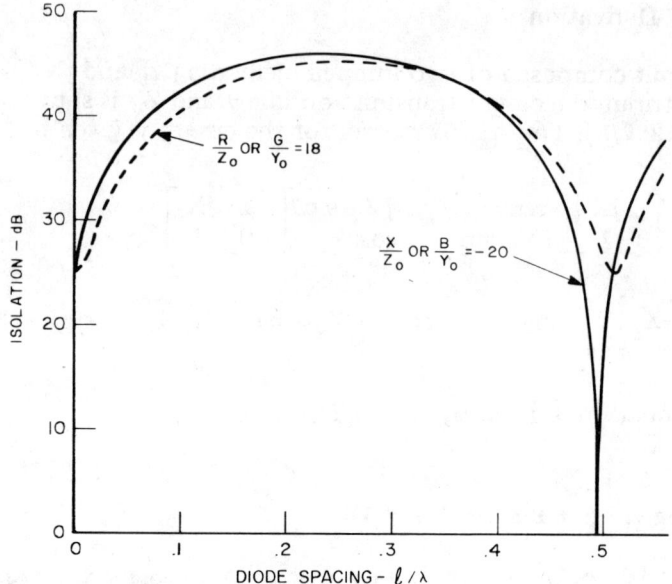

Fig. 6-4   Isolation of two 20-dB diodes

The condition for zero insertion loss is that

$$2 X_N \cos \theta - X_N^2 \sin \theta = 0$$

which gives

$$\theta = \tan^{-1} \left(\frac{2}{X_N}\right) \tag{6-11}$$

when X_N is small (insertion loss state) $\theta \approx 90°$. When X_N is large isolation state), $\theta \approx 180°$.

The condition for maximum isolation is that

$$\frac{\partial}{\partial \theta} (2 X_N \cos \theta - X_N^2 \sin \theta) = 0$$

which gives

$$\theta = \tan^{-1} \left(\frac{X_N}{2}\right) \tag{6-12}$$

when X_N is large $\theta \approx 90°$.

MULTIPLE DIODE SWITCHES

Thus a quarter wavelength between diodes gives both lowest insertion loss and highest isolation with two identical diodes, while half wavelength must be avoided because isolation goes to zero.

Substituting Eq. (6-12) into Eq. (6-10) gives the maximum isolation for 2 diodes when spaced for maximum isolation.

$$\eta_2 = 20 \log (1 + \tfrac{1}{2} X_N^2) \qquad (6\text{-}13)$$

This is to be compared to the isolation of one diode

$$\eta_1 = 10 \log \left(1 + \tfrac{1}{4} X_N^2\right) \qquad (6\text{-}14)$$

For X_N large

$$\eta_2 = 10 \log \left(\tfrac{1}{4} X_N^4\right) = 2 \eta_1 + 6 \text{ dB} \qquad (6\text{-}15)$$

Thus two identical diodes at the optimum spacing give twice the isolation of one diode plus approximately six dB extra. The extra isolation is given more exactly by

$$\eta_{EX} = \eta_2 - 2\eta_1 = 20 \log \left(2 - 10^{-\eta_1/10}\right) \qquad (6\text{-}16)$$

This extra isolation falls to 5 dB when $\eta_1 = 6.5$ dB. Third, fourth, etc. diodes also give this extra attenuation with each one.[1] Thus two 10 dB diodes will give 26 dB (actually 25.5 dB), three will give 42 dB (actually 41 dB), and so on.

The calculations so far have been at one frequency with approximately a quarter-wavelength between diodes. The influence of the second diode on wideband switching may be seen from Fig. 6-5, which is a plot of Eq. (6-10).

2. Insertion Loss

The normalized reactance of a series diode in the low loss state is given by $X_N = \dfrac{\omega L}{Z_0}$ which is positive and proportional to ω. The normalized susceptance of a shunt diode in the low loss state is also positive and proportional to ω: $B_N = \dfrac{\omega C}{Y_0}$. Since ℓ/λ is proportional to ω for a fixed ℓ in TEM transmission line, as frequency is swept from d-c upward, a straight line is drawn on Fig. 6-5 beginning at the origin and traveling into the first quadrant. The insertion loss

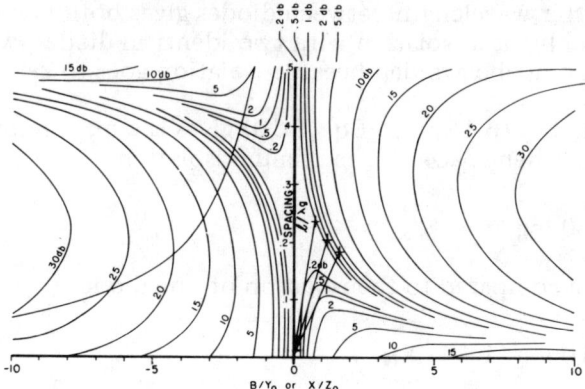

Fig. 6-5   Attenuation of two lumped elements

behaves like a single ripple low pass Tchebyscheff filter having a 3-element prototype. The end of each line gives the maximum X_N (B_N) for the attenuation curve that it is tangent to and stops at. These lines give the "two diodes in trans. line" curve on Fig. 6-2. By increasing the Z_0 of the transmission line between the diodes the connecting section approaches a lumped inductance, and the Tchebyscheff curve for "ends of 3-element filter" on Fig. 6-2 can be approached.

To make a baseband switch using Fig. 6-5, the upper end of the desired straight line would give the spacing between diodes at the highest frequency ℓ/λ and the normalized diode impedance (admittance) X_N (B_N) at the highest frequency. For example, to obtain 1/2 dB maximum insertion loss up to 4 GHz in a 50-ohm transmission line, the 1/2 dB line indicates ℓ/λ = .206 and X_N (B_N) = 1.2. Thus the diodes should be separated by .206 wavelengths at 4 GHz and have inductance less than .41 nH for a series diode or capacitance less than .164 pF for a shunt diode.

3.   Isolation - Baseband

The normalized reactance (susceptance) of a series (shunt) diode in the isolation state is inversely proportional to frequency and is negative. Therefore, as frequency is varied, the diode will generate a hyperbola in the second quadrant of Fig. 6-5. The figure gives little insight into the isolation performance of two identical diodes; however it does show that the same diode spacings that give minimum insertion loss also give maximum isolation.

MULTIPLE DIODE SWITCHES

Figure 6-6 gives a better insight into the isolation performance of two diodes. A diode was assumed to give 30 dB isolation at 1 GHz in 50-ohm transmission line. This isolation would result from a series diode of .05 pF capacitance or a shunt diode of .126 nH inductance. Two diodes in the same electrical plane given an isolation of:

$$\eta = 10 \log \left[1 + X_N^2 \right] \qquad (6\text{-}17)$$

which is derived from Eq. (6-10) by setting $\theta = 0$. At isolations greater than 10 dB this isolation is just 6 dB more than that available from a single diode.

Now suppose the optimum spacing is used at 1 GHz which gives 30 dB isolation. At this frequency $\dfrac{X_N}{2} = 31.6$ which gives $\theta = 88°$ according to Eq. (6-12). The isolation should be 66 dB according to Eq. (6-15). The curves of Fig. 6-6 were generated using Eq. (6-10) with X_N given by

$$X_N = 63.2/f \qquad (6\text{-}18)$$

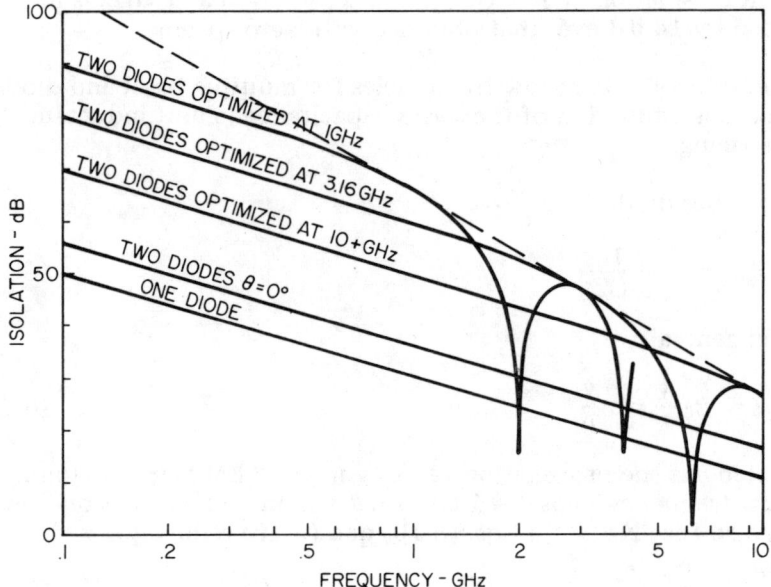

Fig. 6-6   Isolation of two diodes for various spacing

in which f is in GHz. The dashed line in Fig. 6-5 shows the optimum isolation for two of these diodes when properly spaced at each frequency as given by Eq. (6-13). The curve for two diodes optimized at 1 GHz gives the optimum isolation at 1 GHz ($\theta = 88° \times f$). Below 1 GHz the isolation approaches 40 dB greater than the isolation of one diode. Above 1 GHz the isolation falls to zero at 2 GHz, 4 GHz, 6 GHz, etc. when the diodes are separated by multiples of a half-wavelength but peaks again at the optimum at 3 GHz, 5 GHz, 7 GHz, etc. when the diodes are separated by odd multiples of a quarter wavelength. Two diodes which are optimized at the 20 dB point, 3.16 GHz, give 30 dB more isolation than one diode at low frequencies, give zero isolation at even multiples of the design frequency, and optimum isolation at odd multiples of the design frequency as before. The 10 dB isolation point of one diode is slightly above 10 GHz. Two diodes spaced for optimum isolation at this frequency give 20 dB more than the isolation of one diode at lower frequencies, and so on. As shown in Fig. 6-6 the two diode switch retains the baseband characteristics of the single diode with considerably increased isolation over the entire band. The isolation at low frequencies is the isolation of one diode plus the isolation of one diode at the frequency for which optimum spacing was selected plus 10 dB. Even when there is not much room for spacing between diodes, considerable improvement is made possible by the small spacing. The spacing at 1 GHz is only 6.8° when the two diodes are spaced for optimum isolation at 10+ GHz, yet the isolation at 1 GHz is increased by 14 dB over that obtained with zero spacing.

The extra isolation at low frequencies for multiple baseband diode switches is a function of the series capacitance (shunt inductance) and spacing.

For a series diode

$$X_N = -\frac{1}{\omega C Z_0} \qquad (6\text{-}19)$$

and in general

$$\theta = 2\pi \frac{\ell}{\lambda} = \omega \frac{\ell}{v} \qquad (6\text{-}20)$$

in which v is the propagation velocity in the TEM transmission line. At low frequencies $\cos \theta = 1$ and $\sin \theta = \theta$ and when the isolation is high so that $X_N \gg 2$ Eqs. (6-10) and (6-16) reduce to

$$\eta_{EX} = 20 \log \left(\frac{1}{\omega C Z_0} \cdot \frac{\omega \ell}{v}\right) = 20 \log \left(\frac{\ell}{v C Z_0}\right) \qquad (6\text{-}21)$$

MULTIPLE DIODE SWITCHES

This extra isolation is plotted in Fig. 6-7 for Z_0 = 50 ohms. For shunt diodes $C Z_0$ is replaced by $L Y_0$. For Fig. 6-6 it was assumed that C = .05 pF and Z_0 = 50 ohms. Optimum spacing at 1 GHz was at a quarter wavelength which is 7.5 cm in air and $v = 3 \times 10^{10}$ cm/sec in air. Thus Eq. (6-21) provides η_{ex} = 40 dB. Fig. 6-6 shows 1 diode giving 50 dB isolation at .1 GHz while two of them giving optimum isolation at 1 GHz give 90 dB isolation at .1 GHz.

4. Tuned Switching Elements

The insertion loss of resonant diodes is well behaved. The bandwidth and responses in transmission lines will be the same as for baseband but centered at the operating frequency, f_0. If the insertion loss were less than δ from 0 to f_δ, it will be less than δ from $f_0 - \frac{1}{2} f_\delta$ to $f_0 + \frac{1}{2} f\delta$. Fig. 6-4 is cyclic in ℓ/λ; that is, the origin may be put at $\ell/\lambda = .5$ and continue upward, repeating for every multiple of $\ell/\lambda = .5$. A diode designed to resonate at $\lambda/2$ will pass through the $\ell/\lambda = .5$ point at twice the slope it would have at d.c. if it were untuned. Thus the insertion loss at $\lambda/2$ ($\theta = 180°$) spacing between resonant diodes is well behaved. The insertion loss for

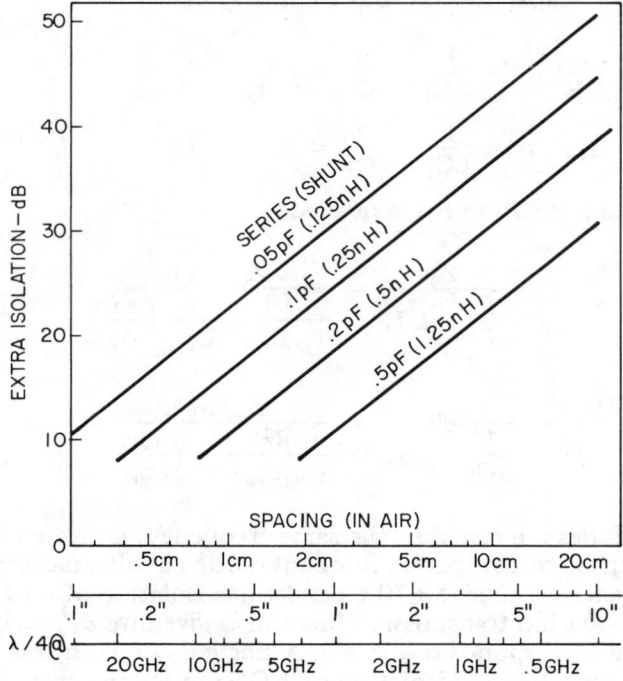

Fig. 6-7 Extra isolation from a second diode

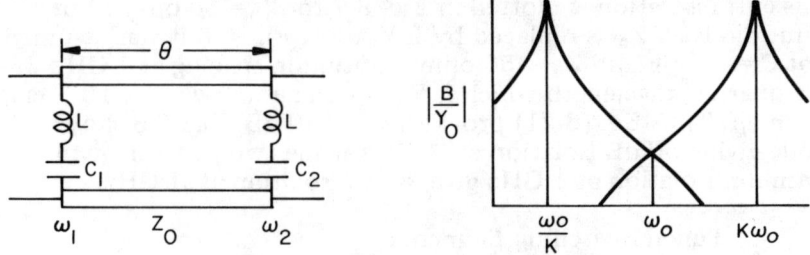

Fig. 6-8 Two diodes tuned to different frequencies

diodes spaced $\lambda/4$ at resonance is much greater than the insertion loss of a single diode. Special filter design methods must be employed with the insertion loss when $\lambda/4$ spacing is used for tuned diodes.

a. Isolation

The circuit for two resonant shunt diodes[2] is shown in Fig. 6-8. The diodes have the same series inductance L but are permitted to have different series capacitors C_1 and C_2 satisfying

$$\omega_1^2 = \left(\frac{\omega_0}{K}\right)^2 = \frac{1}{LC_1} \quad \frac{1}{C_1} = \frac{\omega_0^2 L}{K^2}$$

$$\omega_2^2 = (K\omega_0)^2 = \frac{1}{LC_2} \quad \frac{1}{C_2} = K^2 \omega_0^2 L \quad (6\text{-}22)$$

For calculating isolation B_N is needed.

$$B_{1N} = \frac{B_1}{Y_0} = \frac{Z_0}{\frac{1}{\omega C_1} - \omega L} = \frac{Z_0/\omega_0 L}{\frac{1}{(\omega/\omega_0)K^2} - \left(\frac{\omega}{\omega_0}\right)} \quad (6\text{-}23)$$

$$B_{2N} = \frac{B_2}{Y_0} = \frac{Z_0}{\frac{1}{\omega C_2} - \omega L} = \frac{Z_0/\omega_0 L}{\frac{K^2}{(\omega/\omega_0)} - \left(\frac{\omega}{\omega_0}\right)}$$

When the diodes are tuned to the same frequency, ω_0, then $K = 1$. For the purpose of comparing resonant diode circuits, diode parameters are selected to give $\pm 10\%$ bandwidth isolation of 30 dB at 1 GHz. For 50 ohm transmission line this is given by $Z_0/\omega_0 L = 12.64$ which corresponds to .63 nH. A single diode with this inductance gives 16.1 dB isolation of 1 GHz in the untuned mode. Although the curves are shown for 1 GHz, they will be the same

MULTIPLE DIODE SWITCHES

around any tuned frequency when the resonant elements are properly scaled.

For these values the isolation of one or two resonant diodes is shown in Fig. 6-9. For 0° spacing the isolation of two diodes is 6 dB more than the isolation of one diode. For $\theta_0 = 180°$ (θ_0 is the spacing between diodes at resonance) the isolation is considerably enhanced at lower isolations and slightly improved at higher isolations. For $\theta_0 = 90°$ the enhancement at higher isolations is better than at lower isolations.

A figure of merit can be derived for comparing isolation bandwidths. A standard is selected as Eq. (6-13), the widest bandwidth that can be achieved by optimizing the diode spacing for each isolation. A pair of switches will have a figure of merit of 100% when they are at their frequency of optimized isolation based on their spacing. The switches of Fig. 6-6 optimized at 3.16 GHz give the figure of merit shown as "Baseband" in Fig. 6-10. The curve is characteristic of the baseband circuit and will shift left or right on the figure without changing shape. It peaks at 100% at 46 dB isolation, which results from having been designed for the 20 dB point of a single diode. The diodes selected for Fig. 6-9 give 100% figure of merit

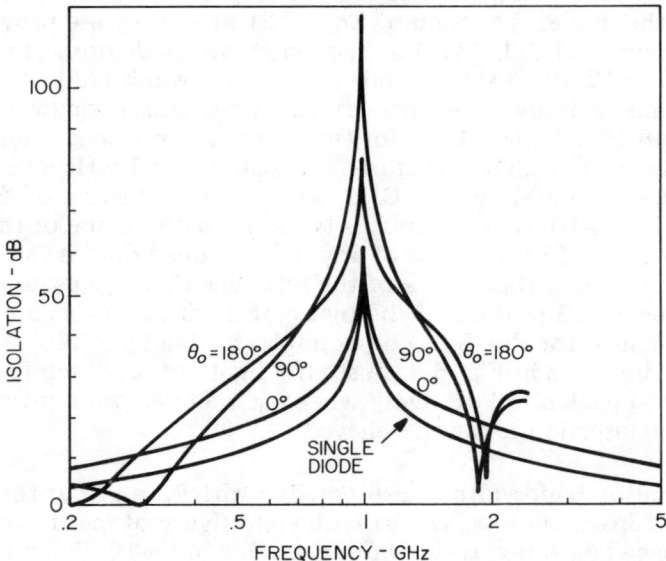

Fig. 6-9   Isolation of two resonant diodes (tuned to the same frequency)

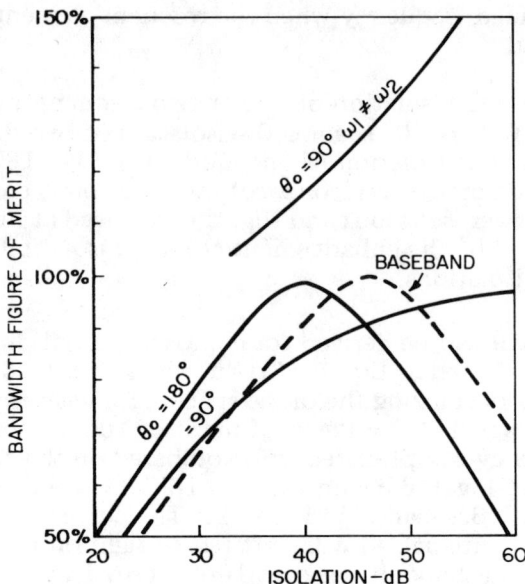

Fig. 6-10 Bandwidth figure of merit for resonant diodes

for 40 dB isolation. This would suggest that the electrical spacing between the diodes is optimum (Eq. 6-12) when they are providing this 40 dB at 1.54 GHz. At this frequency one diode does provide 1/2 (40-6) = 17 dB. A single diode gives 17 dB when $1/2\, B_N = +7$ (plus because it is above resonance) which indicates (using Eq. 6-12) that $\theta = 98.15°$. Adding 180° for the extra half wavelength and dividing by 180° which normalizes the distance to 1 GHz give that the frequency should be 1.54 GHz, which it is. The shape of this curve on Fig. 6-10 for $\theta_0 = 180°$ is the same as the shape of the baseband curve. The response of any pair of tuned diodes can be estimated by using the reverse of the rationale that was used to explain the 40 dB peaking of the figure of merit for the sample diodes assumed for Fig. 6-9. For example, setting $\theta_0 = 360°$ for the same diodes as in Fig. 6-9 causes the figure of merit curve in Fig. 6-10 to peak at 56 dB. Only when θ_0 is an integral multiple of 180° is the insertion loss well behaved.

Good isolation bandwidth is also possible with $\theta_0 = 90°$ at the expense of insertion loss. The isolation and figure of merit for two diodes spaced $\theta_0 = 90°$ is shown in Figs. 6-9 and 6-10. When the diodes are tuned to different frequencies ($k \neq 1$) with $\theta_0 = 180°$, the isolation at ω_0 goes to zero and therefore stagger tuning is not useful. Because it is impossible to get perfectly synchronous tuning

MULTIPLE DIODE SWITCHES

$\theta_0 = 180°$ is not useful for multiple tuned diode switches. However, at $\theta_0 = 90°$ marked improvements are possible at higher isolations by stagger tuning the diodes.[2]

For $\theta_0 \approx 90°$ and high isolation, Eq. (6-9) can be simplified to

$$\eta = 10 \log \tfrac{1}{4} B_{1N}^2 B_{2N}^2 \tag{6-24}$$

Using Eq. (6-23)

$$B_{1N}^2 B_{2N}^2 = \left[\frac{Z_0/\omega_0 L}{\frac{1}{(\omega/\omega_0)K^2} - \left(\frac{\omega}{\omega_0}\right)}\right]^2 \left[\frac{Z_0/\omega_0 L}{\frac{K^2}{(\omega/\omega_0)} - \left(\frac{\omega}{\omega_0}\right)}\right]^2$$

$$= \frac{(Z_0/\omega_0 L)^4}{F^2} \tag{6-25}$$

in which

$$F = \left[\left(\frac{\omega}{\omega_0}\right)^2 + \frac{1}{(\omega/\omega_0)^2}\right] - \left[K^2 + \frac{1}{K^2}\right] \tag{6-26}$$

the isolation will be

$$\eta = 10 \log \tfrac{1}{4} \left(\frac{Z_0}{\omega_0 L}\right)^4 \left(\frac{1}{F^2}\right) \tag{6-27}$$

which when compared to the first part of Eq. (6-15) indicates that the total isolation exceeds the optimum isolation of the two untuned elements at the center frequency by

$$\eta_{EF} = 10 \log \frac{1}{F^2} \tag{6-28}$$

Fig. 6-11A shows the extra isolation from F, η_{EF} when K is selected for octave bandwidth. The function F is shown in Fig. 6-11B. When $|F|<1$ the two diodes give more isolation than if they were untuned and had optimum spacing. The function F is the simple difference between two functions

$$F = f\left(\frac{\omega}{\omega_0}\right) - f(K) \tag{6-29}$$

where

$$f(x) = x^2 + \frac{1}{x^2} \tag{6-30}$$

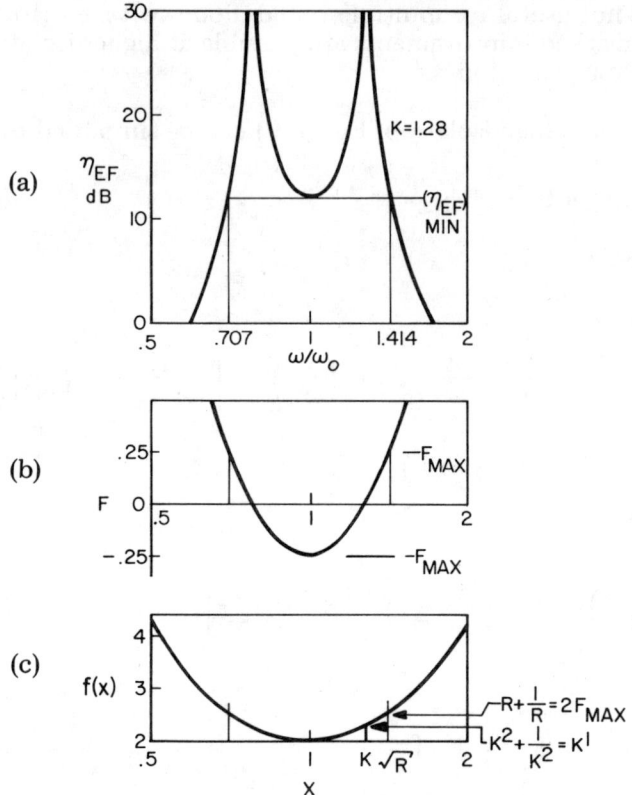

Fig. 6-11 Evolution of extra attenuation of stagger-tuned diodes for octave bandwidth

all of which can be seen by the construction shown in Fig. 6-11C. The constant K should be adjusted so that

$$K^2 + \frac{1}{K^2} - 2 = \tfrac{1}{2}\left(R + \frac{1}{R} - 2\right) \tag{6-31}$$

in which R is the desired bandwidth ratio (R = 2 for octave bandwidth). Bandwidths much larger than an octave will violate the approximation that $\theta \approx 90°$. Solving Eq. (6-31) for K gives

$$K = \tfrac{1}{2}\sqrt{\left(2 + R + \frac{1}{R}\right) + \sqrt{\left(2 + R + \frac{1}{R}\right)^2 - 16}} \tag{6-32}$$

At $\dfrac{\omega}{\omega_0} = 1$ (center of the operating band) and at $R^{\pm \tfrac{1}{2}}$ (ends of the operating band) F is given by

MULTIPLE DIODE SWITCHES

$$F = \tfrac{1}{2}(R + \tfrac{1}{R} - 2) \tag{6-33}$$

which for $R = 2$ gives $F = \tfrac{1}{4}$ and, by Eq. (6-28), $\eta_{EF} = 12$ dB minimum over the operating band.

For the same diodes assumed in Fig. 6-9, the isolation for several stagger tunings is shown in Fig. 6-12. The improvement in isolation is shown in Fig. 6-10 (the curves with figure of merit greater than 100%). For $k = 2$ the bandwidth was worse than what could be achieved by properly spacing the diodes in the baseband mode. The $\omega_1 \neq \omega_2$ curve on Fig. 6-10 crosses below 100%.

In summary, stagger tuning gives an isolation improved by 12 dB over what can be achieved by the best arrangement of two identically tuned diodes for octave bandwidth. The design for stagger tuning is straightforward, while for identically tuned diodes, not enough flexibility is available to obtain maximum bandwidth for any required isolation. The stagger tuned circuit is limited to bandwidths of the order of an octave or less and presents difficulty in obtaining low insertion loss.

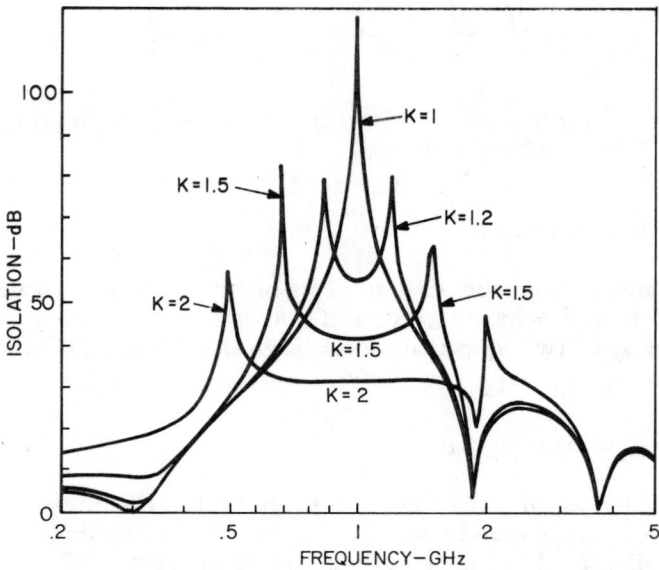

Fig. 6-12 Isolation of two stagger-tuned diodes, $\theta_0 = 90°$

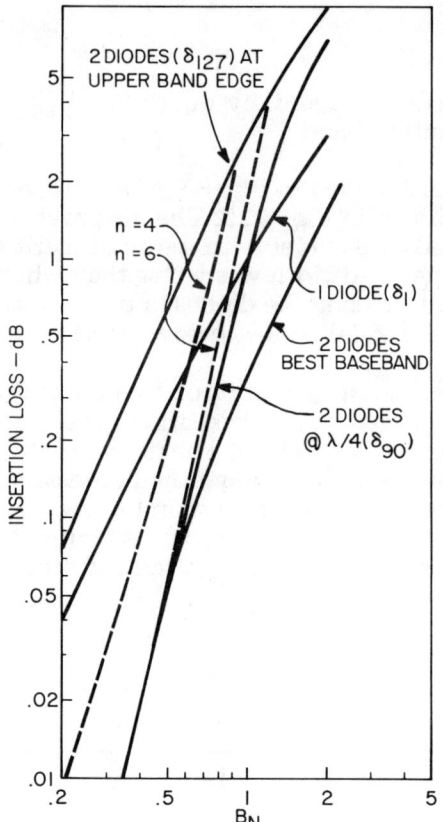

Fig. 6-13 Insertion loss bandwidth comparison, two diodes at $\lambda/4$, octave bandwidth

b. Insertion Loss

As was stated earlier, the insertion loss of tuned diodes spaced $\lambda/4$ is not as well behaved as for $\lambda/2$ spacing. For $\lambda/4$ spacing the insertion loss of two identical diodes is given by Eq. 6-10 with $\theta = 90°$

$$\delta_{90} = 10 \log (1 + \tfrac{1}{4} B_N^4) \qquad (6\text{-}34)$$

The insertion loss of one diode δ_1 is given by Eq. (6-1). For diodes tuned over an octave and spaced $\lambda/4$ at mid band, the angle θ is given by $90\sqrt{2} = 127°$ at the upper end of the band and Eq. (6-10) gives

MULTIPLE DIODE SWITCHES

$$\delta_{127} = 10 \log \left[1 + (.606 B_N + .398 B_N^2)^2\right] \quad (6\text{-}35)$$

The insertion loss at the lower end of the band is about the same because both B_N and the cosine terms change sign, keeping the signs the same. The coefficients are slightly different because the resonant response is geometrically rather than arithmetically symmetrical. Eqs. (6-35), (6-1), and (6-34) are compared in Fig. 6-13. Also shown in Fig. 6-13 is the maximum B_N for various baseband single ripple responses ("2 Diodes Best Baseband"). At the upper band edge and 1 dB insertion loss, the B_N can be only .6 times the B_N for a single diode or for two diodes $\lambda/4$ apart. And the B_N can be only .36 times the B_N that could be used for baseband operation.

Some improvement in insertion loss is afforded by tuning the low loss diode capacitance with a shunt inductance. The susceptance of the parallel inductor and capacitor is given by

$$B = \omega C - \frac{1}{\omega L} \quad (6\text{-}36)$$

Since they will be resonant at mid-band, $\omega_0 C$ then

$$\omega_0 C = \frac{1}{LC} \quad (6\text{-}37)$$

$$B = (\omega - \frac{\omega_0 C^2}{\omega}) C \quad (6\text{-}38)$$

The upper band edge ω_u occurs at $\omega_0 C \sqrt{2}$ for octave bandwidth

$$B_{NC} = \frac{1}{\sqrt{2}} \omega_0 C \, C \, Z_0 \quad (6\text{-}39)$$

A parallel LC combination behaves like a shorted quarter wavelength stub in that the susceptance at resonance is zero and they both have the same sign above and below resonances. The susceptance B_{NS} of a short stub Y_S normalized to a main transmission line Y_0 is given by

$$B_{NS} = -\frac{Y_{oS}}{Y_0} \cot 2\pi \frac{\ell}{\lambda} \quad (6\text{-}40)$$

At the upper band edge $2\pi \frac{\ell}{\lambda} = 120°$ which gives

$$B_{NS} = (.578) \frac{Y_{oS}}{Y_0} \quad (6\text{-}41)$$

Matching LC and stub admittances at the upper band edge[3] of an octave bandwidth gives (equating Eqs. (6-39) and (6-41))

$$B_{NC} = B_{NS} = \frac{\omega_0 C \, C Z_0}{\sqrt{2}} = .578 \frac{Y_0 S}{Y_0} \qquad (6-42)$$

Since comparisons are made at the upper band edge $\omega_u = \sqrt{2}\,\omega_0 C$ Eq. (6-42) can be reduced by

$$C = 1.156 \frac{Y_0 S}{\omega_u} \qquad (6-43)$$

It is important to note that a stub has resonance at 90° and has 60° and 120° length at the upper and lower bands of an octave bandwidth, thus it should be $\lambda/4$ at $(\omega_L + \omega_u)/2$. The LC circuit has resonance at $\omega_0 C^2 = \frac{1}{LC}$ and has equal values over the octave bandwidth at $\frac{\omega_0 C}{\sqrt{2}}$ and $\sqrt{2}\,\omega_0 C$. Thus its resonant frequency should be $\omega_0 C = \sqrt{\omega_L \omega_u}$ which is $\omega_0 C = \frac{\omega_u}{\sqrt{2}}$ while $\omega_0 S = \frac{3}{4}\omega_u$. As the inductance tuning C becomes distributive and thus represented by a stub, the resonant frequencies come closer to each other.

c. Stubs

The typical impedances for shunt stubs using filter synthesis are shown in Fig. 3-11. For low impedance shunt stubs, the series section characteristic impedances become prohibitively high. A more satisfactory design is that of Mumford[4], in which the series section characteristic impedance is the same as the input and output impedances of the filter (refer to Fig. 3-30). The attenuation of Mumford's filter is a maximally flat response given by

$$\alpha = 10 \log \left[1 + K_n \frac{\cos^{2n}\left(2\pi \frac{\ell}{\lambda}\right)}{\sin^2\left(2\pi \frac{\ell}{\lambda}\right)} \right] \qquad (6-44)$$

Since the upper band edge of the octave bandwidth switch occurs at $2\pi\ell/\lambda = 120°$ then the insertion loss at the upper band edge is given by

$$\delta = 10 \log \left[1 + K_n \frac{.25^n}{.75} \right] \qquad (6-45)$$

MULTIPLE DIODE SWITCHES

in which n is the number of shunt stubs in the filter and K_n is a constant dependent on n and given by Mumford.[4] The filters have their lowest characteristic impedance stubs at their center and their highest at the ends. The lower the characteristic impedance of a stub, the higher is the capacitance that can be substituted for it to match it at the band edge. Therefore the equivalent B_{NS} of the center two stubs are calculated using Eqs. (6-45) and (6-41). For example, when n = 4, Eq. (6-45) becomes

$$\delta = 10 \log [1 + .0052 \, K_4] \qquad (6\text{-}46)$$

For $K_2 = 1.109 = \dfrac{Y_{os}}{Y_0}$, K_4 is given by $K_4 = 21.6$. These values give $\delta = .46$ dB and $B_{NS} = .64$ which lies on the curve in Fig. 6-13 for n = 4. An example of a 4 element filter switch is shown in Fig. 7-16.

It can be seen from Fig. 6-13 that the filter structure gives only slight improvement in insertion loss at the 1 dB level but significant improvement at lower levels of insertion loss.

Several additional comments will help to clarify the utility of the filter circuit and Fig. 6-13. The tuned diodes replaced the center two stubs in the filter structure. The end stubs may remain stubs as prescribed by the filter design or be replaced by dummy parallel LC resonant circuits. One of the stubs could be used for bias purposes. The B_N shown in Fig. 6-13 refers to the total B_N of the parallel resonant circuit at the upper band edge. Using Eq. (6-30) the diode capacitance is calculated by

$$C = \frac{\sqrt{2} \, B_N}{\omega_0 \, Z_0} \qquad (6\text{-}47)$$

For a 2-4 GHz switch $\omega_0 = 4\sqrt{2}\pi \times 10^9$. A 0.2 dB insertion loss could be obtained with n = 6 by $B_N = .64$. Thus for a $Z_0 = 50$ ohm transmission line C = 1 pF will be maximum diode capacitance. It is important to keep in mind that Fig. 6-13 relates to octave bandwidth switches. Another bandwidth would require the construction of another figure. All of the upper band edge curves would be shifted toward the other curves for narrower bandwidths, and away from them for wider bandwidths.

Since Fig. 6-13 is for octave bandwidths, the optimum insertion loss arrangement can be obtained by comparing curves. For example, "2 Diodes Best Baseband" gives $B_N = 1.2$ for 0.5 dB insertion loss.

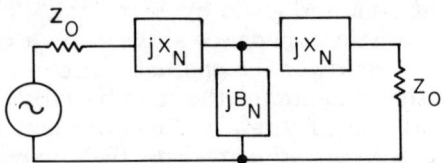

Fig. 6-14   Lumped Element Switch Equivalent Circuit

Any curve representing a diode in a tuned structure that has $B_N > \frac{1.2}{2}$ for .5 dB insertion loss would be better than the baseband structure because the diode capacitance could be larger. The "1 Diode" curve has a value of $B_N = 0.7$ and therefore will be the best giving a capacitance of 1.1 pF for a 2-4 GHz switch.

C. Series-Shunt — Lumped Circuit Elements

In Section VI A the diodes were either the shunt elements or the series elements in the filter structure but not both. Now the filter structure is considered to be made up entirely of diodes. Fig. 6-14 illustrates the 3 element filter structure to be evaluated.

In the low loss state the series diodes are in conduction and the shunt diode is reverse biased. The series diodes are inductive, and represented by

$$X_N = +\frac{\omega L}{Z_0} \tag{6-48}$$

and the shunt diode is capacitive represented by

$$B_N = +\frac{\omega C}{Y_0}$$

The structure has the appearance and response of a low pass filter as shown in Fig. 6-1. The normalized ABCD matrix of the circuit as given by

$$\begin{vmatrix} A & B \\ C & D \end{vmatrix}_N = \begin{vmatrix} 1 & jX_N \\ 0 & 1 \end{vmatrix} \begin{vmatrix} 1 & 0 \\ jB_N & 1 \end{vmatrix} \begin{vmatrix} 1 & jX_N \\ 0 & 1 \end{vmatrix}$$

MULTIPLE DIODE SWITCHES

$$= \begin{vmatrix} 1 - X_N B_N & j(2X_N - X_N^2 B_N) \\ jB_N & 1 - X_N B_N \end{vmatrix}. \quad (6\text{-}50)$$

The attenuation is given by

$$\alpha = 10 \log \tfrac{1}{4} |A + B + C + D|_N^2$$

$$= 10 \log \tfrac{1}{4} \left[4(1 - X_N B_N)^2 + (2X_N + B_N - X_N^2 B_N)^2 \right]. \quad (6\text{-}51)$$

In the high loss state the series diodes are reverse biased and the shunt diode is forward biased. The series diode is represented by a capacitor giving

$$X_N = -\frac{1}{\omega C Z_0} \quad (6\text{-}52)$$

and the shunt diode by an inductor giving

$$B_N = -\frac{1}{\omega L Y_0}. \quad (6\text{-}53)$$

The structure has the appearance and response of a high pass filter as shown in Fig. 6-1. Note that changing the sign of both X_N and B_N in Eq. (6-51) does not change that expression of attenuation. Equation (6-51) is plotted in figure 6-15. Note that normalized shunt reactance ($1/B_N$) is plotted rather than normalized shunt susceptance.

A similar derivation for resistances was made. The normalized series resistance is R_N and the normalized shunt conductance is G_N. The attenuation is given by

$$\alpha = 20 \log \tfrac{1}{2} \left[G_N (R_N + 1)^2 + 2(R_N + 1) \right] \quad (6\text{-}54)$$

which gives curves quite close to those given in Fig. 6-15 for isolations greater than 40 dB.

Assume typical PIN diodes are used in hybrid integrated circuits to make a series-shunt switch as shown in Fig. 6-14. These diodes available in chip or beam lead form have $C \approx .05$ pF and $L \approx .2$ nH when mounted. The magnitude of their normalized reactances (normalized to 50 ohms) is given by

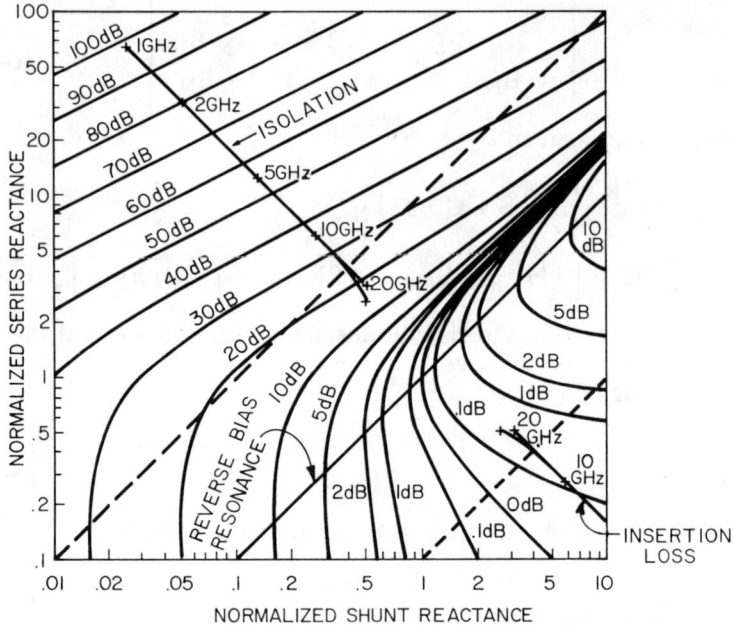

Fig. 6-15 Attenuation of Lumped Element T Switch

	1 GHz	2 GHz	5 GHz	10 GHz	20 GHz
.05 pF	64	32	12.8	6.4	3.2
.2 nH	.025	.05	.125	.25	.5

For the isolation state the .05 pF is in series and the .2 nH is in shunt. These numbers give the line on Fig. 6-15 running from 98 dB at 1 GHz to 18 dB at 20 GHz. For the insertion loss state the .2 nH is in series and the .05 pF is in shunt. In Fig. 6-15 the insertion loss will be less than .1 dB up to 9 GHz and about .3 dB at 20 GHz. The slope of the line generated by changing frequency will always be -45°, and when identical diodes are used in series and in shunt both isolation and insertion loss line will be pieces of the same lines as they approach the "Reverse Bias Resonance" line. For the isolation curve shown, the normalized series reactance goes to zero at 50 GHz while the normalized shunt capacitance is 1.25. For the insertion loss curve the normalized shunt reactance will go to zero at the same frequency while the normalized series reactance will be 1.25. The curves will begin to bend when they enter the region bounded by the dashed lines. The isolation curve is degraded and the insertion loss, lowered compared to their low frequency extrapolations.

MULTIPLE DIODE SWITCHES

The isolation will be limited also by the resistances of the diode. One ohm shunt resistance and 5000 ohms parallel resistance will give a normalized shunt resistance of .02 and a normalized series resistance of 100, which according to Fig. 6-15 should give more than 100 dB isolation. Thus the resistances need not be considered.

How does the isolation of these three diodes compare to the isolation that could be obtained from them if they were separated by the optimum distances derived in Section VI-B? For example, take the point given by a normalized series reactance of 10 and normalized shunt reactance of .1. Fig. 6-15 indicates that the isolation is about 54 dB. Eq. (6-14) indicates that each diode could give 14.15 dB alone. Eq. (6-16) indicates that the extra attenuation from these diodes, when spaced an optimum distance, is 5.58 dB. The total isolation of three of these diodes with optimum spacing would be given by 3 (14.15) + 2 (5.85) = 54.15 dB which is the same isolation given in Fig. 6-15. The slope of the isolation curves indicates that the optimum isolation is obtained for all values of isolation given in Fig. 6-15. There are two series diodes in the circuit of Fig. 6-14 and one shunt diode. A change in the series diodes gives twice the change in isolation that a change in the shunt diode gives in Fig. 6-15. The slope of the isolation lines is thus 0.5. For a pi circuit normalized series reactance is changed to normalized shunt susceptance and normalized shunt reactance is changed to normalized series susceptance.

Suppose more isolation is required than can be obtained from Fig. 6-15. More elements could be added to the structure, alternating between series and shunt elements, and the isolation would continue to be optimum. The insertion loss would have to be calculated using filter theory. Another approach might be to use two T sections. At zero spacing between the T's the isolation would be characteristic of 5 elements because the two series diodes without a shunt diode between them would behave like a single diode. As the spacing between these two diodes is increased, extra isolation would be provided as given by Eq. (6-21). Normal precautions have to be taken when making microwave integrated circuits to avoid waveguide modes in the circuit enclosure or high isolations will not be realized.

REFERENCES

1. R.V. Garver, "Theory of TEM Diode Switching," *IRE Trans. on Microwave Theory and Techniques*, Vol. MTT-9, May 1961, pp. 224-238.

2. J.F. White and K.E. Mortenson, "Diode SPDT Switching at High Power and Octave Microwave Bandwidth," *IEEE Trans. Microwave Theory and Techniques*, Vol. MTT-16, January 1968, pp. 30-36.

3. R.E. Fisher, "Broadbanding Microwave Diode Switches," *IEEE Trans. on Microwave Theory and Techniques*, Vol. MTT-13, September 1965, p. 706.

4. W.W. Mumford, "Tables of Stub Admittances for Maximally Flat Filters Using Shorted Quarter-Wave Stubs," *IEEE Trans. Microwave Theory and Techniques*, Vol. MTT-13, September 1965, pp. 695-696.

Chapter 7
Multiple Throw Switches

Multiple throw diode switches have been made in three basic configurations. The first configuration uses circulators or 3 dB couplers. It has the advantages of utilizing simple diode switches and can provide a variety of switching arrangements. The second configuration utilizes transmission line filter structures. Its advantage is that the diode junctions can be large and yet provide modest bandwidth. A large junction permits the switch to control high power levels. The third configuration utilizes the diode as part of a lumped circuit filter structure. Its advantage is that extreme bandwidths are possible with it.

A. Use of Coupling Devices

 1. Circulators

The simplest double-throw switch uses one basic diode switch and a circulator as shown in Fig. 7-1. The input at port A can be switched to either port B or port C. When the basic switch is ON the power conveyed through the circulator continues on out port B. When the basic switch is OFF the power from the circulator is reflected by it and re-enters the circulator emerging from port C. The isolation to port B will be that which the basic switch provides in the isolation state. The isolation to port C will be limited by the directivity of the circulator and the VSWR of the basic switch and load when ON. Circulators typically provide 20 dB to 30 dB isolation. A basic switch with a 1.2 VSWR in the ON state will have a reflection coefficient given by

$$|\Gamma| = \frac{\rho - 1}{\rho + 1} = .091 \qquad (7-1)$$

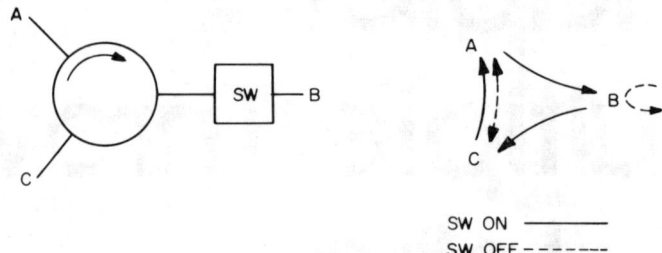

Fig. 7-1   Double Throw Switch Using a Circulator

which provides reflected power reduced by

$$\eta = 20 \log \frac{1}{|\Gamma|} = 20.8 \text{ dB} \qquad (7\text{-}2)$$

To obtain 40 dB isolation the isolator must provide greater than 40 dB isolation and the basic switch must have a VSWR less than 1.02. It is possible to put a tuner between the ON basic switch and the circulator to obtain any desired isolation even though the circulator and the basic switch have limited isolation capacity. The higher the isolation obtained with the tuner, the narrower will be its bandwidth.

By adding another circulator to the structure as shown in Fig. 7-2, the circuit operates as a switchable four-port circulator. Inputs at A or C can be switched to outputs at B or D. The circuit is a one-directional transfer switch.

2.   3-dB Couplers

There are two classes of 3 dB couplers, 90° and 180°, as shown in Fig. 7-3. The ports are labeled so that power into port A or C is equally divided between ports B and D, and similarly power into ports B or D is equally divided between ports A and C. The TEM 90° coupler is a quarter wavelength coupled directional coupler. A single quarter-wavelength coupled section provides octave bandwidth and much greater bandwidths are possible by using more coupled sections. The waveguide short slot hybrid junction typically covers the waveguide bandwidth. All of the 180° couplers shown and the square 90° coupler have 10% to 20% bandwidth. Wider Wider bandwidth 180° couplers are possible using more complex structures.

MULTIPLE THROW SWITCHES

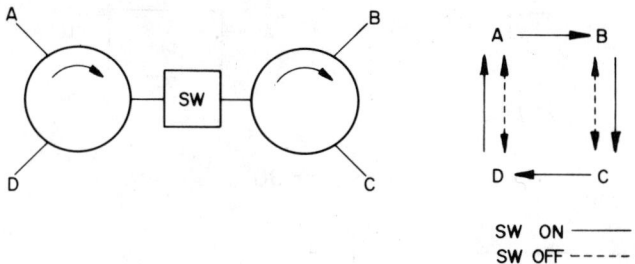

Fig. 7-2 Unidirectional Transfer Switch

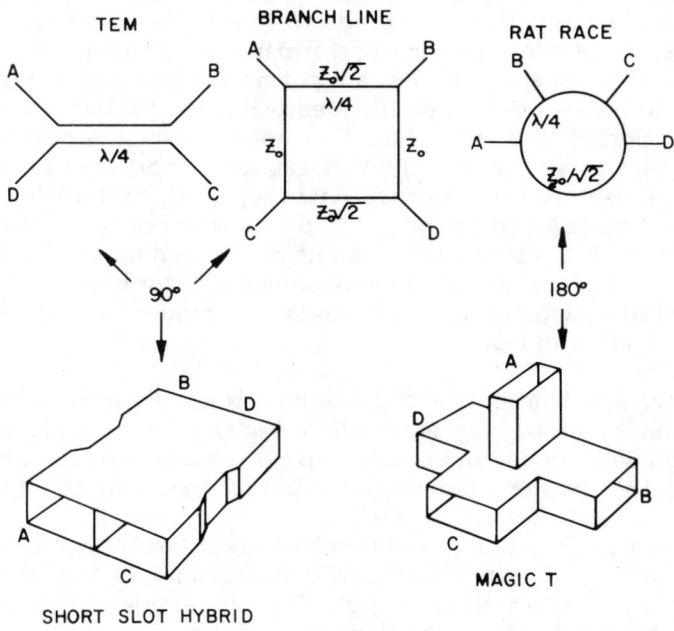

Fig. 7-3 3 dB Couplers in TEM Transmission Line and Waveguide

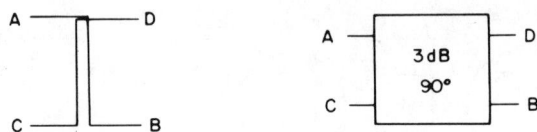

Fig. 7-4 Cross-over 90° dB Coupler

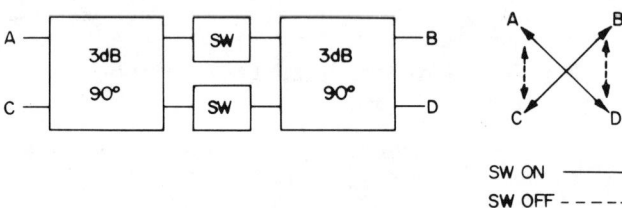

Fig. 7-5 DPDT Switch with 3 dB 90° Couplers

An important difference exists between the 180° and 90° couplers when used in conjunction with diode switches. This difference can be seen by considering their performance when identical reflectors (simulating off diodes) are put on the two output ports. Taking A to be the input port, the identical perfect reflectors are put on ports B and D. In the 90° couplers the phases of the reflected waves are such that the voltages add at port C and cancel at port A, thus power into port A exits via port C. In the 180° couplers the voltages add at port A and cancel at port C, so that power into port A is reflected back out port A and none emerges from port C. The 90° couplers prove convenient for making multiple throw switches while additional techniques have to be used with 180° couplers.

The quarter wavelength TEM 3 dB coupler is usually made with the coupling lines crossed over, as shown in Fig. 7-4, to facilitate interconnection with diodes and other 3-dB 90° couplers. As shown in the figure, the coupler will be represented in circuits by the square.

When two of these couplers and two basic switches are combined as shown in Fig. 7-5, a double throw switch results. In fact, the operation is that of a transfer switch. The connections are A-D and B-C or A-C and B-D. When the basic switches are ON, power from port A goes past them, recombines in the right hand 3 dB coupler, and emerges from port D. When the basic switches are OFF, they reflect incident power which recombines in the left hand 3 dB

MULTIPLE THROW SWITCHES

coupler and emerges from port C. The diodes must be identical and in the same electrical plane to satisfy the reflection phase requirements for power in A to exit via C. The same scheme could work with 3-dB 180° couplers if one diode were a series switch and the other were a shunt switch or if the distance to the identical reflectors were different by a quarter wavelength as shown in Fig. 7-6.

As shown in Fig. 7-7 the switch of Fig. 7-5 is especially suited to be a T-R switch in a radar. When the basic switches are OFF, the transmitted power into port A is routed to the antenna on port C. When the switches are ON, the power from the antenna in port C is conveyed to the receiver on port B. The circuit has two very attractive features; the transmitter power that gets past the basic switches in the OFF state (due to their finite isolation) is conveyed to the matched load on port D instead of to the receiver on port B. The transmitter-to-receiver isolation is 20 dB to 40 dB greater than the isolation of the basic switches because of the directivity of the 3 dB couplers. The switches must have identical transfer coefficients, S_{21}, in the isolation state, to obtain the maximum transmitter-to-receiver isolation. The second attractive feature of the application shown in Fig. 7-7 is that when the basic switches are shunt diodes, the maximum transmitter power they can control is not limited by their maximum reverse breakdown voltage. They are always in the conduction state when high power is impinging on them.

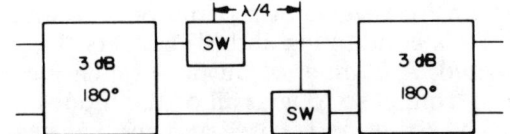

Fig. 7-6   DPDT Switch with 3 dB 180° Couplers

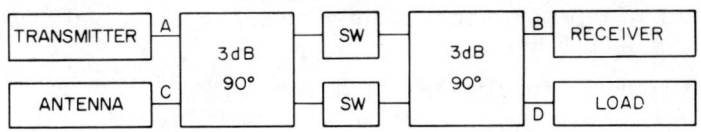

Fig. 7-7   RADAR Duplexer

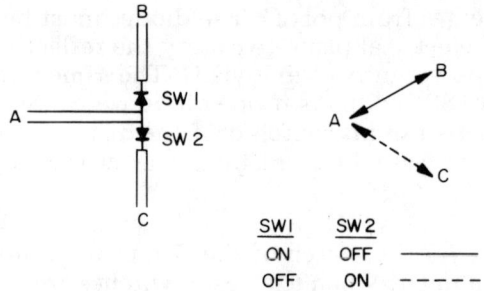

Fig. 7-8 Lumped Element Double Throw Switch

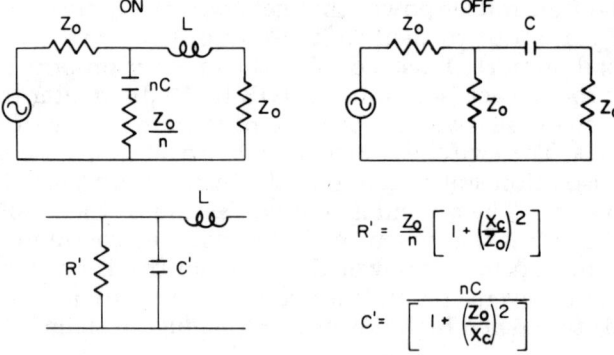

Fig. 7-9 Equivalent Circuit of Lumped Element Multiple Throw Switch (n = no. of OFF diodes)

B. Lumped Circuit Element Structure

The simplest multiple throw switch is shown in Fig. 7-8. When SW1 is in conduction SW2 is reverse biased and ports A and B are connected. When SW2 is conducting and SW1 reverse biased, ports A and C are connected. A number of output branches can be connected to a common terminal so long as all of the diodes are reverse biased but one. The structure is base-band and has been made to work up to 20 GHz[1] using hybrid integrated circuit techniques. This simple circuit is represented by the equivalent circuits given in Fig. 7-9, in which L represents the inductance of the conducting diode and nC represents the capacitance of the n reversed biased diodes to the OFF lines.

The ABCD matrix for the OFF diode is given by

$$\begin{vmatrix} A & B \\ C & D \end{vmatrix} = \begin{vmatrix} 1 & 0 \\ Y_0 & 1 \end{vmatrix} \cdot \begin{vmatrix} 1 & jX_c \\ 0 & 1 \end{vmatrix} = \begin{vmatrix} 1 & jX_c \\ Y_0 & 1+jX_c Y_0 \end{vmatrix}. \quad (7\text{-}3)$$

MULTIPLE THROW SWITCHES

The isolation η is given by

$$\eta = 10 \log \frac{1}{4} \left[A + BY_0 + CZ_0 + D \right]^2 = 10 \log \left[\frac{9}{4} + \left(\frac{X_c}{Z_0} \right)^2 \right] \qquad (7\text{-}4)$$

Note that the series diode has $\eta = 10 \log [1 + \frac{1}{4} (X_c/Z_0)^2]$ when the incident power is reflected. The factor $1/4$ is missing because the open circuit voltage of the generator is not across the diode. The ON line impedance halves the voltages and thus causes the power to be reduced by a factor of four.* For 20 dB isolation $X_c/Z_0 \approx 10$ which causes R' in Fig. 7-9 to be given by $R' \approx 100 Z_0$. This value of R' will introduce about .043 dB insertion loss by itself and C' will introduce about .001 dB insertion loss by itself. Therefore, an OFF branch with 20 dB isolation or more does not add significant insertion loss to the ON branch. An .05 pF diode (a typical capacitance readily available in a PIN diode) will provide 20 dB isolation up to 6.36 GHz in a 50 ohm transmission line multiple throw switch.

Multiple diodes in each branch are necessary to obtain good performance at higher frequencies or to use larger capacitance diodes at the lower frequencies as discussed in Chapter VI.

If the isolation is kept high by a second series diode spaced a distance ℓ down the transmission line from the first diode, then the equivalent circuit shown in Fig. 7-10A can be used to calculate the insertion loss of the ON branch. (The OFF branch will give 6 dB more isolation than the diode circuit would in the normal single throw structure.) C_p is the capacitance of the next reverse biased diode as transformed by the transmission line length ℓ separating it from the first diode. This length should be less than a quarter wavelength at the highest frequency. Since the capacitive reactance of an OFF series diode is quite a bit higher than the characteristic impedance of the transmission line in which it is mounted, the capacitance C_p is primarily that of an open circuit transmission line of length ℓ. When the electrical length approaches a quarter wavelength, C_p approaches infinity, but doesn't become

---

* This same ratio influences the power limitations of double throw and multiple throw switches. The peak power rating of a series diode multiple throw diode switch is four times eq. (4-63) which is the same as for a shunt diode single throw or multiple throw rating, eq. (4-64). The average power rating of a shunt diode multiple throw switch is four times eq. (4-70) which is the same as for a series diode single throw or multiple throw rating, eq. (4-69). Thus eq. (4-64) and eq. (4-69) prescribe the peak and average power ratings of multiple-throw switches whether series or shunt diodes are used.

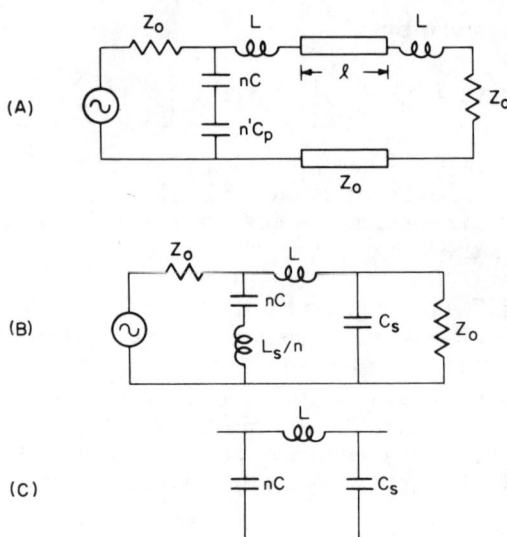

Fig. 7-10 Equivalent Circuits of Lumped Element Multiple Throw Switch with Multiple Diodes for Each Batch

inductive unless the line is longer than a quarter wavelength. Thus, nC_p will have the effect of reducing the capacitance of the branch containing nC and nC_p in Fig. 7-10A. Since nC_p cannot become inductive the maximum capacitance will be nC.

Alternatively, the isolation can be kept high by a shunt diode immediately following the series diode at the junction. The equivalent circuit shown in Fig. 7-10B can then be used to calculate the insertion loss of the ON branch. At reverse bias the shunt diode has capacitance C_s and at forward bias it has inductance L_s. Since the inductive reactance is much smaller than Z_0 while the series diode reactance is much larger than Z_0, the series resonance of nC and L_s/n is much higher than the maximum operating frequency of the switch. The branch can be represented quite accurately by nC alone. The circuit takes on the character of the pi circuit shown in Fig. 7-10C.

In physically relizable circuits C is more than the capacitance of the OFF diode. Because several diodes are being crowded into a common junction, it is necessary to extend each center conductor for a finite distance to each diode junction. The capacitance of these center conductors add to C. A 50 ohm air dielectric transmission line has 2/3 pF per centimeter. Diode chips for hybrid integrated circuits are about .020" (0.5mm) in diameter, which would give .1 pF on an

MULTIPLE THROW SWITCHES

alumina substrate. Thus the very good diode with .05 pF will, in fact, have at least .15 pF as an OFF arm when mounted in a multiple throw switch. The problem of making multiple throw switches having low insertion loss and VSWR reduces to matching into this augmented capacitance of the OFF arms of the switch. The problem is the same as that of overcoming the capacitance of shunt switching diodes, as discussed in Section VI A1 and as portrayed in Fig. 6-2. The Bode limit for .15 pF in a 50 ohm transmission line to 20 GHz is .005 dB or a VSWR of 1.07. A .1 dB insertion loss ripple or 1.3 VSWR up to 20 GHz would be caused by .265 pF in a 50 ohm transmission line. Broader bandwidth can be obtained by having the capacitance in the center of a filter structure.* Thus a filter circuit surrounding a capacitance nC, as shown in Fig. 7-11, will be required to obtain substantial bandwidth. The elements L' and C' are selected from filter theory to give the desired maximum insertion loss or VSWR ripple and upper cut-off frequency.

The inductance L from Fig. 7-10 may be included in L' if C_s is small. L' can be made no smaller than L. The capacitance C_s may also be included in C' and C' can be no smaller than C_s. (Normally C' will be on the order of half of nC). An example of a double throw switch made by using these filter techniques[1] is shown in Figs. 7-12 and 7-13. The series PIN diodes No. 3 and No. 4 were two mesa diodes on the same semiconductor chip (.04 pF) and the shunt diodes were planar junctions (.1 pF). The OFF diode is the first capacitance in the filter structure. The strap to the first shunt diode is the first series inductor. The first shunt diode is the middle shunt capacitor in the filter structure. And the strap to the second shunt diode and the shunt diode form the final series L and shunt C in the filter structure. In order to make the theory fit the performance, the distributed properties of the straps have to be taken into account at the higher frequency, since they really are short sections of transmission line. The switch gives less than 2 dB insertion loss and more than 55 dB isolation up to 18 GHz. Bias is supplied to the diodes via watch spring inductors (which have no upper frequency limit) built into the connectors.

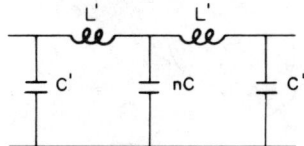

Fig. 7-11   Circuit for Wide Insertion Loss Bandwidth of Multiple Throw Switch

* See methods in appendix C for meeting or exceeding the Bode Limit.

192 MICROWAVE DIODE CONTROL DEVICES

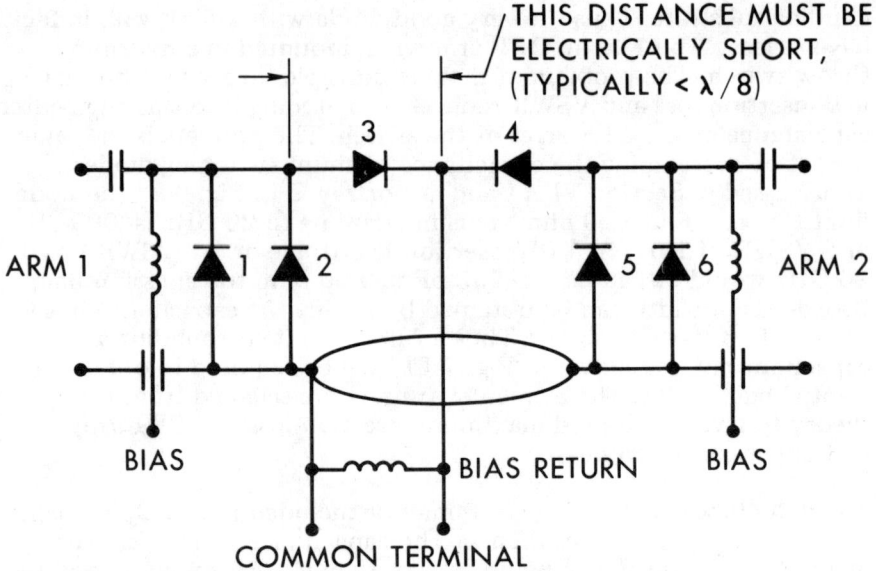

Fig. 7-12 SPDT Switch Circuit (courtesy Hewlett-Packard Co.)

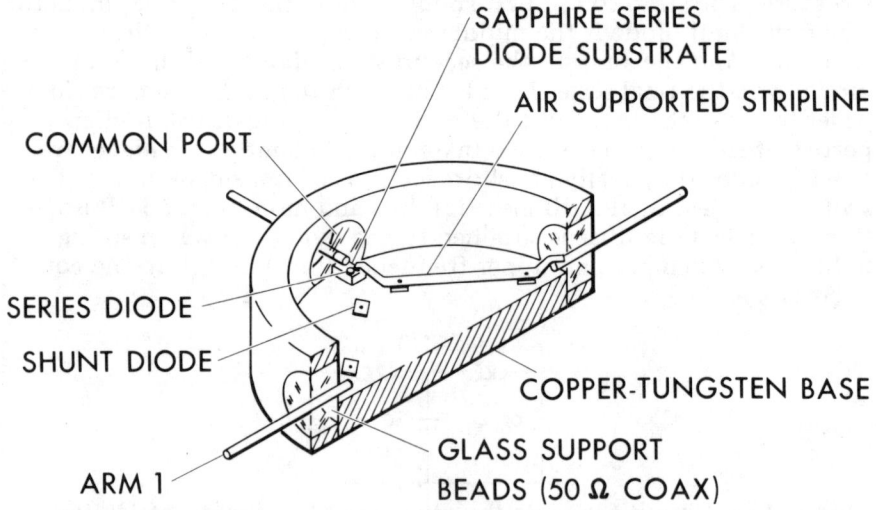

Fig. 7-13 SPDT Switch Mechanical Structure: Cross Section with One Stripline Removed (courtesy of Hewlett-Packard Co.)

MULTIPLE THROW SWITCHES

C. Use of Stubs

The series diodes at the common junction in the previous section had to have low capacitance in order to provide wide bandwidth. The amount of RF power they can control is limited to several watts because of the small size of the junction and because it is difficult to dissipate heat generated in series diodes. Shunt diodes are permitted to have higher capacitance, and because one end is grounded they can also dissipate more heat. Thus a multiple throw switch using shunt diodes can control substantially more power than one using series diodes. The shunt diodes must be mounted a quarter wavelength from the common junction to keep the OFF branches from disturbing the ON branch, as shown in Fig. 7-14. As many diodes as needed can be put after the first diode to obtain any desired isolation. The OFF branch is a shorted quarter wavelength transmission line having the same characteristic impedance as the input line. As discussed in Chapter III and as illustrated in Fig. 3-3 ($Z_g/Z_0 = 1$) the insertion loss introduced by this line at the band edges is about .3 dB for an octave bandwidth and 1 dB for a 3:1 bandwidth. It will provide .1 dB insertion loss and a VSWR of 1.3 at the edges of a 1.4 to 1 bandwidth. The circuit of Fig. 7-14 is quite useful for waveguide double throw switches since OFF diodes are equivalent to a short circuit across the waveguide in the plane of the diode and bandwidth is necessarily limited to that of the waveguide.

1. Muehe & Young Structure

Even better low VSWR performance can be achieved by using the structure of Fig. 3-7A and Fig. 3-8. When $Z_Q/Z_0 = Z_0'/Z_0$, then the structure can be used as a double throw switch as illustrated in Fig. 7-15.

It can be seen from Fig. 3-8 that $Z_Q/Z_0 = Z_0'/Z_0 = .75$ gives a bandwidth ratio of 2.0 and maximum VSWR of 1.06. And $Z_0'/Z_0 = Z_Q/Z_0 = .6$ gives a bandwidth ratio of about 3.0 giving a maximum VSWR of about 1.35. This bandwidth ratio is substantially greater than the 1.4 given by the Z_0 stubs alone for the same maximum VSWR.

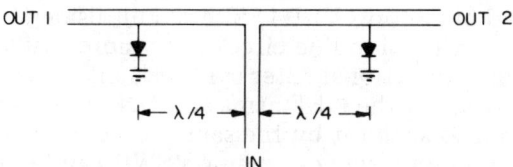

Fig. 7-14 SPDT Switch Using λ/4 Stubs

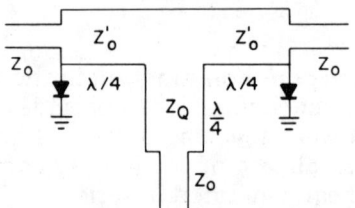

Fig. 7-15 SPDT Switch Using Muehe & Young Structure

In a similar fashion the Tchebyscheff filter section of Fig. 3-11 could be used by requiring that $Z_{02} = Z_0'$. I.e., the Z_{02} section would be dropped and another $Z_0' - Z_{01}'$ section would come off of the filter at the middle point. The shunt diodes would be at the two output $Z_{01}' - Z_0'$ junctions. The VSWR of the switch would be equal to that using the 37.5 ohm stubs prescribed above and the structure would now be bigger and more difficult to realize. Usually, characteristic impedances out of the range from 25 ohms to 70 ohms are to be avoided.

The discussion so far has been concerned only with the contribution of the OFF arm stub to insertion loss. The ON diode also contributes to the insertion loss. Assuming the stub insertion loss is low enough, the insertion loss bandwidth of a single shunt capacitive ON diode can be determined from Fig. 2-12. For example, assume a shunt diode is to be used in each arm of a 2-4 GHz switch using the 37.5 ohm stubs as described above. A single untuned .5 pF diode in each arm would give .5 dB maximum insertion loss at 4 GHz. A tuned 1 pF diode would give .5 dB maximum insertion loss over the 2-4 GHz band. According to the Bode Limit (Fig. 6-2), a maximum insertion loss of .5 dB allows a maximum B_N of 2.83, which indicates a capacitance of 2.25 pF at the center of the low pass filter. The end capacitance which would be controlling the maximum incident RF power would be lower according to the filter design requirements, and thus would not have the full power rating that might be associated with 2.25 pF.

2. Mumford Structure

As was discussed in Section VI B4, filters built using Mumford's maximally flat transmission line circuits are more easily realized than those using conventional filter synthesis methods. Using this type filter, the stubs of the OFF arms and ON diode capacitance can all be taken into account by the same design parameters, and a prescribed maximum insertion loss or VSWR can be obtained.[2] A typical Mumford filter has series 50 ohm stubs and shunt stubs

MULTIPLE THROW SWITCHES

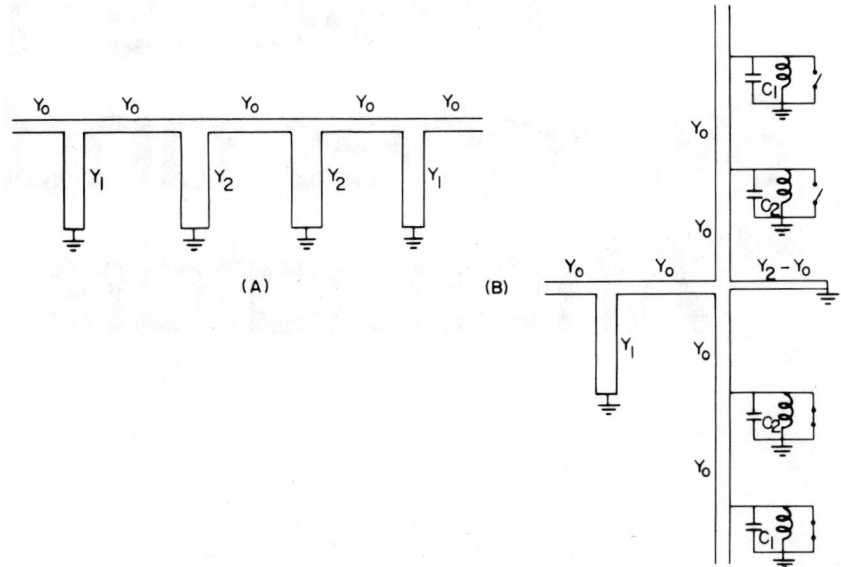

Fig. 7-16 SPDT Switch Using Mumford Filter

with $Z_0 < 50$ ohms as shown in Fig. 7-16. The structure may be rearranged as shown in Fig. 7-16B to provide a double throw switch. When the switches are in the positions shown, the lower stub is short circuited at the first diode in the lower branch. The lower Y_0 stub and the $Y_2 - Y_0$ stub combine in parallel to give an admittance of Y_2. Capacitors C_2 and C_1 in the upper (and lower) branch(es) are selected to match the admittance of the Y_2 and Y_1 stubs at the band edges as in Section VI B4. The structure may be used for n-way switches in which the combined admittance of the OFF stubs is $(n-1) Y_0$ and the $Y_2 Y_0$ stub is $Y_2 - (n-1)Y_0$.

REFERENCES

1. W.W. Nelson, "Hybrid Circuits for Multioctave Multiple-Throw Microwave Switches," *IEEE Journal of Solid-State Circuits*, Vol. SC-3, no. 3, Sept. 1968, pp. 243-246.

2. J.F. White and K.E. Mortenson, "Diode SPDT Switching of High Power with Octave Microwave Bandwidth," *IEEE Trans. on Microwave Theory and Techniques*, Vol. MTT-16, no. 1, January 1968, pp. 30-36.

Matched Switches and Attenuators

Chapter 8

An ideal diode switching element either reflects or transmits incident RF power. In applications requiring match, the simplest means for obtaining it is to put attenuators or isolators in front of and behind the diode switching element. A diode shunting a line will have attenuation given by

$$\alpha_{SHUNT} = 20 \log (1 + \tfrac{1}{2} \alpha_{RF} I Z_0) \qquad (8\text{-}1)$$

in which α_{RF} is the coefficient of RF conductivity ($\alpha_{RF} = 80$ is typical for PIN diodes), I is the diode current in amperes, and Z_0 is the characteristic impedance of the transmission line the diode is shunting. This equation was derived in Section II D as given in Eq. (2-70) and Fig. 2-18. It is assumed that all of the normal diode parasitics are tuned out or negligible. For a series diode the attenuation is given by

$$\alpha_{SERIES} = 20 \log \left(1 + \frac{1}{2\alpha_{RF} I Z_0}\right) \qquad (8\text{-}2)$$

These equations are plotted in Fig. 8-1.

A. Coupling Devices

In Chapters II and VII circuits were described using circulators and 3 dB couplers which provided matched input characteristics. Fig. 2-17 shows the incident power being dissipated in the diodes in the isolation state. These circuits can be augmented by having a matched load behind the diodes with the diodes either series mounted or shunt mounted. The attenuation for the diode alone as the termination is given in Eq. (2-65) which reduces to

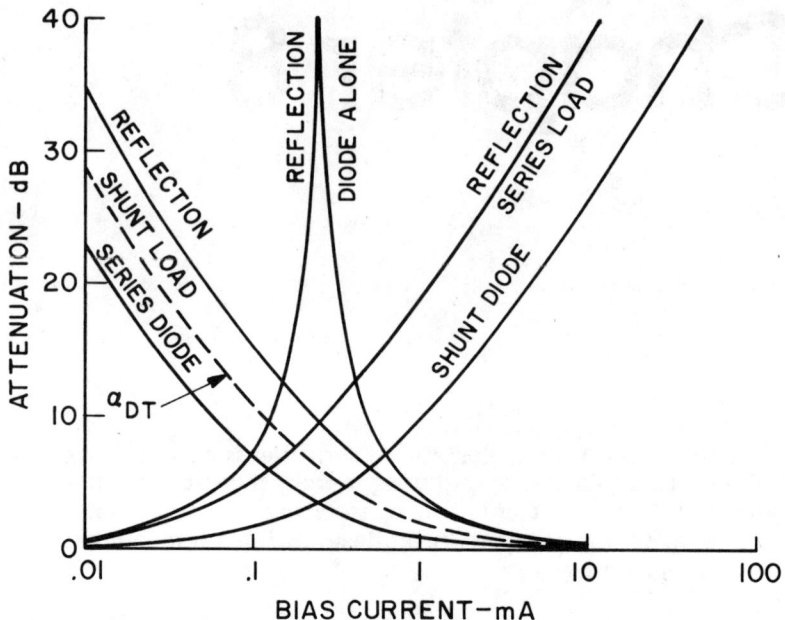

Fig. 8-1 PIN Diode Attenuation: $Z_0 = 50$, $\alpha_D = 80$

$$\alpha_{DIODE} = 20 \log \left[\frac{1 + \alpha_{RF} I Z_0}{1 - \alpha_{RF} I Z_0} \right] \tag{8-3}$$

under the conditions assumed above. This equation gives the curve for reflection diode alone[1] as shown in Fig. 8-1.

When a matched load is placed in shunt with diode in a reflection circuit the attenuation is given by

$$\alpha_{\substack{REFL \\ SH\ LOAD}} = 20 \log \left[1 + \frac{2}{\alpha_{RF} I Z_0} \right] \tag{8-4}$$

and when the load is in series with the diode

$$\alpha_{\substack{REFL \\ SER\ LOAD}} = 20 \log \left[1 + 2 \alpha_{RF} I Z_0 \right] \tag{8-5}$$

These two equations are also plotted in Fig. 8-1.

MATCHED SWITCHES AND ATTENUATORS

When the circulator is used for a reflection type switch the input VSWR of the switch is determined by the circulator. Normally only 20 dB to 30 dB isolation can be obtained over any useful bandwidth because the circulator must have unusually low output VSWR's to obtain higher isolation. An isolation of 20 dB corresponds to an output VSWR of about 1.2 and 30 dB, 1.06. If more isolation is required, additional circulator-diode circuits must be added in tandem.

When a 3 dB coupler is used, the input VSWR is determined by the quality of the directional coupler and how much the two diodes are alike. The same arguments hold about isolation.

More isolation can be obtained with coupling structures by using the circuits shown in Figs. 7-2 and 7-5. The input VSWR properties are similar to those for the reflection devices while the attenuation vs. current is the same as that for the simple series or shunt diodes. Any amount of isolation can be obtained using the methods of Chapter VI. These circuits are useful in TEM transmission line and waveguide.

One additional circuit giving matched input uses two pairs of power splitters[2] as shown in Fig. 8-2. Power which enters the two balanced arms of the power splitter cancels when the two waves are equal in magnitude and out of phase. Since the thru paths are identical, the waves add in phase to come out the common arm of the second power splitter (acting as a power combiner). Any power reflected by the diodes will be out of phase at the input power splitter since the round trip reflection path of one diode is one half wavelength longer than that of the other. A single stage power splitter has about 20% bandwidth.

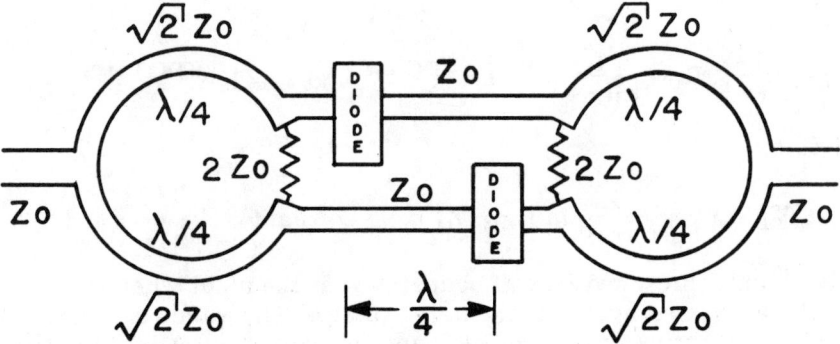

Fig. 8-2 Matched Attenuator Using Dent Coupler

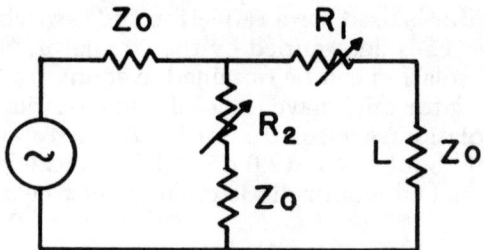

Fig. 8-3   Matched Input Variable Attenuator

B. Multiple Diode Absorbing

1. Match Using Standard Double Throw Switch

The simplest absorbing diode switches are matched in only one direction. The double throw switch shown in Fig. 7-8 can be made to be matched looking into the A terminal by taking the output from B and putting a matched load on C. The bias on SW2 is adjusted so that the admittance of the two arms always gives a matched input. Fig. 8-3 represents the switching circuit. The condition for matched input is that

$$\frac{1}{Z_0} = \frac{1}{R_2 + Z_0} + \frac{1}{R_1 + Z_0} \tag{8-6}$$

which reduces to

$$R_2 = Z_0^2/R_1 \tag{8-7}$$

Letting

$$R_1 = \frac{1}{\alpha_{RF} I_1}, \quad R_2 = \frac{1}{\alpha_{RF} I_2}, \quad I_1 = I_0 e^{\alpha_{DC} V_1}, \text{ and } I_2 = I_0 e^{\alpha_{DC} V_2}$$

gives

$$V_1 + V_2 = -\frac{2}{\alpha_{DC}} \ln (\alpha_{RF} Z_0 I_0) = \text{constant} \tag{8-8}$$

The double throw switch will be matched at the input when the diodes are biased by complimentary voltages. The attenuation of this double throw switch can be found by writing the ABCD matrix for Fig. 8-3.

MATCHED SWITCHES AND ATTENUATORS

$$\begin{bmatrix} A & B \\ C & D \end{bmatrix} = \begin{bmatrix} 1 & 0 \\ \frac{1}{R_2 + Z_0} & 1 \end{bmatrix} \begin{bmatrix} 1 & R_1 \\ 0 & 1 \end{bmatrix} = \begin{bmatrix} 1 & R_1 \\ \frac{1}{R_2 + Z_0} & 1 + \frac{R_1}{R_2 + Z_0} \end{bmatrix} \quad (8\text{-}9)$$

which gives

$$\alpha = 10 \log \tfrac{1}{4} \left[1 + \frac{R_1}{Z_0} + \frac{Z_0}{R_2 + Z_0} + 1 + \frac{R_1}{R_2 + Z_0} \right]^2 \quad (8\text{-}10)$$

Substituting Eq. (8-7) into Eq. (8-10) gives

$$\alpha = 20 \log \left[1 + \frac{R_1}{Z_0} \right] \quad (8\text{-}11)$$

or using the substitutions as before

$$\alpha_{DT} = 20 \log \left[1 + \frac{1}{\alpha_{RF} I_1 Z_0} \right] \quad (8\text{-}12)$$

which is plotted as the dashed line in Fig. 8-1.

2. Match Using Complementary Double Throw Switch

Another simple circuit that is matched[3] is shown in Fig. 8-4. Assuming both outputs 1 and 2 have matched loads, the input admittance is given by

$$Y_{IN} = \frac{1}{Z_1 + Z_0} + \frac{Y_0^2}{Y_2 + Y_0} \quad (8\text{-}13)$$

which reduces to $Y_{IN} = Y_0$ when $Z_1 = 1/Y_2$. Using identical diodes for Z_1 and Y_2 causes the input to be matched for all diode impedances. When the diodes are biased in series, so that the same bias current passes through both of them, both have the same impedance

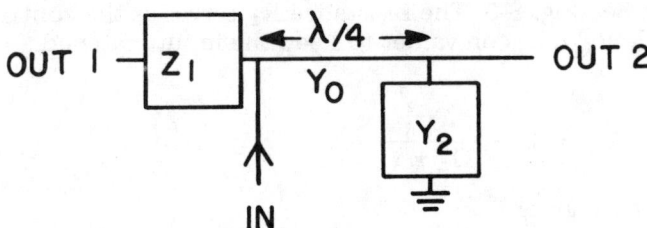

Fig. 8-4 Matched Input Variable Attenuator with $\lambda/4$

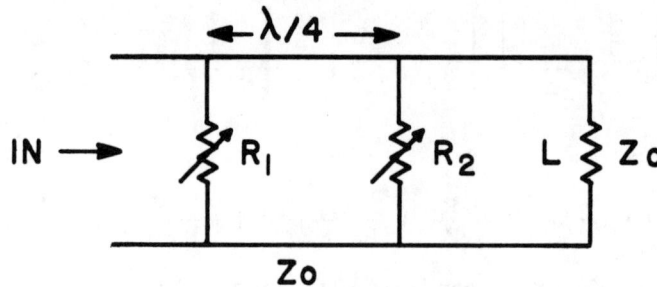

Fig. 8-5 Matched Input Variable Attenuator with Two Shunt Diodes

and complex bias circuits are not necessary to keep the variable attenuator matched. The attenuation through OUT 1 is the same as Eq. (8-12). The attenuation through OUT 2 is given by

$$\alpha = 20 \log [1 + \alpha_{RF} I Z_0] \qquad (8\text{-}14)$$

which lies midway between Shunt Diode and Reflection Series Diode on Fig. 8-1. A dummy load is put on the unused output to make a single output variable attenuator. Either output may be used as the variable attenuator output. Output 2 is used when the diode will give more isolation in the shunt mode and output 1 is used when the diode works better in series. (Note that in double throw switching structures the ½ is deleted from the attenuation equations: $\alpha = 10 \log [(R_N + 1)^2 + X_N^2]$ and so on).

3. Match Using Two Shunt Diodes (One Direction Only)

A simple method for making variable attenuators that avoids the use of coupling devices but provides octave bandwidth is that of multiple shunt variable resistance diodes. In its simplest form it consists of two variable resistance diodes and is matched from only one input. See Fig. 8-5. The resistance R_2 serves as the controlling diode while R_1 takes on values to keep the input matched.

$$Y_{IN} = Y_0 = G_1 + \frac{Y_0^2}{G_2 + Y_0} \qquad (8\text{-}15)$$

Solving for R_1 gives

$$R_1 = Z_0 + R_2 \qquad (8\text{-}16)$$

Solving for I_1 gives

$$I_1 = \frac{1}{\alpha_{RF} Z_0 + \frac{1}{I_2}} \qquad (8\text{-}17)$$

which is plotted in Fig. 8-6. I_1 is made a constant multiplier of I_2 by biasing both diodes from the same high voltage source through different series resistors. Thus I_1 is defined by

$$I_1 = k I_2 \qquad (8\text{-}18)$$

and the input VSWR is calculated

$$|\Gamma| = \left| \frac{Y_0 - Y_{IN}}{Y_0 + Y_{IN}} \right| \qquad (8\text{-}19)$$

$$\rho = \frac{1 + |\Gamma|}{1 - |\Gamma|} = \left[\frac{Y_0}{Y_{IN}} \right]^{\pm 1} \quad \text{(greater than one)} \qquad (8\text{-}20)$$

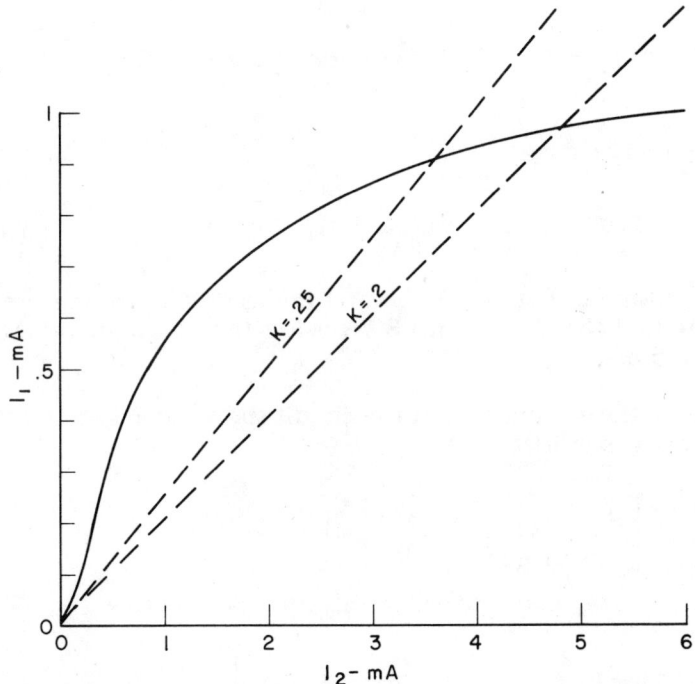

Fig. 8-6 Current Conditions for Matching Two Shunt Diodes $\lambda/4$ Apart $\alpha_{RF} = 16.6$, $Z_0 = 50\Omega$

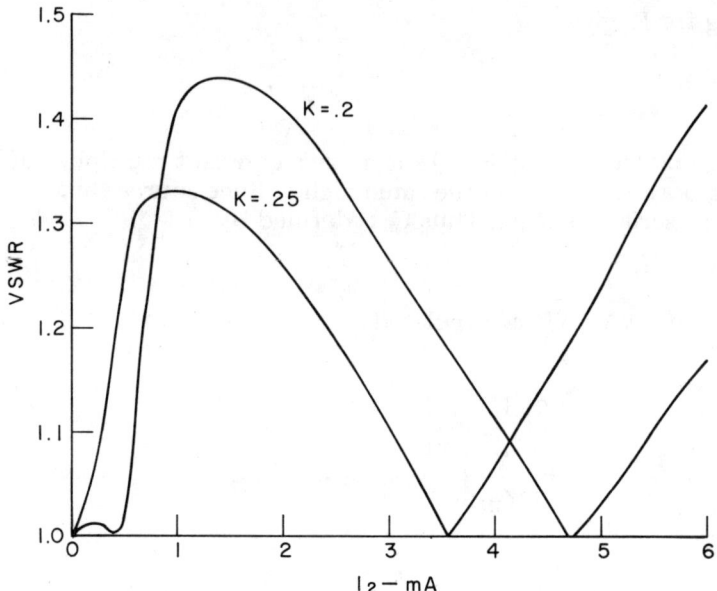

Fig. 8-7   VSWR for Various Current Ratios

Using Eq. (8-15) gives

$$\rho = \left[\frac{\alpha_{RF} k I_2}{Y_0} + \frac{Y_0}{\alpha_{RF} I_2 + Y_0} \right]^{\mp 1} \tag{8-21}$$

which is plotted in Fig. 8-7 for several values of k. Selecting k = .25 will insure that the maximum VSWR is less than 1.33 for values of I_2 up to 5.5 mA.

To calculate the attenuation of the circuit shown in Fig. 8-5, the ABCD matrix is written

$$\begin{bmatrix} A & B \\ C & D \end{bmatrix} = \begin{bmatrix} 1 & 0 \\ G_1 & 1 \end{bmatrix} \cdot \begin{bmatrix} 0 & jZ_0 \\ jY_0 & 0 \end{bmatrix} \cdot \begin{bmatrix} 1 & 0 \\ G_2 & 1 \end{bmatrix} \tag{8-22}$$

Letting $G_1 = k G_2$ and solving for the attenuation gives

$$\alpha = 20 \log \tfrac{1}{2} \left[2 + (1+k) \frac{G_2}{Y_0} + k \frac{G_2^2}{Y_0^2} \right] \tag{8-23}$$

MATCHED SWITCHES AND ATTENUATORS

For k = .25 the circuit of Fig. 8-5 gives 15 dB attenuation for 5 mA = I_2. At some increase in complexity, more control diodes can be added for greater attenuation range. The method of current ratios continues to provide a low VSWR for more diodes. The circuit has its narrowest matched bandwidth at maximum isolation. In this condition $R_1 = Z_0$ and $R_2 \approx 0$ ohm. The circuit has the same input impedance as a matched load with a shunt Z_0 stub. This stub has a 1.3 VSWR over a 1.4 to 1 bandwidth, and 1.8 over an octave bandwidth as shown in Fig. 8-8 for $G_{IN} = 1$.

4. Match Using Two Input Matching Diodes

Fig. 8-9 shows the circuit used to calculate the input matching of a high isolation switch having two matching PIN diodes at the input. The input admittance is given by the following sequence of equations:

$$Y_T = G_1 - jY_0 \cot\theta \tag{8-24}$$

$$\frac{Y_1}{Y_0} = \frac{Y_T + jY_0 \tan\theta}{Y_0 + jY_T \tan\theta} = \frac{G_1 + jY_0(\tan\theta - \cot\theta)}{2Y_0 + jG_1 \tan\theta} \tag{8-25}$$

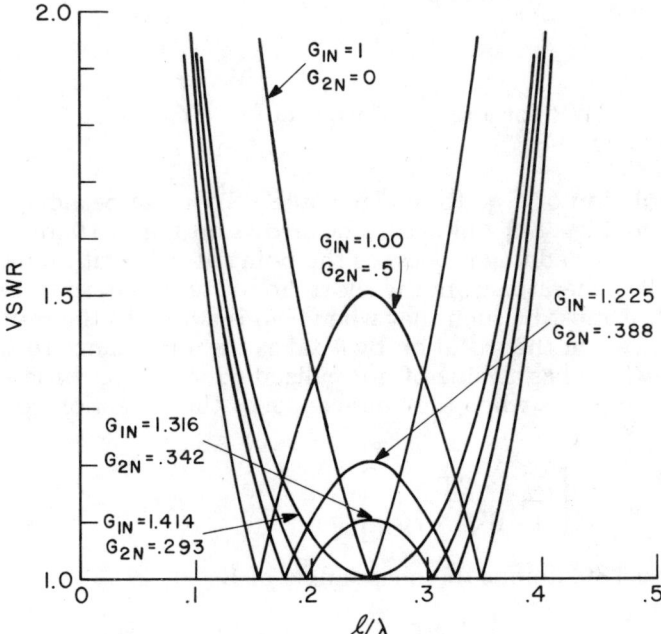

Fig. 8-8 VSWR of Two Shunt Diodes Matching into a Short

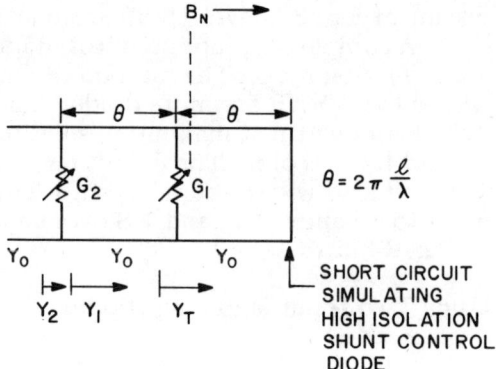

Fig. 8-9 Circuit of Two Diodes Matching into a Short

$$\frac{Y_2}{Y_0} = \frac{Y_1}{Y_0} + \frac{G_2}{Y_0} \qquad (8\text{-}26)$$

The input VSWR, ρ, is derived using

$$\Gamma = \frac{1 - Y_2/Y_0}{1 + Y_2/Y_0} \quad \text{and} \quad \rho = \frac{1 + |\Gamma|}{1 - |\Gamma|} \qquad (8\text{-}27)$$

The input VSWR for a range of values of $G_1/Y_0 = G_{1N}$ (and G_{2N}) is given in Fig. 8-8.

The calculation of G_{2N} for equal ripple VSWR can be aided by considering Figs. 8-9 and 8-10. The arrows on Fig. 8-10 indicate the admittance adding process as the point of observation moves toward the generator from the short circuit termination in Fig. 8-9. The angle θ must be such that when G_{1N} is added to the resulting susceptance, another rotation by θ takes the admittance to the real axis which has a value of normalized conductance of less than one. This happens when the phase angle of the reflection coefficient of Y_{TN},

$$\theta = \tan^{-1}\left[\frac{2 B_N}{1 - B_N^2 - G_{1N}^2}\right] \qquad (8\text{-}28)$$

is equal to $180°$ minus the phase angle of B_N

$$180° - \theta = \tan^{-1}\left[\frac{2 B_N}{1 - B_N^2}\right] \qquad (8\text{-}29)$$

MATCHED SWITCHES AND ATTENUATORS

Recalling that $\tan\theta = -\tan(180° - \theta)$ gives

$$\frac{2 B_N}{1 - B_N^2 - G_{1N}^2} = -\frac{2 B_N}{1 - B_N^2} \tag{8-30}$$

which reduces to

$$B_N = \sqrt{1 - G_{1N}^2}/2 \tag{8-31}$$

The VSWR of $G_{1N} - B_N$ can be calculated using Eqs. (8-27) and G_{2N} is found using

$$G_{2N} = 1 - \frac{1}{\rho} = 1 - \frac{G_{1N}}{2} \tag{8-32}$$

Having selected the value of G_1, G_2 and θ to obtain a match at the design frequency, the frequency dependence of this circuit may be calculated. The dashed curve on Fig. 8-10 shows the input admittance as ℓ/λ varies over a 3:1 ratio.

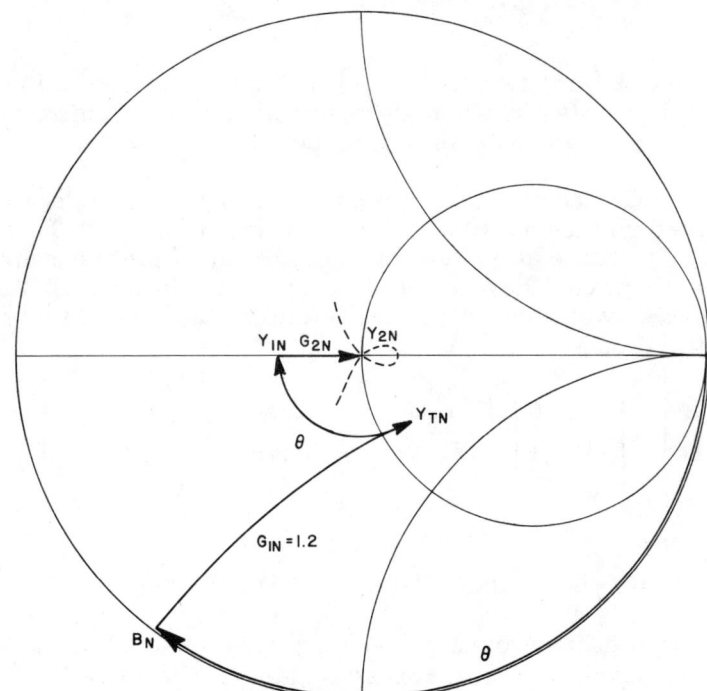

Fig. 8-10 Admittance Adding Process of Two Diodes Matching into a Short

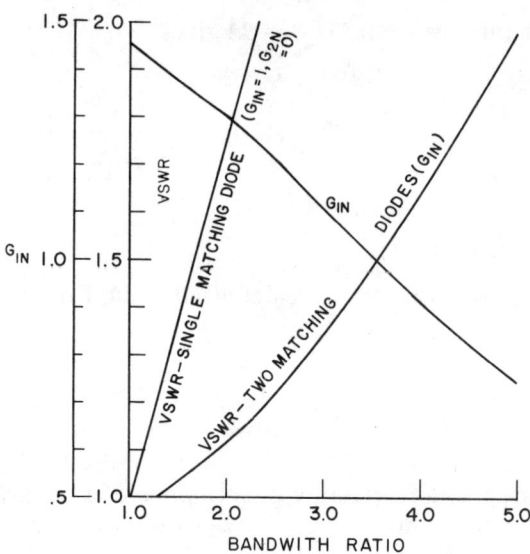

Fig. 8-11 Bandwidth and Conductance of One and Two Diodes Matching into a Short

The bandwidth using two PIN diodes in this fashion is shown in Fig. 8-11. There is quite a substantial improvement in bandwidth over what can be achieved with a single diode.

To make a matched variable attenuator, the matching diodes would be mounted on the ends of an array as shown in Fig. 8-12. To obtain the desired isolation as many control diodes are put in the center section as required. The attenuation is calculated from the ABCD matrix of the two central diodes to determine the bandwidth of the control diodes.

$$\begin{bmatrix} A & B \\ C & D \end{bmatrix} = \begin{bmatrix} 1 & 0 \\ G_c & 1 \end{bmatrix} \cdot \begin{bmatrix} \cos\theta & jZ_0 \sin\theta \\ jY_0 \sin\theta & \cos\theta \end{bmatrix} \cdot \begin{bmatrix} 1 & 0 \\ G_c & 1 \end{bmatrix}$$

(8-33)

$$\alpha = 10 \log \left[(1 + G_{CN})^2 + G_{CN}^2 (1 + \tfrac{1}{2} G_{CN})^2 \sin^2\theta \right] \quad (8\text{-}34)$$

This equation differs from Eq. (6-10) because the shunting elements are real rather than reactive. This attenuation is shown in Fig. 8-13 for $G_{CN} = 50$. The maximum attenuation drops by only 1.3 dB over an octave bandwidth and by 4.6 dB over a 4:1 bandwidth.

MATCHED SWITCHES AND ATTENUATORS

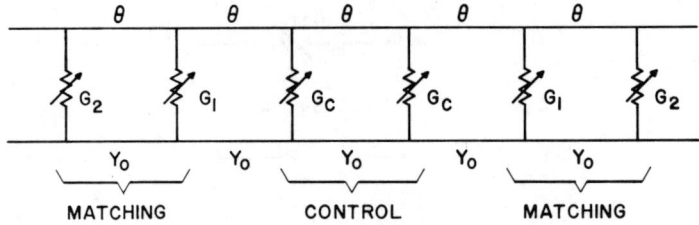

Fig. 8-12   Circuit of a Bilaterally Matched Variable Attenuator Having Two Control Diodes and Two Matching Diodes on Each End

More matching diodes on each end should enable matching over a wider bandwidth. Commercially available attenuators[4] of this type maintain matched input by having separate series bias resistors for each diode, as was done for Fig. 8-7.

C. Lumped Element Variable Attenuators

Consider conventional T and Pi circuit resistive attenuators as shown in Fig. 8-14. When the fixed resistors are replaced by variable resistance PIN diodes, the attenuator can be made variable and matched by properly controlling the diode biases. Matched conditions for the T circuit are found by setting up the following relationship:

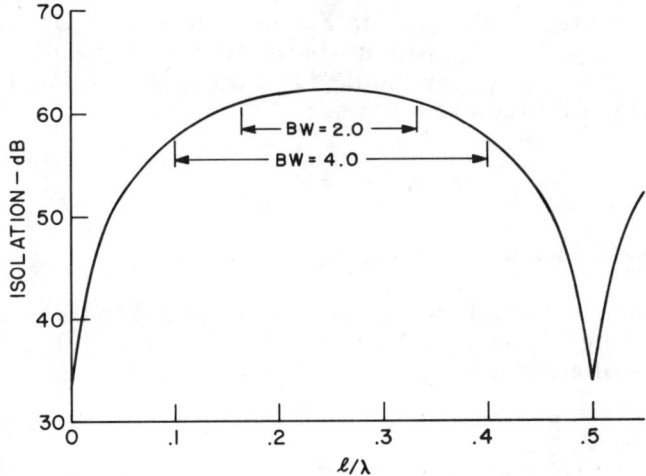

Fig. 8-13 Isolation of Two 1-ohm Control Diodes Shunting a 50-ohm Transmission Line Separated by ℓ/λ

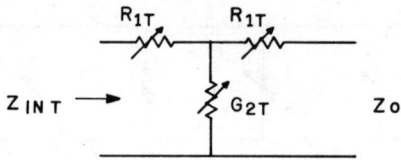
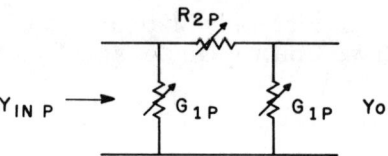

Fig. 8-14   Equivalent Circuit of T and Pi Attenuators

$$Z_{IN\ T} = R_{1T} + \cfrac{1}{G_{2T} + \cfrac{1}{R_{1T} + Z_0}} = Z_0 \qquad (8\text{-}35)$$

$$2(R_{2T})_N = \frac{1}{(R_{1T})_N} - (R_{1T})_N \qquad (8\text{-}36)$$

For low loss $(R_{1T})_N \approx 0$, the $\dfrac{1}{(R_{1T})_N}$ term dominates and the complementary voltage circuit used in Fig. 8-3 can be used. But for large values of attenuation $(G_{2T})_N$ must be allowed to vary while $(R_{1T})_N$ remains at $(R_{1T})_N \approx 1$. Handbooks[6] give the values of $(R_{1T})_N$ and $(G_{2T})_N$ for the T circuit as follows:

$$(R_{1T})_N = \frac{k-1}{k+1} \text{ and } (G_{2T})_N = \frac{k^2-1}{2k} \qquad (8\text{-}37)$$

in which $\alpha = 20 \log k$. $\qquad (8\text{-}38)$

The same equations and arguments may be applied to the pi circuit by substituting $(G_{1P})_N$ for $(R_{1T})_N$ and $(R_{2P})_N$ for $(G_{2T})_N$. For large values of attenuation

$$(G_{2T})_N = \frac{k}{2} \qquad (8\text{-}39)$$

which gives

$$k = (G_{2T})_N = 2\alpha_{RF} l Z_0 = 2\alpha_{RF} Z_0 I_0 e^{\alpha_{DC} V} \qquad (8\text{-}40)$$

MATCHED SWITCHES AND ATTENUATORS

$$\alpha = 20 \log k = [\, 20 \log (2\alpha_{RF}\, Z_0\, I_0)\,] + [8.7\, \alpha_{DC}] \cdot V \quad (8\text{-}41)$$

The attenuation is linearly related to the voltage applied to the control diode.

Another very useful circuit for lumped circuit variable attenuators is the bridged T[5] as shown in Fig. 8-15. The values of R_{1N} and G_{2N} can be given by

$$R_{1N} = G_{2N} = k - 1 \quad (8\text{-}42)$$

in which k is the voltage attenuation ratio as before. Since $G_{1N} G_{2N} =$ a constant for all values of attenuation, complementary voltage drive will assure match just as in Fig. 8-3. The attenuation is also linearly related to the voltage for high values of attenuation as in Eq. (8-41) except for the factor 2. The dc circuit for obtaining this complimentary voltage bias is shown in Fig. 8-16.

The circuit of Fig. 8-3 can be modified to make a bilaterally matched variable attenuator as shown in Fig. 8-17. Also shown is the dual to the T circuit. The normalized input admittance is given by

$$Y_{INN} = \frac{1}{1 + R_{1N}} + \frac{1}{R_{2N} + \dfrac{1}{1 + \dfrac{1}{1 + R_{1N}}}} \quad (8\text{-}43)$$

Letting $R_{1N} = k_c/R_{2N}$ (complementary voltage bias) causes the normalized input impedance to be given by

$$Z_{INN} = \frac{2 R_{2N} + 3 k_c + 1 + 2 k_c/R_{2N} + k_c^2/R_{2N} + k_c^2/R_{2N}^2}{2 R_{2N} + k_c + 3 + 4 k_c/R_{2N} + k_c^2/R_{2N}^2} \quad (8\text{-}44)$$

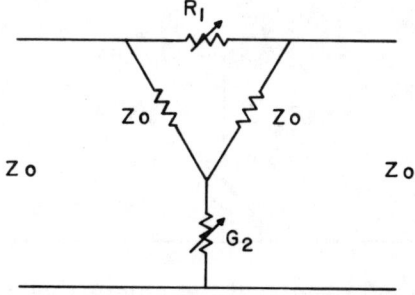

Fig. 8-15 Bridged T Attenuator

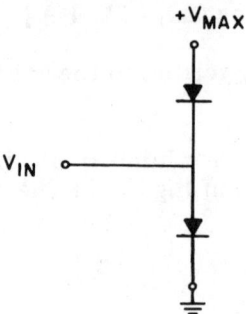

Fig. 8-16 Circuit for Complementary Bias of Diodes

When $R_{2N} = 0$ then $Z_{INN} = 1$ and when $R_{2N} = \infty$, $Z_{INN} = 1$; therefore the attenuator will be matched at both attenuation extremes. k_c remains free to be adjusted for equal ripple VSWR over the attenuation range. For example, to make $Z_{INN} = 1$ for $R_{2N} = 1$, k_c must be given by $k_c = \sqrt{2}$. The normal dual substitutions may be made in Eq. (8-44) to describe the T circuit.

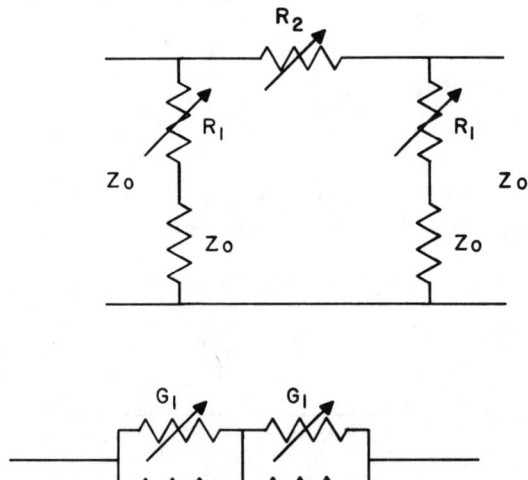

Fig. 8-17 Bilaterally Matched Variable Attenuators Based on Unilaterally Matched Attenuator of Fig. 8-3

MATCHED SWITCHES AND ATTENUATORS

When using PIN diodes for variable attenuators, harmonics and inter-modulation products can be generated. These are most prevalent at low bias currents, low frequencies, and high RF power. When all three of these conditions can be avoided the harmonics and intermodulation products will be minimum. To work at lower frequencies, diodes should be used that have larger than normal carrier life-times and I regions. The longer life-time and I region will reduce self-rectification and the longer I region will decrease α_{RF} requiring more dc current for a given conductance.

REFERENCES

1. D.A.E. Roberts and S.J. Robinson, "A PIN Diode Modulator for the Frequency Band 2.5 - 7.5 Gc/s," *The Microwave Journal*, Vol. 6, No. 12, December 1963, pp. 74-78.

2. R. Ekinge and T. Hedstrom, "A New Variable Microwave Attenuator, *IEEE Trans. on Microwave Theory and Techniques*, Vol. MTT-18, No. 9, September 1970, pp. 661-662.

3. J.C. Hoover, "A Mono-Control Microwave Semiconductor Switch," 1964 *IEEE PTGMTT International Symposium Digest*, May 19-21, 1964, pp. 204-207.

4. J.K. Hunton and A.G. Ryals, "Microwave Variable Attenuators and Modulators Using PIN Diodes," *IRE Trans. on Microwave Theory and Techniques*, Vol. MTT-10, No. 4, July 1962, pp. 262-273.

5. C.A. Prufer, "Constant-Impedance PIN-Diode Attenuators," *Microwaves*, August 1967, pp. 46-49.

6. Anon. *Reference Data for Radio Engineers*, 4th ed. IT&T Corp., New York, N.Y. 1956, pp. 252-254.

Limiters

Chapter 9

An ideal limiter has no attenuation when low power is incident upon it but starting at some threshold has an attenuation that increases with increasing power to maintain the output power constant as shown by the solid line in Fig. 9-1. The main use of a microwave limiter is to prevent transmitter power in a radar from burning out its own receiver mixer. When designed for wider bandwidth the limiter can also protect the mixer from other nearby radar transmitters. Limiters are also useful for reducing the amplitude modulation of swept frequency oscillators and for reducing amplitude modulation in phase detection systems.

Realizable limiters give responses similar to the broken line curves in Fig. 9-1. They are characterized by a finite insertion loss δ, a finite isolation η, and vacillation Δ, about the limit power level P_{LIM}. The insertion loss and isolation are derived from the diode circuit considered as a switch, at zero bias for the insertion loss state and in conduction for the isolation state. All information on insertion loss and isolation will be found in other chapters. This chapter covers output characteristics, recovery time, PIN diode limiting, and useful circuits.

A. Output Characteristics

Fig. 9-2 shows the basic circuit for a limiter. Two diodes are mounted in opposite polarity across a load resistor. The voltage-current characteristic of the two diodes combined is also shown on the figure. Each diode conducts very little current in the forward direction until the "contact" potential V_ϕ is exceeded. This contact voltage is typically .3V for germanium diodes, .7V for silicone diodes, and 1.1V for gallium arsenide diodes. For sine wave voltage

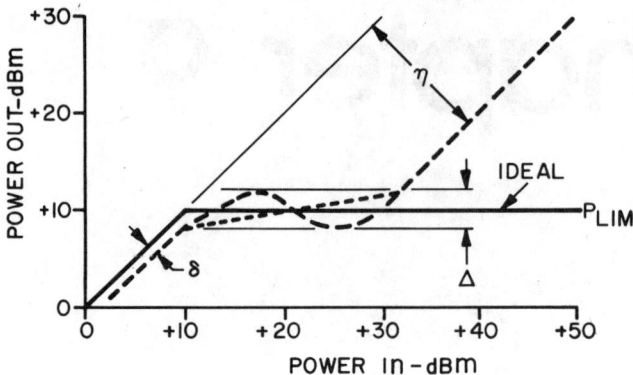

Fig. 9-1 Definition of Limiter Parameters.

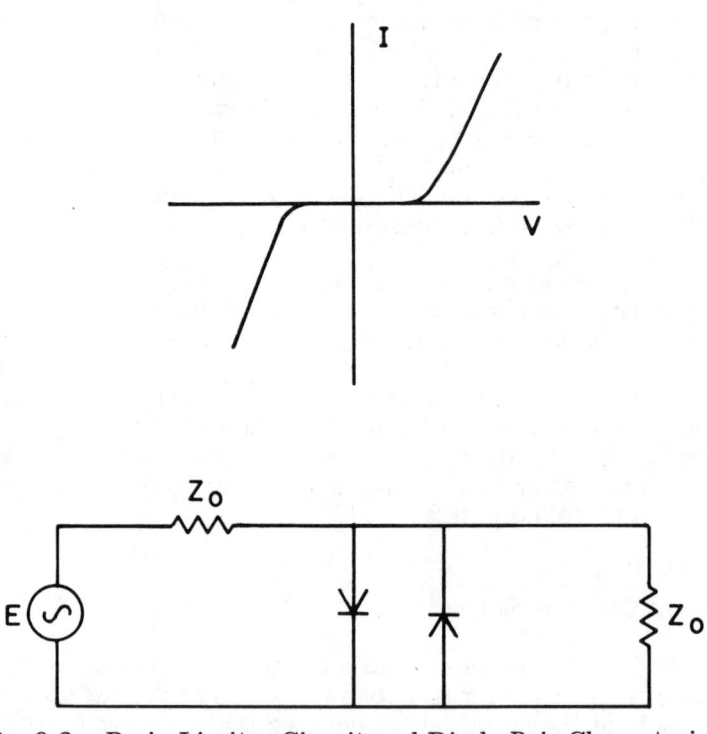

Fig. 9-2 Basic Limiter Circuit and Diode Pair Characteristic.

LIMITERS

less than V_ϕ, no current flows through the diodes, and the maximum available generator power is delivered to the load L. When the voltage exceeds V_ϕ, the diode resistance $\left(\frac{\partial V}{\partial I}\right)$ rapidly drops below Z_0, causing the sine wave to be clipped as shown in Fig. 9-3. In the limit the power in the load reaches the value

$$P_{LIM} = V_\phi^2 / Z_0 \qquad (9\text{-}1)$$

as a square wave. When the generator and load are replaced by matched transmission lines having characteristic impedances of Z_0, the limiter can be visualized in its microwave circuit. In the transmission line, a sine wave is incident on the limiter, a clipped sine wave emerges from it traveling toward the load, the clipped peaks are reflected back into the generator.

The square wave is rich in odd harmonics. The fundamental power getting to the load is given by

$$P_{LIM} = \frac{8}{\pi^2} V^2 / Z_0 \qquad (9\text{-}2)$$

When a very fast diode is used for the limiter (hot carrier diode) the RF waveform would be as shown in Fig. 9-3, and a low pass filter would have to be placed after the limiter to eliminate the harmonics if they disturb the system. The waveform of Fig. 9-3 would give the outputs shown in Fig. 9-4. Various factors moderate the output as follows. Varactors are normally used as limiters because they provide higher isolation at high power, and can dissipate more peak and average power than hot carrier diodes. They do not rectify the RF as well as the hot carrier diodes

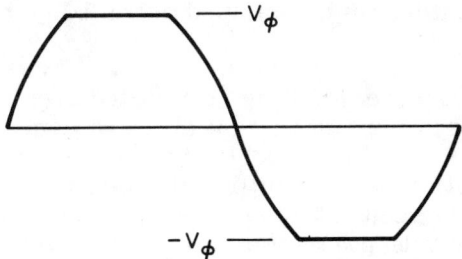

Fig. 9-3   Clipped Sine Wave.

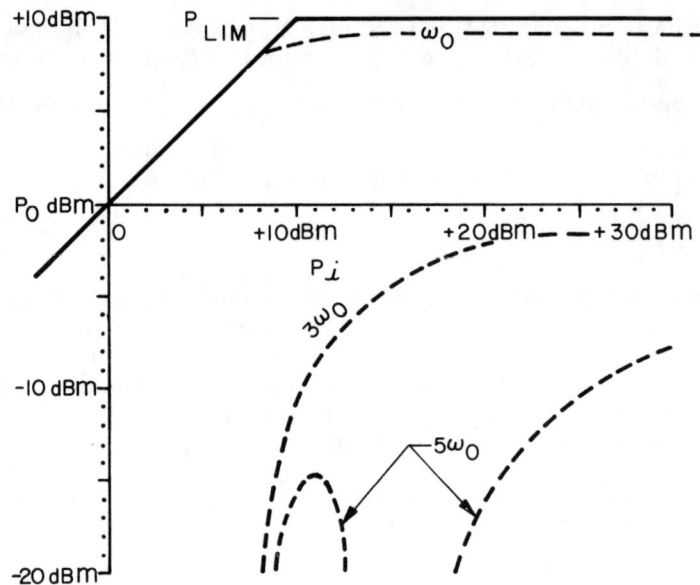

Fig. 9-4 Spectral Output for the Waveform given in Figure 9-3.

can and therefore are not as rich in harmonics. They do, however, tend to interact parametrically with each other and with the circuits surrounding them thereby producing even harmonics as well.

When a low pass filter is used to stop the harmonics, the electrical distance to the filter influences the limiting characteristics as portrayed in Fig. 9-1. The deviation Δ about P_{LIM} is typically less than 2 dB. As the low pass filter is moved away from the limiter, the characteristic varies about the P_{LIM} line, within the envelope given by the broken lines. At the proper distance to the low pass filter, very flat limiting can be obtained over a 10 - 15 dB range of input power levels.

The limited output power level can be adjusted over a modest range by placing it in a circuit that magnifies or demagnifies the voltage appearing across the diode. In the extreme, this could be a high Q cavity with input and output loops. A more restricted but more practical circuit is to place the shunt diode(s) in the middle of a half wavelength section of transmission line having characteristic impedance Z_{0H}. The input quarter wavelength transforms the generator impedance to Z_{0H}^2/Z_0 and the output quarter wavelength section transforms the load to this same impedance level. Therefore, the limiter diode is effectively mounted in a Z_{0H}^2/Z_0 transmission line. Taking the practical range of

LIMITERS

transmission line characteristic impedances to be 12.5 ohms to 100 ohms, the diode can be mounted in character impedances from 3 ohms to 200 ohms providing the range of limited power outputs as shown in Fig. 9-5. Care must be taken to use this same magnified Z_0 for calculating insertion loss and isolation. Using a higher Z_0 lowers the limited output power at the cost of increased insertion loss, but on the other hand, isolation is increased.

B. Recovery Time

The close-in range of a radar is restricted by the slow recovery of the limiter protecting its receiver. It will be restricted by other things such as recovery from receiver saturation and transmitter waveform, but the limiter had the slowest recovery when TR tubes were used. Recovery time is defined as the time between the end of the pulse being limited and the time when the attenuation has decreased to within 3 dB of the low loss attenuation.

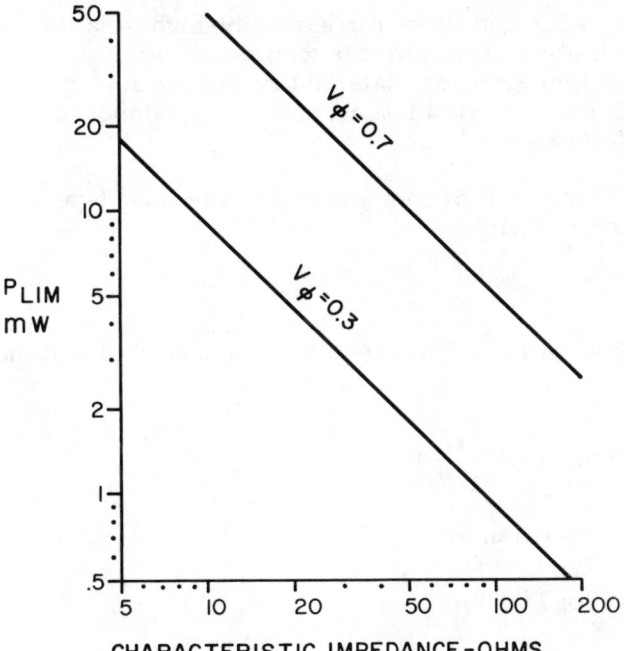

Fig. 9-5 Output Limit Levels for Germanium (.3v) and Silicon (.7v) Diodes.

Diode limiter recovery is caused by the finite time it takes injected carriers to recombine. High isolation is caused by the low resistance of a depletion region filled with injected carriers. The bulk conductivity of the depletion region is given by

$$\sigma = ne\mu \tag{9-3}$$

in which n is the carrier density, e is the carrier charge and μ is the mobility. The conductance of the depletion region is given by

$$G = \frac{A}{h}\sigma \tag{9-4}$$

in which A is the area and h is the height of the depletion region. The carriers recombine such that after the source is extinguished their quantity is described by

$$n = n_s \epsilon^{-t/\tau} \tag{9-5}$$

in which n_s is the density of carriers in the high isolation state and τ is carrier lifetime (typically a few microseconds in intrinsic material at room temperature). Material that is neutral because of equal doping will have a lower lifetime, and raising temperature will shorten lifetime.

Eqs. (9-3) through (9-5) combine to give the time dependent diode junction conductivity.

$$G = k\epsilon^{-t/\tau} \tag{9-6}$$

in which $k = An_s e\mu/h$. The attenuation of a single shunt diode is given by

$$\alpha = 20 \log\left(1 + \frac{G}{2Y_0}\right) \tag{9-7}$$

which can be written as

$$\alpha = 20 \log\left(1 + k_N \epsilon^{-t/\tau}\right) \tag{9-8}$$

in which $k_N = \dfrac{k}{2Y_0}$, the constant that defines the isolation of the limiter with the pulse power incident on it. At high isolation the attenuation is given by

LIMITERS

$$\alpha = 20 \log k_N \epsilon^{-t/\tau} = 20 \log k_N - 8.7 \frac{t}{\tau} \quad (9-9)$$

The attenuation falls at 8.7 dB per lifetime. As the attenuation approaches 0 dB ($t=t_0$)*, the "1+" in Eq. (9-8) becomes important, causing the attenuation to decrease more slowly, as shown in Fig. 9-6. The 3-dB attenuation occurs about one lifetime after the 0-dB time using Eq. (9-9), and the 1-dB attenuation, at 2 lifetimes.

The calculations were made for one diode because two diodes in the same plane would give the same result. When two diodes are used and are not in the same electrical plane, the diode closest to the generator has the highest power incident on it and will dominate the recovery process. Frequently the diode closest to the generator is also made to handle higher power than the others, which leads to a larger junction and a longer lifetime.

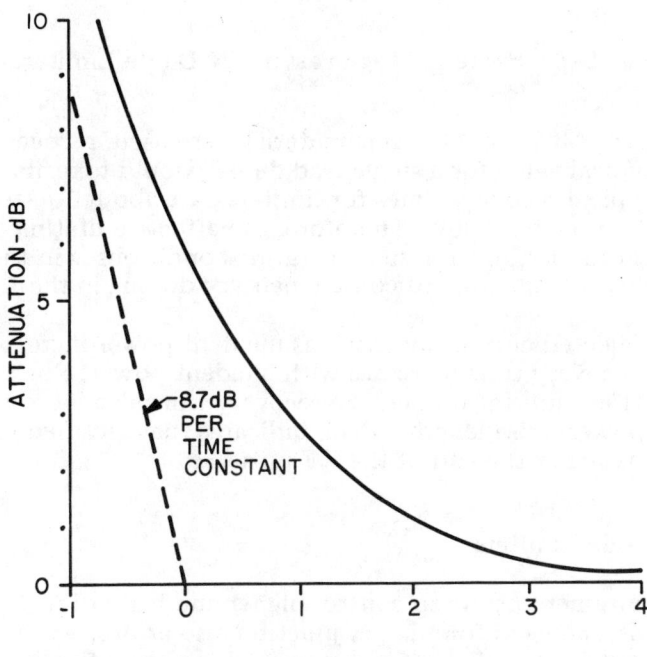

Fig. 9-6   End of Limiter Recovery.

*t_0 is defined as that time when the attenuation would drop to zero if the high attenuation approximation were valid there.

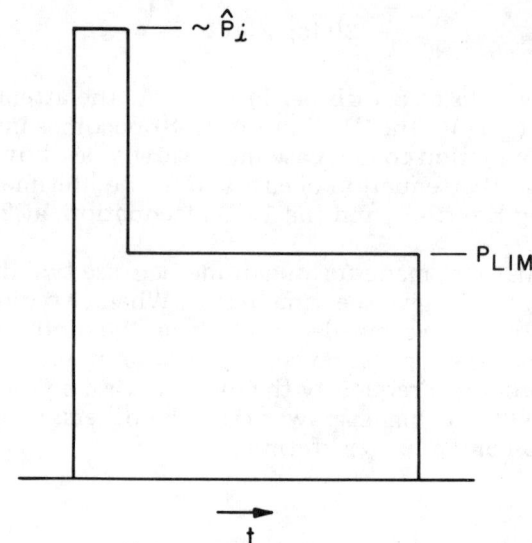

Fig. 9-7 Spike Leakage Past a PIN Diode Limiter.

If intrinsic ($\tau \approx 3.7\,\mu$sec) lifetime material were used, recovery from 20 dB (a typical value for a single lead diode) would take about 11 μsec. Typical recovery times for limiters[1] are about 1 μsec for 1 kW and 50 nsec for 100W. Therefore, the effective lifetimes of these limiter diodes under actual operating conditions is in the 300 nsec to 20 nsec range (due to complementary doping in the I region).

Note that injected current increases as incident power increases; therefore, recovery time increases with incident power. For the poorer rectifier limiting diodes, the recovery time also increases with high power pulse length, which indicates that rectified current is still increasing at the end of the RF pulse.

C. PIN Diode Limiters

As the requirement grows to control higher and higher incident power levels, the need for a larger junction also grows. As the junction gets bigger, detection efficiency goes down until finally in the megawatt region the limiters have to be switched to the isolation state by a driver and cease to be limiters. In the several kilowatt range, the medium size junctions still rectify well enough to provide self-activation, but turn-on is accompanied by spike leakage as illustrated in Fig. 9-7. The output of the limiter has a leading

LIMITERS

edge spike[2] power nearly equal to the incident power level. When the diode builds up self-rectified bias, the output is constant at P_{LIM}.

As the rectified current is building up for a high power pulse, the diode impedance is moving continuously from its low level value ($\sim 1\,k\Omega$) to its high level value ($<1\,\Omega$). In the process, it passes relatively slowly to the impedance where 50% of the incident power is absorbed in the diode (refer to Fig. 4-18) but then suddenly because of the lower impedance of the diode compared to the generator impedance, the rectified current increases rapidly causing a sudden end of the spike.

Even though a PIN diode is not a very good rectifier at higher frequencies, it continues to provide flat limiting if it is correctly mounted. The RF resistance R of a PIN diode as derived from rectification[3] is given by

$$\frac{1}{G} = R = \frac{h}{\beta \sqrt{D/\omega}} \sqrt{\frac{Z_0}{P_i}} \tag{9-10}$$

in which h is I region thickness (in cm), $\beta = q/kT = 40$, D is diffusion coefficient (15.6 cm^2/sec), and P_i is the incident pulse power. A single diode shunting a transmission line provides attenuation given by

$$\alpha = 20 \log \left[1 + \frac{G}{2Y_0} \right] \approx 20 \log \left[\frac{\beta \sqrt{D/\omega}}{2hY_0} \sqrt{\frac{P_i}{Z_0}} \right] \tag{9-11}$$

$$\alpha = 10 \log \left[\frac{\beta^2 D Z_0 P_i}{4\omega h^2} \right] = 10 \log \left[\frac{P_i}{P_{LIM}} \right] \tag{9-12}$$

$$P_{LIM} = \left[\frac{4h^2 \omega}{\beta^2 D Z_0} \right] = .13\, h'^2\, f \text{ Watts} \tag{9-13}$$

in which h' is in mills and f is in GHz.

Eq. (9-13) gives the output limit levels shown in Fig. 9-8 for $Z_0 = 50\,\Omega$ transmission line. At very low frequencies the P_{LIM} is 10 mW. These curves are substantiated by the data of Leenov[3] and Brown[4]. Leenov built a limiter with a .002" thick I region which had $P_{LIM} = +29$ dBm ± 1 dB at 1 GHz which compares favorably with $+27.5$ dBm predicted here. The limiter also had 1 μsec spikes. Fig. 9-8 can be used to predict spike leakage. Based on observations it is estimated that all PIN diodes limiting to 1 Watt will have 1 μsec spike leakage (and 100 mW, .1 μsec, etc.). No simple model has yet

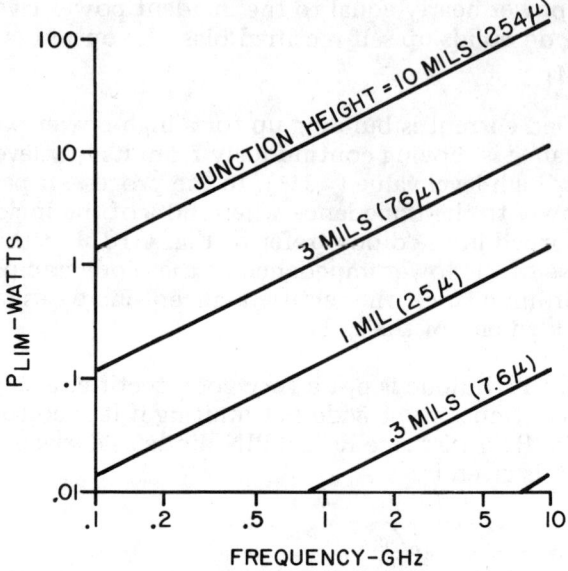

Fig. 9-8 Output Limit Levels for PIN Diode Limiters (Z_0 = 50 ohms).

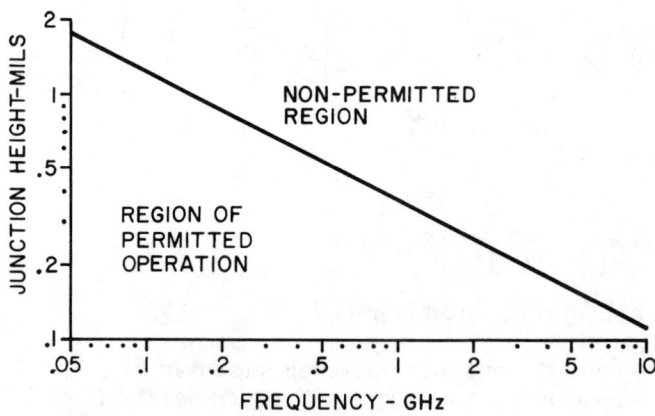

Fig. 9-9 Useful Range of PIN Diodes as Limiters from Brown[4].

LIMITERS

been derived that satisfactorily describes PIN diode limiter spike leakage. Brown studied the spike leakage and limiting of a range of diodes as shown in Fig. 9-9. When frequency and thickness combine above the curve then spike leakage, higher threshold, and eventually burnout, occur. The data points fall neatly near the bottom of Fig. 9-8 where P_{LIM} begins to increase.

When the spike leakage and increased threshold of the PIN diode become problems, the PIN limiter can be followed by a varactor limiter as shown in Fig. 9-10. The PIN diode protects the varactor from high incident power, and the varactor eliminates the spike and lowers the limited output power to 10 mW.

Flat limiting occurs when the PIN diodes are mounted in the same electrical plane. When like PIN diodes are $\lambda/4$ apart, their input-output characteristic can take on a severe S shape if no isolator is inserted between them.

To use PIN diodes at Megawatt levels, they must be operated as switches to get them into their very low loss state. A method that accomplishes this without using complex driving circuits is shown in Fig. 9-10B. A high power diode detector is loosely coupled to the input line which does the rectifying for the PIN diode.

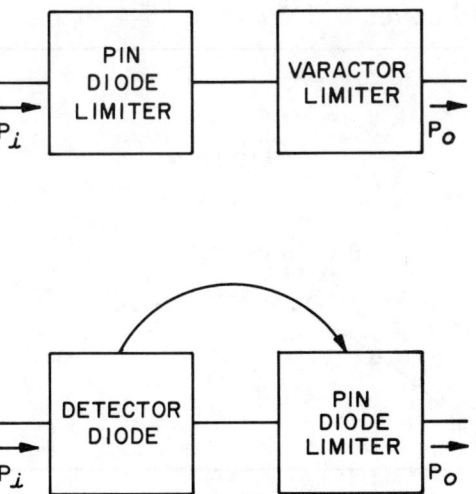

Fig. 9-10 PIN Diode Limiters with Varactor Limiter Cleanup (A) and with Detector Diode Driver (B).

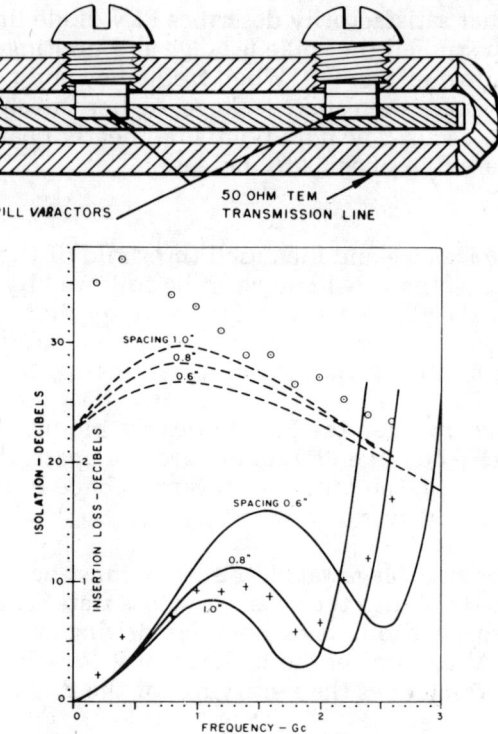

Fig. 9-11 Insertion Loss and Isolation of a Two Varactor Diode Baseband Limiter.

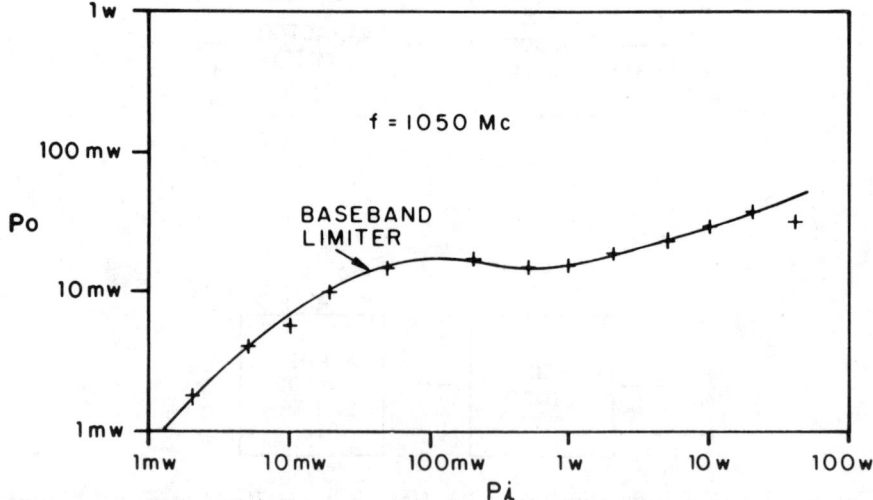

Fig. 9-12 Input-Output Characteristic of a Two Varactor Diode Baseband Limiter.

LIMITERS

D. Useful Circuits

Very wideband limiters have been made by using broad band switch design principles. Fig. 9-11 shows the insertion loss and isolation of a limiter[5] made in rectangular air-filled coax with pill type varactor diodes. The device limits from DC to 2.5 GHz with an input-output characteristic as shown in Fig. 9-12, and power rating as shown in Fig. 4-16. Hewlett Packard extended the frequency range up to 18 GHz using hybrid integrated circuits, as shown in Fig. 5-23.

Practically any varactor or PIN diode can be made into a limiter by using the circuits disclosed in Chapter V. The basis for selecting a switch circuit for a limiter is only that it be in the isolation state when the diode is drawing current. If a diode happens to have the wrong switching sense, the switching sense can be corrected by using it as a reflection switch. One of the first limiters made[6] used X-band diode switches that had the wrong switching sense. The switches were mounted on a 3-dB coupler as shown in Fig. 9-13. The output was very flat over a fair bandwidth as shown in Fig. 9-14. This limiter was used for eliminating the AM on a klystron being swept through a mode. Subsequent models of this limiter were made at higher powers to clean up the spike leakage past a ferrite limiter[7].

Limiters are very useful in duplexing circuits, where a transmitter and receiver share an antenna. The circuit is a double throw switch as discussed in Chapter VIII and in particular in Fig. 7-7.

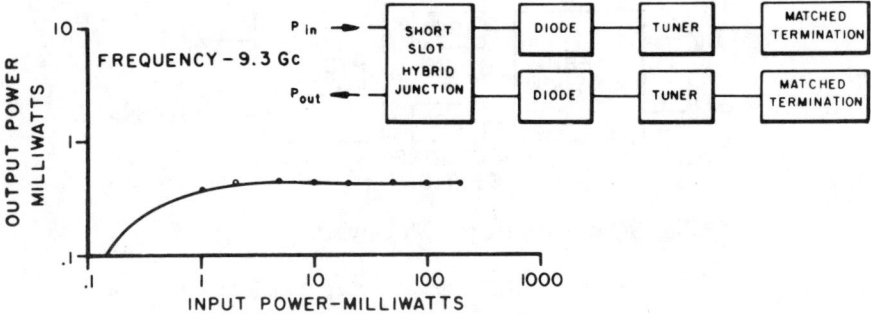

Fig. 9-13 Block Diagram for X-Band Matched Limiter.

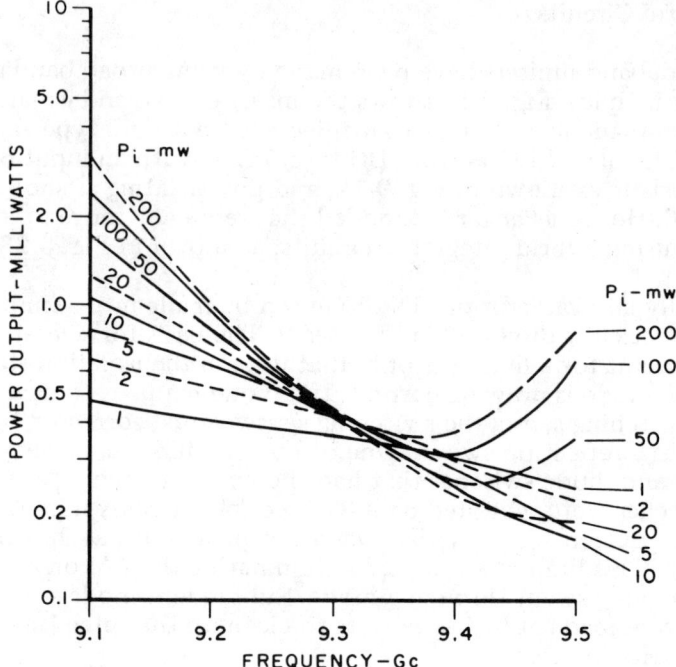

Fig. 9-14 X-Band Matched Limiter Output. 1N263 Diodes used for Diode and Tuner (spaced $3\lambda/4$).

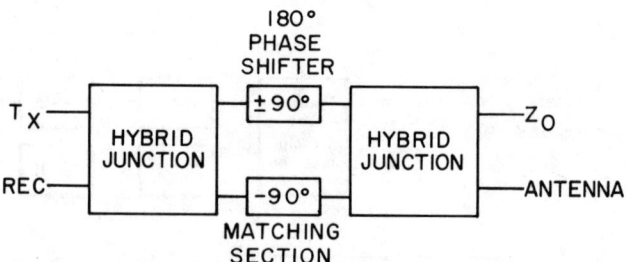

Fig. 9-15 Circuit for Megawatt Duplexer.

LIMITERS

An extremely high power duplexer[8] has been made (2 MW @ C band) using the circuit of Fig. 9-15. The 180° phase shifter is of the loaded line type. When both paths have the same phase shift, the transmitter is connected to the antenna and the receiver to the load. When their phase shifts are different by 180°, the antenna is connected to the receiver and the transmitter to the load. By using a loaded line phase shifter for the switching elements, power can be shared (almost) equally by the diodes (the phase shift section has low insertion loss) and, being loosely coupled, each diode can handle a significantly higher power than a single shunt diode. By referring to Eqs. (10-127) and (10-87) improvement in power handling capability can be calculated. By using 10 diodes 5 36° bits will be required for 180° phase shift. The improvement in power handling capability is a factor of 40. (A 180° reflection phase shifter using two diodes on a 3-dB coupler is being compared to a 180° loaded line phase shifter.) The circuit permits the application of very high power duplexing and switching techniques at the higher frequencies.

A programmable leveler[9] can be made with back-to-back hot carrier diodes, as shown in Fig. 9-16(A). Both diodes are biased into conduction. When RF exceeds bias current, the diode stops conducting and blocks RF power. The output is a clipped sine wave just like the shunt limiter, except that the output (Fig. 9-3) is a current waveform clipped at I_B instead of a voltage waveform clipped at V_0. The limited power level is settable by adjusting the bias current I_B. 10 mA bias would give 5 mA (=I_B) per diode. The limited output (assuming the output is filtered and Z_0 = 50 Ω) is given by

$$P_{LIM} = \frac{8}{\pi^2} I_B^2 Z_0 = 1 \text{ mW} \tag{9-14}$$

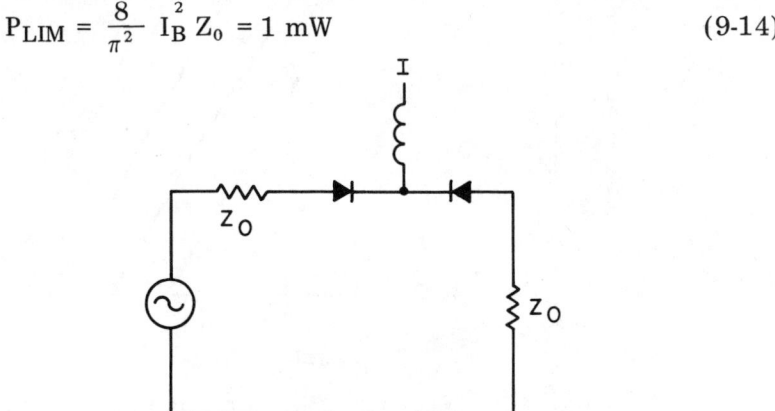

Fig. 9-16(a) Hot Carrier Diode Leveler.

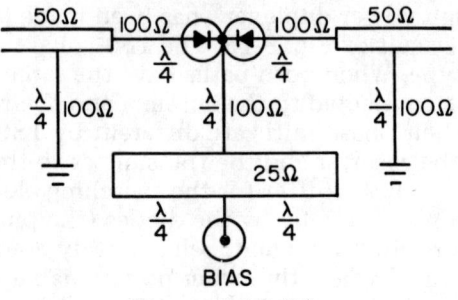

Figure 9-16 (b).

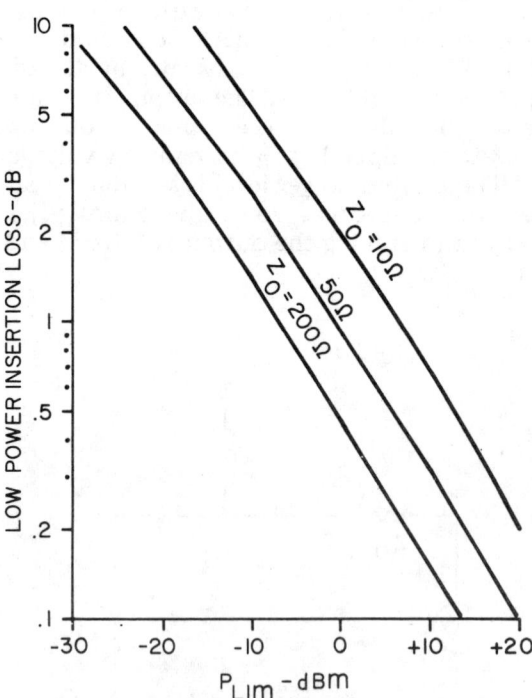

Fig. 9-17 Insertion Loss of the Hot Carrier Diode Leveler for Sub-Level Signals.

LIMITERS

The insertion loss is given by the series resistance of the two biased diodes.

$$R_S = \frac{1}{\alpha_D I_B} = \frac{1}{40 I_B} = 5\,\Omega \tag{9-15}$$

$$\delta = 20 \log\left(1 + \frac{2R_s}{2Z_0}\right) = .83\,\text{dB} \tag{9-16}$$

The limiter shown in Fig. 9-16(B) has 100-Ω $\lambda/4$ transformers around it transforming the diodes into a characteristic impedance of 200 Ω. A bias of I_B = 2.5mA per diode would give limiting at 1mW and .43dB insertion loss. Other insertion losses and limit levels are given in Fig. 9-17. The limiter can be made to limit at very low power levels at a sacrifice in insertion loss. Care must be taken at high incident powers that the breakdown voltage of the diodes is not exceeded. The maximum isolation will be dominated by the capacitance of a single back biased diode as would be given by a single series diode switch (see Fig. 2-12 for Z_0 = 50 Ω).

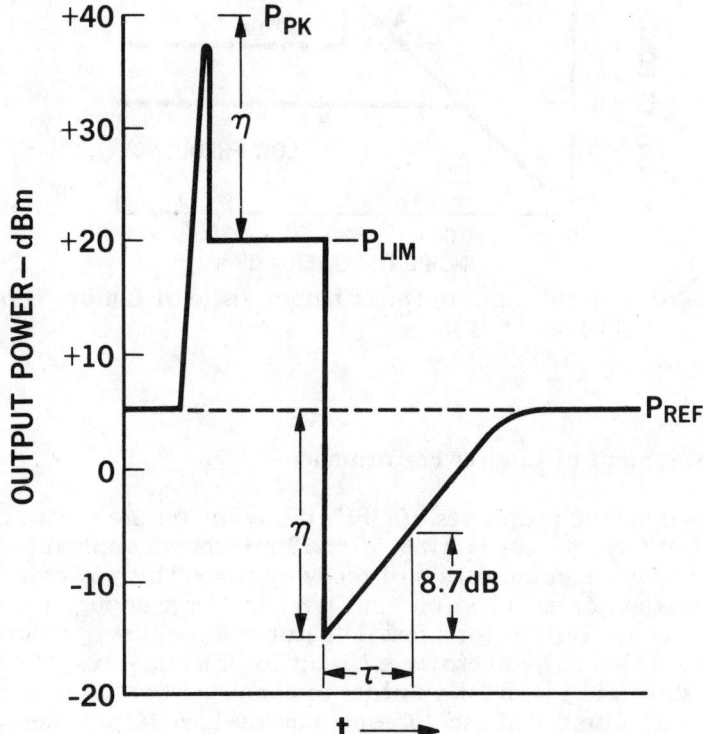

Fig. 9-18 PIN Diode Limiter Output Response.

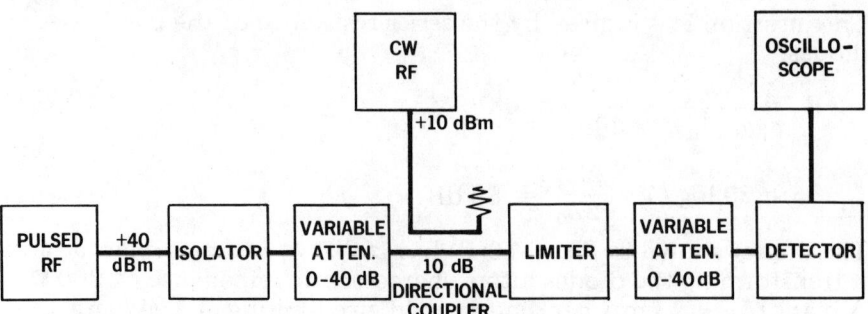

Fig. 9-19 Waveguide Circuit for Measuring Limiter Response.

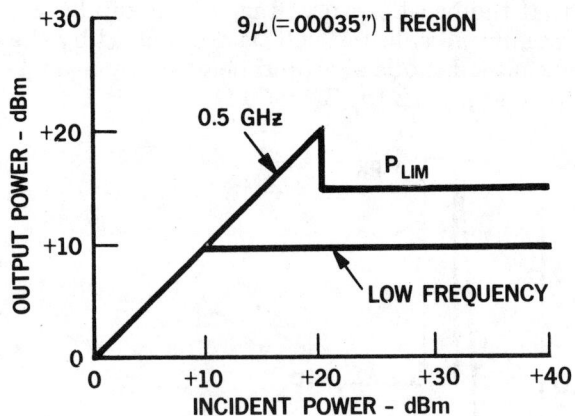

Fig. 9-20 Typical Input-output Characteristic of Limiter Diode (Plateau Only).

E. Measurement of Limiter Performance

All of the dynamic properties of a PIN diode limiter are shown in Fig. 9-18. A CW reference power of +5dBm is shown applied to the limiter to provide an indication of recovery time. This power level is less than the normal +10dBm limit level and large enough to provide a detectable output for a fast detector and oscilloscope during the recovery time. The microwave circuit for injecting this reference power is shown in Fig. 9-19. In order to construct the full picture of Fig. 9-18 a number of oscilloscope pictures have to be taken with different settings of the detector input attenuator and of the oscilloscope gain.

LIMITERS

The input-output characteristic of a typical PIN diode is shown in Fig. 9-20. Frequently P_{LIM} is not flat (increases gradually with incident power) and doesn't exactly fit Fig. 9-8. This is normally found in diodes with a short lifetime τ and quick recovery. The discontinuity in the input-output characteristic is normally found in PIN diodes at frequencies for which P_{LIM} is higher than +10dBm. As P_{LIM} becomes higher the step discontinuity in the input-output curve becomes larger. When P_{LIM} is high the diode also requires more time before the onset of limiting — a longer RF pulse if required — the leakage spike is longer. For the purpose of measurement, thermal dissipation should be taken into account by lowering the pulse repetition rate or by using a special heat sink diode mount.

REFERENCES

1. J.J. Brule, "X-Band Solid-State Detector Amplifier," Contract Da 36-039 AMC 03274E, March 1965, AD 617 003.

2. R. Tenenholtz, "Solid-State S and X-Band Limiters," Contract NObsr-87307, 4th Q, May 1963, AD 405 976.

3. D. Leenov, "The Silicon PIN Diode as a Microwave Radar Protector at Megawatt Levels," *IEEE Trans. on Electron Devices*, Vol. ED-11, February 1964, pp. 53-61.

4. N.J. Brown, "Design Concepts for High-Power PIN Diode Limiting," *IEEE Trans. on Microwave Theory and Techniques*, Vol. MTT-15, December 1967, pp. 732-742.

5. R.V. Garver and J.A. Rosado, "Broad-Band TEM Diode Limiting," *IRE Trans. on Microwave Theory and Techniques*, Vol. MTT-10, September 1962, pp. 302-310.

6. R.V. Garver and D.Y. Tseng, "X-Band Diode Limiting," *IRE Trans. on Microwave Theory and Techniques*, Vol. MTT-9, March 1961, p. 202.

7. W.F. Krupke, T.S. Hartwick, and M.T. Weiss, "Solid-State X-Band Power Limiter," *IRE Trans. on Microwave Theory and Techniques*, November 1961, Vol. MTT-9, pp. 472-480.

8. P. Baskin, "Application of Semiconductor Devices to High Power Duplexers," Contract AF 30(602)-3278, August 1966, AD 502 454.

9. H. Oelke, Harry Diamond Laboratories, private communication.

Phase Shifters

Chapter 10

The diode switch is the basic element for digital diode phase shifters and the varactor is the basic element for continuous phase shifters. The simplest digital phase shifter consists of two SPDT switches that permit either of two paths to be selected for the transmission path,[1,2] as shown in Fig. 10-1. Next in complexity is the reflection type phase shifter,[3,4] as shown in Fig. 10-2. When the switch is closed, the plane of reflection is in the plane of the switch. When the switch is opened, the plane of reflection is at the short and the output wave is delayed by $\Delta \ell$. (A 90° 3 dB coupler may be used instead of the circulator requiring two identical switch-transmission line circuits.) The periodically loaded transmission line[5,6] is more complex than the first two digital phase shifters. As shown in Fig. 10-3, when the shunt inductors are switched in, phase velocity is increased. The $\lambda/4$ between the elements causes a partial cancellation of their mismatches. When the capacitors are switched in, the phase velocity is reduced causing additional phase delay. The fourth type of digital diode phase shifter,[7,8] as shown in Fig. 10-4, is the most complex. The circuit elements are selected so that the T circuit is exactly matched whether the switches are in position to form the low-pass circuit or the high-pass circuit. The circuit provides phase delay in the low-pass position and phase advance in the high-pass position.

Continuous phase shift may be obtained from all but the first digital phase shifter. A varactor terminating a circulator will give a high reflection coefficient and continuously variable phase as the capacitance is varied. Parallel L-C circuit pairs shunting a line a quarter wavelength apart will be fairly well-matched near resonance, but the loading can be changed from inductive to capacitive by varying C. A range of phase from phase delay thru phase advance will be available. The T circuit of Fig. 10-4 may be

236 MICROWAVE DIODE CONTROL DEVICES

SWITCHED LINE:

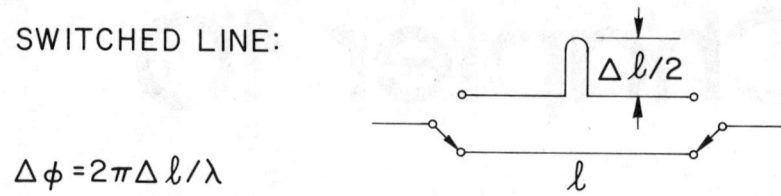

$\Delta\phi = 2\pi\Delta\ell/\lambda$

Fig. 10-1 Switched Line Phase Shifter Circuit

REFLECTION:

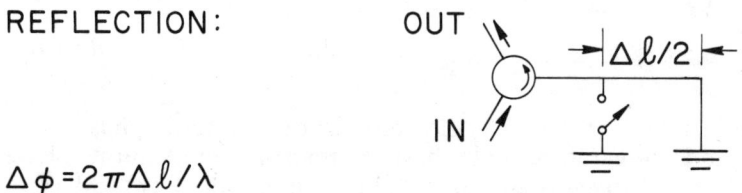

$\Delta\phi = 2\pi\Delta\ell/\lambda$

Fig. 10-2 Reflection Phase Shifter Circuit

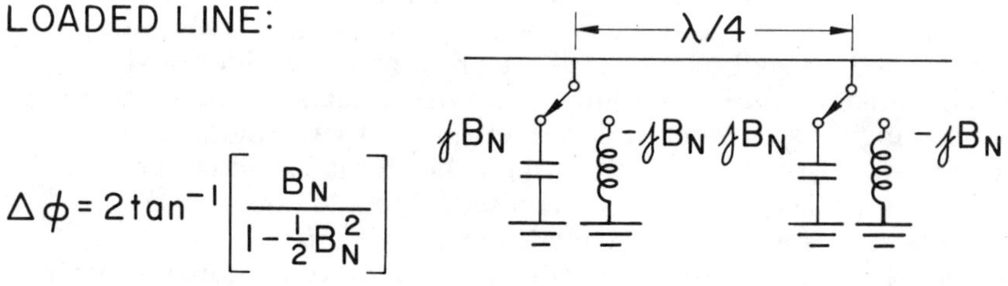

LOADED LINE:

$\Delta\phi = 2\tan^{-1}\left[\dfrac{B_N}{1-\frac{1}{2}B_N^2}\right]$

Fig. 10-3 Loaded Line Phase Shifter Circuit

HI-LOW PASS:

$X_N = \tan\dfrac{\Delta\phi}{4}$

$B_N = \sin\Delta\phi/2$

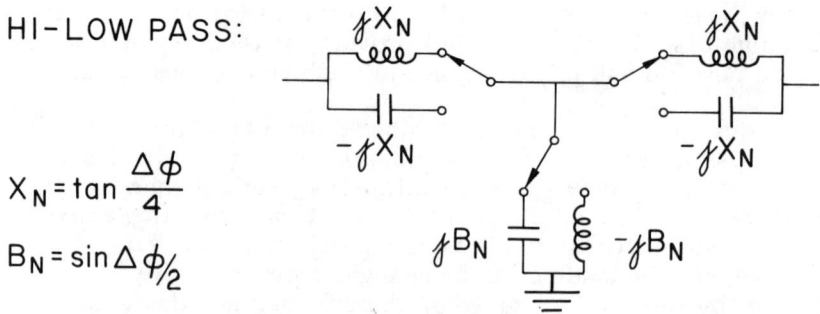

Fig. 10-4 Hi-Lo Pass Phase Shifter Circuit

PHASE SHIFTERS

varied by having a series L-C circuit for each X_n pair and a parallel L-C circuit for the B_N pair. By varying the values of C separately, the circuit can be kept matched and the phase will be variable.

Digital phase shifters have found wider application than variable phase shifters for a number of reasons. Digital computers are being used more widely for automatic control and for information processing. A digital computer can drive a digital phase shifter directly, while a digital-to-analog converter is required to drive a variable phase shifter. The variable capacitance of a varactor diode used in variable phase shifters is a non-linear function of voltage for both dc and RF. This nonlinearity normally requires the drive voltage to be predistorted to give a linear relationship between control voltage and phase. This variable capacitance also causes harmonic generation of the input RF power and produces intermodulation products when multiple RF signals are present. The capacitance of a varactor changes most rapidly around zero bias; thus, greater phase shift occurs around zero bias. The bias must be substantially larger than RF voltage to control the phase shift. Thus, the varactor operating at zero bias can control only low RF power levels. When the bias is a low voltage and high RF power is incident on the varactor phase shifter, it generates harmonics, generates intermodulation products, does not have the same phase shift that it has at low power, and has additional losses (due to harmonic generation). Varactor phase shifters are made up of a varactor and some form of tuning inductance and therefore are usually more frequency sensitive than digital phase shifters. Varactor phase shifters are also difficult to make because the capacitance of varactor diodes varies from diode to diode. But varactor phase shifters have three important advantages over digital phase shifters. (1) They work entirely in the reverse bias region and therefore require bias voltage, but practically no bias current, which makes them very efficient. (2) Since they are in reverse bias at all times, no minority carriers become involved in changing phase states; thus, they respond instantaneously to changes in modulation voltage, which makes them extremely fast. Very high modulation rates can be applied to them. (3) All phases are available permitting the use of relatively few diodes to obtain smooth and complete phase coverage. Thus, where high efficiency, high speed, or number of diodes is important, the varactor variable phase shifter is useful.

To be compatible with the digital computer and to give the greatest selection of phase states with the lowest number of elements, a binary array of phase shift bits is normally used, as shown in Fig.

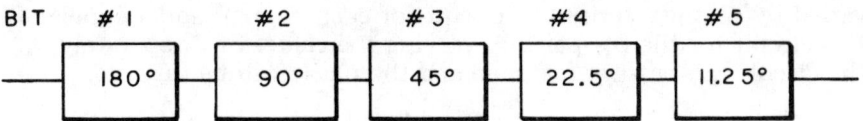

Fig. 10-5 Binary Bits for Digital Phase Shifters

10-5. Each bit is half the size of the bit preceeding it. The bits as shown may be added in any combination to come within 5.625° of any desired phase. Cumulative system errors usually determine what minimum bit size is useful. For example, if the cumulative system phase error were ±10% (out of 360°), bits smaller than 36° would contribute to loss and not to phase accuracy. Thus, most practical systems use only 4 bits.

Phase is not always the most important parameter to be considered in the description of the performance of phase shifters.[9] For example, a phased array radar will not necessarily have a good short pulse response if it has wide bandwidth phase shifters. The important parameter is time delay from common feed to each radiating element. Thus, the bits should be described in units of time delay rather than in units of phase delay, and the phase shift elements should have the same frequency-phase shift properties as time delay elements. The phase shifters of Figs. 10-1 and 10-2 as shown are time-delay phase shifters.

A. Digital Phase Shifters

   1. Switched Line Phase Shifters

   a. General

As shown in Fig. 10-1, the switched line phase shifter uses 2 SPDT switches (discussed in Chapter VII). The lower path has transmission length ℓ while the upper path has transmission length $\ell + \Delta \ell$. The upper path has a phase delay longer than the lower path given by

$$\Delta \phi = 2 \pi \Delta \ell / \lambda \qquad (10\text{-}1)$$

   b. The Spacing ℓ

The insertion loss[2] is influenced by ℓ. If ℓ and $\ell + \Delta \ell$ are approximately multiples of $\lambda/2$ according to Fig. 6-5, then the insertion loss (due to interacting reflections) of the switches will add up to

PHASE SHIFTERS

give the highest value. If, on the other hand, ℓ and $\ell + \Delta\ell$ are odd multiples of $\lambda/4$ according to Fig. 6-5, then the reflections causing the insertion loss will tend to cancel. If each ON switch has normalized reactance X_N, two of them will introduce phase shift as shown in Eq. (10-2) even though matched out.

$$\Delta\phi = 2 \cot^{-1} \frac{X_N}{2} \qquad (10\text{-}2)$$

Therefore the switches must be designed to have low VSWR, and ℓ should be adjusted so both paths have the same error bias.

These errors in phase shift and insertion loss may be calculated exactly using the circuit shown in Fig. 10-6. The superposition theorem is used to determine the voltage at B. All impedances are normalized to the generator impedance, which is also equal to all transmission line characteristic impedances. The upper generator voltage vectors are for even excitation of the circuit, as labelled +. For even excitation of the circuit, open circuit terminations may be substituted at the plane of symmetry for the other half of the circuit. The lower generator voltage vectors are for odd excitation of the circuit and are labelled -. For odd excitation of the circuit, short circuit terminations may be substituted at the plane of symmetry for the other half of the circuit.

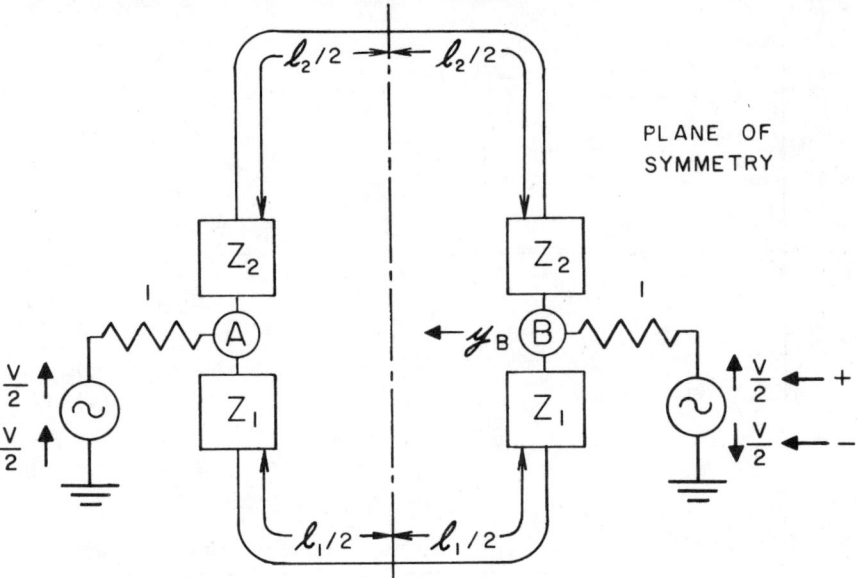

Fig. 10-6 Detailed Equivalent Circuit of the Switched-Line Phase Shifter

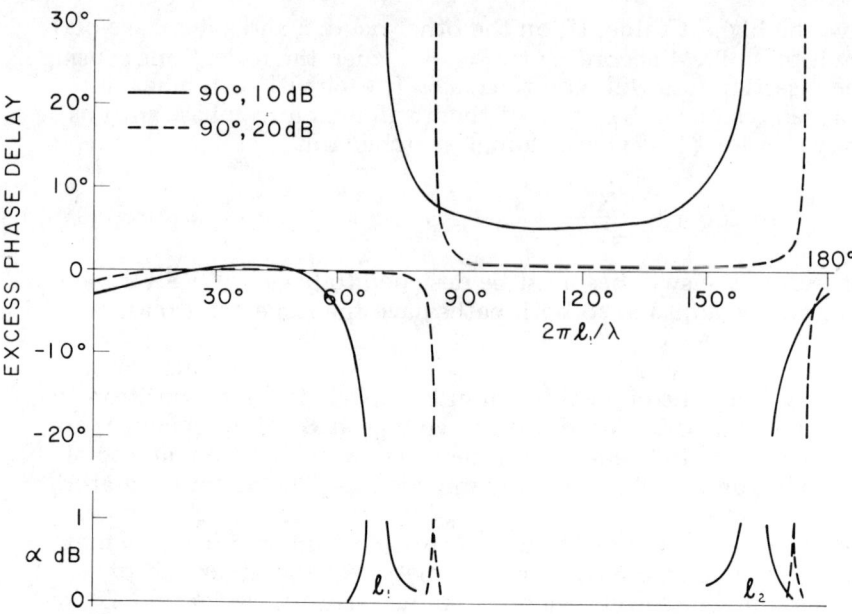

Fig. 10-7 Phase Error and Insertion Loss of a 90° Switched-Line Phase Shifter

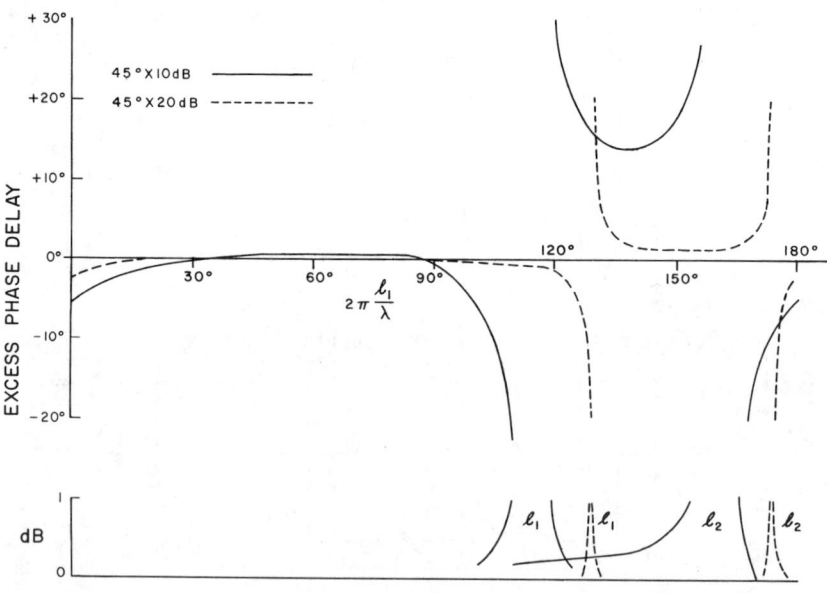

Fig. 10-8 Phase Error and Insertion Loss of a 45° Switched-Line Phase Shifter

PHASE SHIFTERS

For even excitation, the normalized admittance looking at the two transmission lines from point B, y_B^+ is given by

$$y_B^+ = \frac{1}{Z_1 + \dfrac{1}{j\tan(2\pi\ell_1/2\lambda)}} + \frac{1}{Z_2 + \dfrac{1}{j\tan(2\pi\ell_2/2\lambda)}} \quad (10\text{-}3)$$

The normalized admittance for odd excitation y_B^- is given by

$$y_B^- = \frac{1}{Z_1 + j\tan(2\pi\ell_1/2\lambda)} + \frac{1}{Z_2 + j\tan(2\pi\ell_2/2\lambda)} \quad (10\text{-}4)$$

The voltages at B, V_B are given by

$$V_B^+ = \frac{\dfrac{V}{2}\dfrac{1}{y_B^+}}{1 + \dfrac{1}{y_B^+}} \qquad V_B^- = \frac{-\dfrac{V}{2}\dfrac{1}{y_B^-}}{1 + \dfrac{1}{y_B^-}} \quad (10\text{-}5)$$

The transmission coefficient thru the two path network S_{21} is given by adding the voltages for even and odd excitation and dividing by the voltage associated with maximum delivered power $V/2$. Then the left generator generates full voltage and the right generator generates nothing.

$$S_{21} = \frac{V_B^+ + V_B^-}{V/2} = \frac{1}{y_B^+ + 1} - \frac{1}{y_B^- + 1} \quad (10\text{-}6)$$

$$S_{21} = \frac{1}{1 + \dfrac{1}{Z_1 + \dfrac{1}{j\tan(2\pi\ell_1/2\lambda)}} + \dfrac{1}{Z_2 + \dfrac{1}{j\tan(2\pi\ell_2/2\lambda)}}} - \frac{1}{1 + \dfrac{1}{Z_1 + j\tan(2\pi\ell_1/2\lambda)} + \dfrac{1}{Z_2 + j\tan(2\pi\ell_2/2\lambda)}} \quad (10\text{-}7)$$

When $Z_1 = 0$

$$S_{21} = \frac{1}{1 + j\tan(2\pi\ell_1/2\lambda) + \dfrac{1}{Z_2 + \dfrac{1}{j\tan(2\pi\ell_2/2\lambda)}}} - \frac{1}{1 + \dfrac{1}{j\tan(2\pi\ell_1/2\lambda)} + \dfrac{1}{Z_2 + j\tan(2\pi\ell_2/2\lambda)}} \quad (10\text{-}8)$$

or when $Z_2 = 0$, all 1 and 2 subscripts on the right hand side are interchanged.

Integrated circuit switches can be made with very low insertion loss, but high isolation is difficult to obtain with single diodes. Thus, the assumption that an ON diode can be represented by zero ohms is quite reasonable. The errors for diodes which provide capacitively limited SPST isolations and resistively limited SPST isolations have been calculated. For these calculations, it is assumed that path ℓ_1, is shorter than path ℓ_2 and that path ℓ_1 provides the reference phase. Thus, path ℓ_2 will provide a phase delay with respect to path ℓ_1.

When path ℓ_2 is a half-wavelength longer than path ℓ_1 switching from path ℓ_1 to path ℓ_2 introduces an increased phase delay of 180°. Comparison of the phase of S_{21} in Eq. 10-8 for path ℓ_1 and path ℓ_2 indicates that the phase shift is exactly 180° for all values of ℓ_1 as long as all four switching diodes are the same. This holds true only for phase shifts of 180°. The insertion loss is also the same for all values of ℓ_1 and for both phase states. A SPST resistive isolation of 20 dB gives a .5 dB insertion loss; a resistive 10 dB gives 1.8 dB; and a capacitive 10 dB gives 0.1 dB.

When path ℓ_2 is a quarter wavelength longer than path ℓ_1 the phase shift is 90°. The error for various lengths of path ℓ_1 is calculated using Eq. 10-8 and is plotted in Fig. 10-7. Errors become quite significant at two lengths of path ℓ_1. At these two lengths the attenuation in one of the paths becomes infinite and the phase error goes thru a full 360°. The OFF diode capacitance adds to the effective length of the OFF path. When the effective length of the OFF path is a half wavelength (or multiple thereof) it is resonant, and the phases add up in such a manner to reflect all incident power back to the generator. For example, 10 dB capacitive isolation (SPST) is given by a normalized series capacitive reactance of $X_N = -6$. This reactance corresponds to an open circuit line 9.45° long. One of these capacitors on each end of a line will add 18.9° to the effective length of the line. Therefore, the line will have an effective half-wavelength when it is 161.1° long. This length is the ℓ_1 path that gives the insertion loss spike and cycle of phase error for a capacitive 10 dB (SPST) isolation in Fig. 10-7. The other high error length occurs when ℓ_2 is a half wavelength long, which occurs when ℓ_1 is 161.1° - 90° = 71.1°. The SPST 20 dB reactance is $X_n = -20$ which gives an effective line length of 2.88°, which in turn gives large errors of 174.24° and 84.24° for $2\pi\ell_1/\lambda$.

PHASE SHIFTERS

The curves of phase error are cyclic; they repeat every 180°. Note that even when the 10 dB isolation (SPST) diodes are used, phase errors are very low when $2\pi \ell_1/\lambda$ is between 20° and 50°. The insertion loss in this region for the 10 dB (SPST) switch is 0.1 dB in both phase states. The 20 dB switch gives far less insertion loss. (Insertion loss will be dominated by the resistive component of the OFF diode.) Higher isolation of the diodes only narrows the region of large phase error. The error persists.

The reason the phase errors are large when the OFF path is a half wavelength long or multiple thereof is that the isolation of two reactive switches a half-wavelength apart goes to zero (ref. Eq. (6-11) and Fig. 6-4). The phase shift through the OFF path goes through a rapid change near this resonance, reaching a value that upsets the phase and amplitude of the transmission through the ON path.

Fig. 10-8 shows the error curves for a 45° phase shifter. The same rationale as used above gives the lengths for which large errors occur. Note that the region of small error now extends from 20° to 90°. The insertion loss for the 10 dB (SPST) switch is again 0.1 dB, etc.

The calculated length of ℓ_1 that causes large insertion loss and phase errors may be found easily using Fig. 10-9.

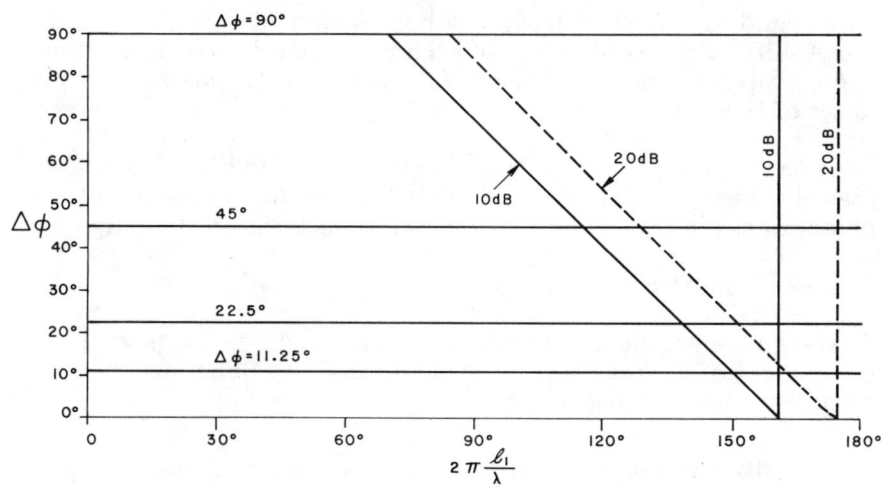

Fig. 10-9 Regions of Maximum Error and Insertion Loss for Switched-Line Phase Shifters

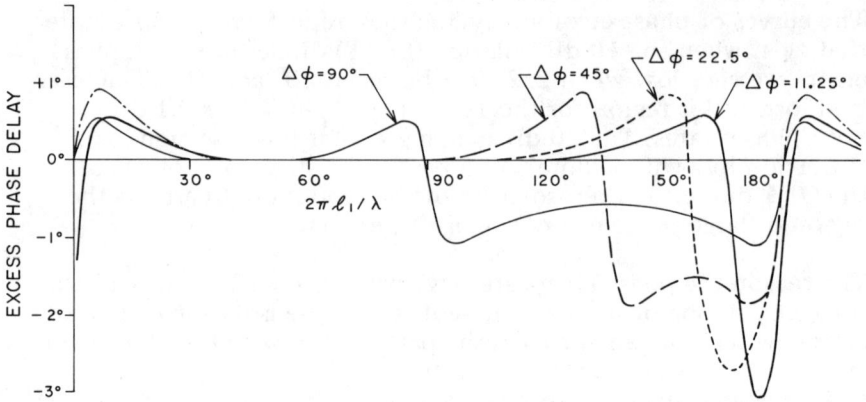

Fig. 10-10 Phase Errors Induced by a Resistively Limited 20 dB Switching Diode

The phase errors were also calculated for a resistively limited diode switch. Fig. 10-10 shows these errors for a 20 dB (SPST) resistive switch. Note the expanded scale. The regions of small phase error are approximately the same as for reactively limited switches. The insertion loss in this region of small phase error was 0.5 dB for the 20 dB (SPST) switch and 1.8 dB for the 10 dB (SPST) switch. The insertion loss of path ℓ_1 decreased at the first negative phase error length, and the insertion loss of path ℓ_2 decreased at $0°$. For small phase shifts, the insertion loss of each path approached zero at the appropriate lengths. In the region of small error, the insertion losses of both phase states were equal.

In general, the selection of $2\pi \ell_1/\lambda$ between $20°$ and $50°$ will insure that phase errors are a minimum, that the insertion losses in both phase states are equal, and that the phase shifter is not too large.

c. The Insertion Loss $\delta\phi$

As shown above, the insertion loss is dominated by the resistive component of the OFF switch. A 20 dB resistive SPST switch has a series resistance given by

$$20 \text{ dB} = 10 \log (1 + \tfrac{1}{2} R_N)^2$$

$$R_N = 18 \qquad (10\text{-}9)$$

PHASE SHIFTERS

For the ON switch the $R_N = 18$ is shunting the line giving $G_N = \frac{1}{R_N} = \frac{1}{18}$. The insertion loss for a shunting resistor is given by

$$\delta = 10 \log (1 + \tfrac{1}{2} G_N)^2 \tag{10-10}$$

which gives $\delta = .24$ dB for $G_N = \frac{1}{18}$. Two of these switches in series would give .48 dB total, which is exactly what was calculated in conjunction with Fig. 10-10. The total insertion loss of each switched line phase bit is thus given by

$$\delta = 2 \times 10 \log (1 + \tfrac{1}{2} G_N)^2 = 40 \log (1 + \tfrac{1}{2} G_N) \tag{10-11}$$

which is the same for all phase bits.

Recall from Fig. 4-1 that the parallel resistance of a reverse biased diode is given by

$$R_P = R_S (1 + Q^2). \tag{10-12}$$

When the resistance is shunting a line then

$$G_N = \frac{Z_0}{R_P} = \frac{Z_0}{R_S (1 + Q^2)} \tag{10-13}$$

and the insertion loss is given by

$$\delta = 40 \log \left[1 + \frac{Z_0}{2 R_S (1 + Q^2)} \right]. \tag{10-14}$$

When 100 GHz diodes are used at 10 GHz in 50-ohm transmission line and $R_s = 2$ ohms, the insertion loss is 2 dB. The ON diodes would contribute insertion loss given by

$$\delta = 2 \times 10 \log \left(1 + \tfrac{1}{2} \frac{R_S}{Z_0} \right)^2 = 40 \log \left(1 + \frac{2}{100} \right) = .34 \text{ dB}. \tag{10-15}$$

Combining Eqs. (10-14) and (10-15) gives the total insertion loss δ_ϕ as approximately

$$\delta_\phi = 40 \log \left(1 + \frac{Z_0}{2 R_S Q^2} + \frac{R_S}{2 Z_0} \right). \tag{10-16}$$

Raising Q in the previous example from 10 to 100 would lower the insertion loss from 2.34 to .36 dB. Lowering R_s from 2 ohms to 1 ohm would make the insertion loss 4.0 dB for Q = 10 and 0.2 dB for Q = 100. There is obviously an optimum value of R_s for each value of Q that provides minimum insertion loss.

Differentiating Eq. (10-16) with respect to R_s/Z_0 and equating to zero shows that the lowest insertion loss is given by

$$R_S/Z_0 = 1/Q. \tag{10-17}$$

Substituting Eq. (10-17) into Eq. (10-16) gives the minimum insertion loss for a given diode Q

$$\delta_{\phi}\text{MIN} = 40 \log \left(1 + \frac{1}{Q}\right). \tag{10-18}$$

d. Power Limitations

The average power each diode can control will be a maximum when the insertion loss is a minimum. When R_s/Z_0 satisfies Eq. (10-17) equal power is dissipated in the OFF diode and in the ON diode. Using Eq. (4-69), the series diode can control incident average power given by

$$\overline{P}_i = \frac{\overline{P}_D Z_0}{R_s} \left(1 + \frac{R_s}{2Z_0}\right)^2. \tag{10-19}$$

Using Eq. (10-17) in Eq. (10-19) yields

$$\overline{P}_i = \overline{P}_D Q \left(1 + \frac{1}{2Q}\right)^2 \tag{10-20}$$

The fact that the insertion loss performance of the diodes could be calculated by assuming the OFF diodes were connected to ground is used to calculate the peak power the diode can control. The OFF diodes thus have the same voltages on them as if they were shunted to ground. Eq. (4-64) gives the peak power a shunt diode can control as

$$\hat{P}_i = \frac{E_B^2}{8Z_0} \tag{10-21}$$

PHASE SHIFTERS

These equations for power ratings are valid in the region of ℓ_1 giving minimum phase error. When half wavelength resonances (which give large attenuation and large phase errors) are encountered, switches which were designed using Eqs. (10-20) and (10-21) will not be able to control the power.

e. Broad Bandwidth

The switched line phase shifter is a time delay device. Phase shift will be inversely proportional to frequency. Wide bandwidth may be achieved by using a Schiffman phase shifter in one of the transmission paths[10] as shown in Fig. 10-11. The phase shift of the Schiffman coupled section is a linear function of frequency plus a sinusoidal function of frequency. When ℓ_1 is selected so that it is parallel to the center portion of the Schiffman curve on the ϕ-ω plot, a relatively constant phase shift is available over a significant bandwidth. Care must be taken that neither line intersects multiples of 180° over the band of interest, or large phase errors and insertion loss will occur as discussed in Section a. above, unless special precautions are taken. These half-wavelength resonances of the OFF line have been successfully eliminated by Podell and Lynes[16] by spoiling the Q of the OFF line. Podell switched the ends of the OFF line to matched loads by using transfer switches at the ends instead of double throw switches and Lynes connected any resistance (close to Z_0) to the OFF line with another switch.

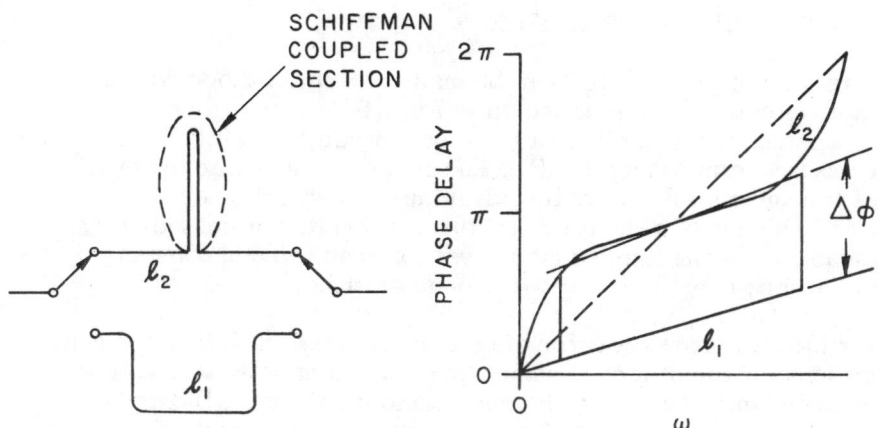

Fig. 10-11 Constant-Phase-Shift Switched-Line Phase Shifter

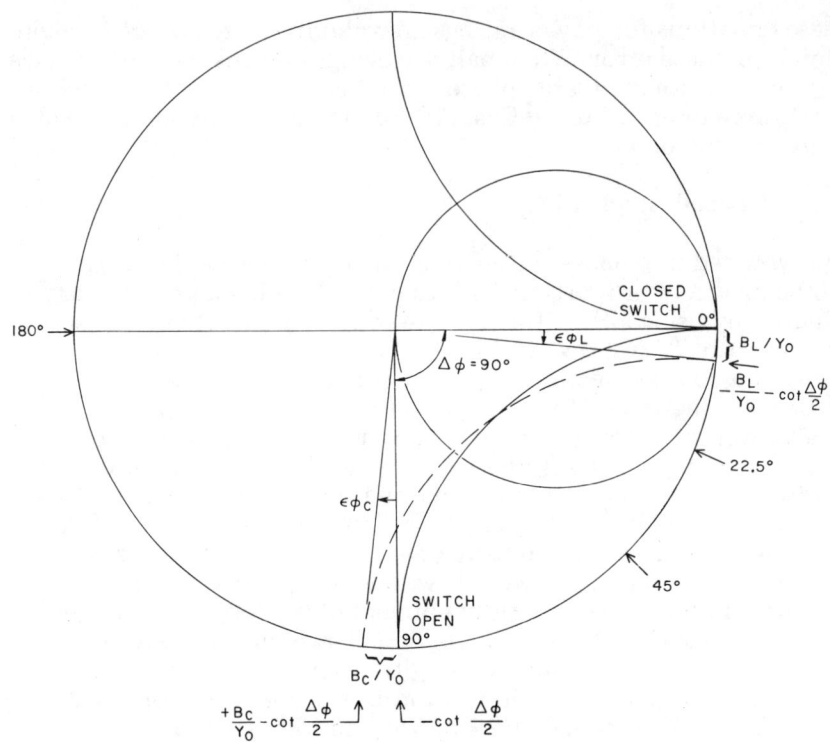

Fig. 10-12   Admittance of a Switching Diode with Parasitics in a 90° Reflection Phase-Shifter

2. Reflection Phase Shifters

A reflection phase shifter can be made of a shunt diode with a short circuit behind it as shown in Fig. 10-2, a series diode with an open circuit behind it, or a lumped circuit including diode parasitics terminating the line. The switches backed up by lengths of transmission line have the advantage of being time delay devices which give wide instantaneous bandwidth for phased-array radars, while the lumped circuit versions can provide constant phase shift over octave or wider bandwidths.

A source of error exists for the reflection phase shifter which is not present for the others. Mismatches intervening between the terminating impedance and the perfect circulator or 3-dB coupler contribute large phase errors. The maximum phase error ϵ_ϕ for an intervening mismatch of Γ_M in front of a reflector Γ_X is given by[11]

PHASE SHIFTERS

$$\epsilon_\phi = \pm\, (1 + 3\, |\Gamma_X| - \tfrac{1}{4} \sin \pi |\Gamma_X|) \sin^{-1} \left| \frac{\Gamma_M}{\Gamma_X} \right|. \tag{10-22}$$

For low loss phase shifters $|\Gamma_X| = 1$, and

$$\epsilon_\phi = \pm 4 \sin^{-1} |\Gamma_M| \tag{10-23}$$

An intervening mismatch having a VSWR of 1.2 will give $\epsilon_\phi = \pm 20.8°$. Therefore, it is very important to have no mismatches between the circulator (or 3 dB coupler) and the reflection element(s). The error is very closely given by

$$\epsilon_\phi = \pm 100°(\rho - 1) \tag{10-24}$$

Caution must also be used in selecting a circulator or 3-dB coupler, because the finite isolation is caused by internal reflections in the coupling device. For example, a circulator having 20 dB isolation has an internal voltage reflection coefficient of 0.1 which gives $\pm 22.8°$ maximum phase error; 30 dB isolation gives $\pm 7.2°$ maximum phase error; and 40 dB, $\pm 2.3°$. By carefully arranging phases, the maximum phase error may be avoided.

a. Time Delay Phase Shifter

The time delay phase shifter shown in Fig. 10-2 has the advantage of having a tendency to have the instantaneous wide bandwidth needed for pulsed phased array radars. The normalized admittance of the termination may be considered as shown in Fig. 10-12. The admittance is shown as it would be measured in the plane of the diode switch. A perfect switch has infinite admittance when closed and gives $\infty + j\infty$ on the Smith Chart for all lengths of line behind it. When the perfect switch is opened, all that is seen is the admittance of the length of line behind it, $\Delta \ell/2$, which is given by

$$Y/Y_0 = -j \cot [2\pi (\Delta\ell/2)/\lambda] = -j \cot \frac{\Delta\phi}{2}. \tag{10-25}$$

The circular arc shown in Fig. 10-12 shows the admittance that would be traced out by a perfect PIN diode switch as current is varied from 0 to ∞ for 90° phase shift. The switch open admittances for other phase shifts are indicated on the figure.

A typical non-perfect diode switch will be inductive for conduction and capacitive for reverse bias. The inductive susceptance will cause the error shown in Fig. 10-12 as $\epsilon_{\phi L}$ while the capacitive susceptance will cause the error shown as $\epsilon_{\phi C}$. These errors will be calculated and it will be shown that proper selection of diode parasitics can permit these errors to cancel out (as illustrated in the figure).

Reflection coefficient is given by

$$\Gamma = \frac{Y_0 - Y}{Y_0 + Y} . \tag{10-26}$$

The line is terminated by a susceptance

$$Y = jB \tag{10-27}$$

$$\Gamma = \frac{Y_0 - jB}{Y_0 + jB} = \frac{(Y_0^2 - B^2) - j2BY_0}{Y_0^2 + B^2} . \tag{10-28}$$

The angle of the reflection coefficient ϕ is given by

$$\phi = \tan^{-1}\left[\frac{-2B/Y_0}{1 - (B/Y_0)^2}\right] = -2\tan^{-1}\left(\frac{B}{Y_0}\right) . \tag{10-29}$$

For diode conduction the normalized diode admittance Y_{DC}/Y_0 is

$$\frac{Y_{DC}}{Y_0} = -j\frac{Z_0}{\omega L} . \tag{10-30}$$

Combining Eqs. (10-25) (10-29), and (10-30) gives the phase for conduction ϕ_C.

$$\phi_C = -2\tan^{-1}\left[-\frac{Z_0}{\omega L} - \cot\frac{\Delta\phi}{2}\right] . \tag{10-31}$$

Similarly the diode at reverse bias has normalized admittance

$$\frac{Y_{DR}}{Y_0} = j\omega C Z_0 \tag{10-32}$$

and gives phase ϕ_R

$$\phi_R = -2\tan^{-1}\left[\omega C Z_0 - \cot\frac{\Delta\phi}{2}\right] . \tag{10-33}$$

PHASE SHIFTERS

For a perfect diode $\phi_C = +180°$ for all $\Delta\phi$. For imperfect diodes $\phi_C < +180°$. For a perfect diode $\phi_R = +90°$ for $\Delta\phi = 90°$ and $\phi_R < 90°$ for imperfect diodes. The phase shift $\Delta\phi'$ including diode parasitics is given by

$$\Delta\phi' = \phi_C - \phi_R \qquad (10\text{-}34)$$

The inductive reactance of the switching diode in forward conduction reduces $\Delta\phi'$ and thus produces negative error $\epsilon_{\phi L}$ while the capacitive reactance of the switching diode in reverse bias increases $\Delta\phi'$, thus producing positive error $\epsilon_{\phi C}$. The errors contributed by various values of B_C/Y_0 and B_L/Y_0 are shown in Fig. 10-13. The SPST insertion loss and isolation of these parasitics are shown along the axes.

The errors can be made to cancel each other. For example, $B_C/Y_0 = .1$ gives $+6°$ error for $\Delta\phi = 90°$. This error can be cancelled by $B_L/Y_0 = 18$ which gives $-6°$.

To calculate the insertion loss from this structure, consider the diode impedance R_s at forward bias and $R_P = R_S(1 + Q^2)$ at reverse bias giving

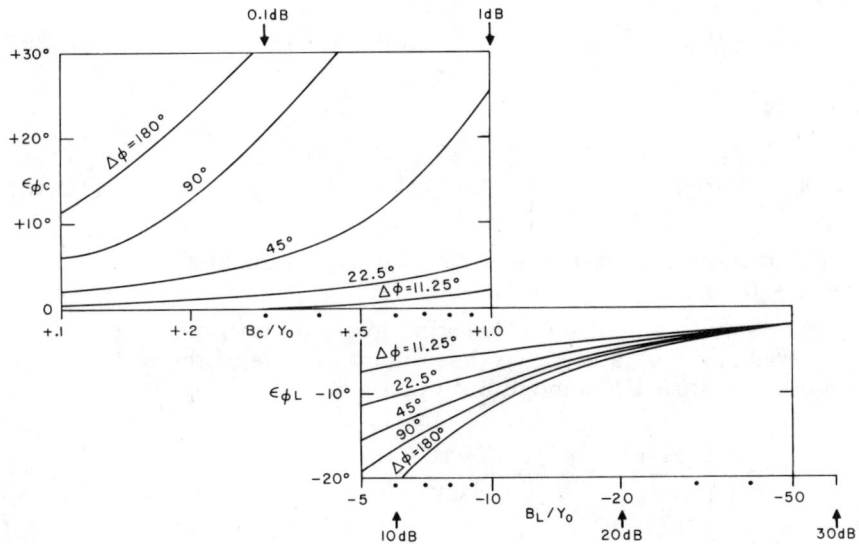

Fig. 10-13 Contribution of Shunt Switching-Diode Parasitics to Reflection Phase-Shifter Errors

$$\frac{Y_c}{Y_0} = \frac{Z_0}{R_S} + j0 \qquad (10\text{-}35)$$

and

$$\frac{Y_R}{Y_0} = \frac{Z_0}{R_P} + jB \qquad B = -\cot\frac{\Delta\phi}{2} \qquad (10\text{-}36)$$

Reverse bias condition usually causes a higher VSWR than forward bias (giving unequal insertion loss for the two phase states). This can be equalized by putting a resistance in parallel with the diode which lowers the reverse bias VSWR, but does not change the forward bias VSWR very much. When this is done the insertion loss is given by

$$\delta = 20\log\left|\frac{1}{\Gamma}\right| \qquad (10\text{-}37)$$

in which $|\Gamma|$ is given by

$$|\Gamma| = \frac{\rho - 1}{\rho + 1} \qquad (10\text{-}38)$$

and ρ, the VSWR is given by

$$\rho = Z_0/R_S \qquad (10\text{-}39)$$

all giving

$$\delta = 20\log\left[\frac{Z_0 + R_S}{Z_0 - R_S}\right] \approx 20\log\left(1 + 2\frac{R_S}{Z_0}\right) \qquad (10\text{-}40)$$

An R_S of 2 ohms terminating a $Z_0 = 50$ ohm transmission line gives $\delta = 0.7$ dB.

Another way to equalize the insertion losses in both phase states is to arrange Z_0, Q, and R_S so that the magnitude of the reflection coefficient is the same in both phase states.

$$|\Gamma| = \left|\frac{1 - Y_N}{1 + Y_N}\right| = \sqrt{\frac{(1 - G_N)^2 + B_N^2}{(1 + G_N)^2 + B_N^2}} \qquad (10\text{-}41)$$

in which the normalized admittance is

$$Y_N = Y/Y_0 = G_N + jB_N \qquad (10\text{-}42)$$

PHASE SHIFTERS

Using Eqs. (10-35), (10-36), and (10-41) the reflection coefficient magnitudes are equal when

$$\sqrt{\frac{\left(1 - \frac{Z_0}{R_S}\right)^2}{\left(1 + \frac{Z_0}{R_S}\right)^2}} = \sqrt{\frac{\left(1 - \frac{Z_0}{R_S Q^2}\right)^2 + B^2}{\left(1 + \frac{Z_0}{R_S Q^2}\right)^2 + B^2}} \qquad (10\text{-}43)$$

$$\frac{Z_0^2}{R_S^2} = \frac{(1 + 2B^2)Q^2 - 1}{1 - Q^{-2}} \approx (1 + 2B^2)Q^2 \qquad (10\text{-}44)$$

$$Z_0 = QR_S \sqrt{1 + 2B^2} = QR_S \sqrt{1 + 2\cot^2 \frac{\Delta\phi}{2}} \qquad (10\text{-}45)$$

Using the same rationale as that for Eq. (10-40) gives

$$\delta = 20 \log \left[\frac{Q\sqrt{1 + 2\cot^2 \frac{\Delta\phi}{2}} + 1}{Q\sqrt{1 + 2\cot^2 \frac{\Delta\phi}{2}} - 1} \right]$$

$$\approx 20 \log \left[1 + \frac{2/Q}{\sqrt{1 + 2\cot^2 \left(\frac{\Delta\phi}{2}\right)}} \right]. \qquad (10\text{-}46)$$

For $\Delta\phi = 180°$, $\delta = 20 \log \left[1 + \frac{2}{Q}\right]$ and $Z_0 = QR_S$. A Q of 25 would give equalized loss for $R_S = 2$ ohms and $Z_0 = 50$ ohms, and 0.7 dB insertion loss as before. A Q of 100, however, and $R_S = .5$ ohm would also give equalized loss in $Z_0 = 50$ ohms and 0.18 dB insertion loss. For $\Delta\phi = 90°$ $\delta = 20 \log \left[1 + \frac{2}{Q\sqrt{3}}\right]$ and $Z_0 = Q\sqrt{3}R_S$. Insertion loss is down to 0.1 dB for $Q = 100$ and Z_0 is higher.

To calculate the power rating of a reflection-type diode phase shifter, reference is made to Fig. 10-14. The peak power limit is calculated using Fig. 10-14(a). The current I is given by

$$I = \frac{E}{Z_0 + jX} \qquad (10\text{-}47)$$

The voltage across the switch is the same as the voltage across the reactance E_X of the short circuit terminated line length.

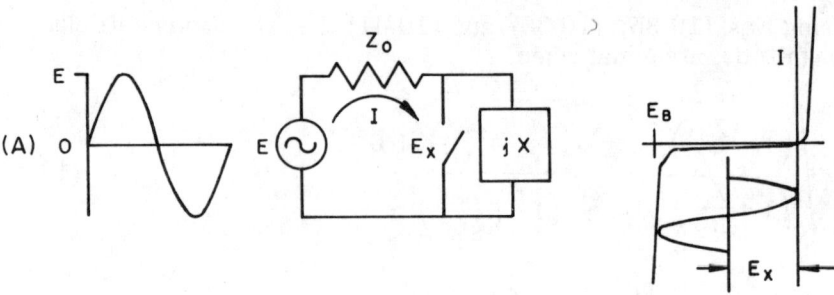

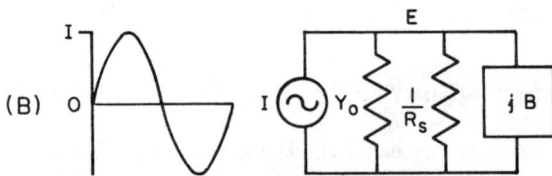

Fig. 10-14   Equivalent Circuits for Calculating the Power Ratings of Reflection Diode Phase Shifters. Under most conditions reverse bias is 10-100V for PIN diodes and $E_X \approx E_B$ as discussed in Chapter IV C 1

$$|E_X| = |(I)(jX)| = \frac{E\, X_N}{\sqrt{1 + X_N^2}} \qquad (10\text{-}48)$$

in which

$$X_N = X/Z_0 = \tan\left(\frac{\Delta\phi}{2}\right). \qquad (10\text{-}49)$$

Eqs. (10-48) and (10-49) reduce to

$$E_X = E \sin\left(\frac{\Delta\phi}{2}\right). \qquad (10\text{-}50)$$

The maximum available power $\hat{P}_i$ from the generator shown in Fig. 10-14(a) is given by

$$\hat{P}_i = \tfrac{1}{2} \left(\frac{E}{2}\right)^2 \frac{1}{Z_0} \qquad (10\text{-}51)$$

and the maximum value of E_X is given by

$$E_X = E_B/2. \qquad (10\text{-}52)$$

PHASE SHIFTERS

(Recall that $E_X \approx E_B$ is possible when harmonics and intermodulation products will not be troublesome, ref. Chapter IV C1.) Equations (10-50), (10-51), and (10-52) combine to give

$$\widehat{P}_i = \frac{E_B^2}{32 \, Z_0 \, \sin^2\left(\frac{\Delta\phi}{2}\right)} \tag{10-53}$$

the maximum peak power the reflection phase shifter can control.

The average power limit of the reflection phase shifter is calculated by Fig. 10-14(b). The voltage E is given by

$$E = \frac{I}{Y_0 + \frac{1}{R_S} + jB} \tag{10-54}$$

The maximum available power, $\overline{P}_i$ from the generator is given by

$$\overline{P}_i = \tfrac{1}{2}\left(\frac{I}{2}\right)^2 \frac{1}{Y_0} \tag{10-55}$$

The power $\overline{P}_D$ dissipated in R_S is given by

$$\overline{P}_D = \tfrac{1}{2} \frac{E E^*}{R_S} = \tfrac{1}{2} \frac{I^2/R_S}{\left(Y_0 + \frac{1}{R_S}\right)^2 + B^2} \tag{10-56}$$

Combining Eqs. (10-36), (10-55) and (10-56) gives

$$\overline{P}_i = \frac{\overline{P}_D}{4}\left[\frac{Z_0}{R_S} + 2 + \frac{R_S/Z_0}{\sin^2\left(\frac{\Delta\phi}{2}\right)}\right] \tag{10-57}$$

the maximum average power the reflection phase shifter can control. For most practical values of R_S/Z_0 and $\Delta\phi$, Eq. (10-57) can be approximated by

$$\overline{P}_i = \frac{\overline{P}_D Z_0}{4 R_S} \tag{10-58}$$

When $\widehat{P}_i = \overline{P}_i = P_{CW}$ (cw operation) then Eqs. (10-53) and (10-58) yield

$$Z_0 = \frac{E_B}{2 \sin\left(\frac{\Delta\phi}{2}\right)} \sqrt{\frac{R_S}{2\bar{P}_D}} \qquad (10\text{-}59)$$

and

$$P_{CW} = \frac{E_B}{8 \sin\left(\frac{\Delta\phi}{2}\right)} \sqrt{\frac{\bar{P}_D}{2R_S}} \qquad (10\text{-}60)$$

b. Wide-Band Reflection Phase Shifter

For wide-band phase shifting, the diodes(s) is (are) connected to the circulator (3 dB coupler) without a length of transmission line behind it (them). A perfect diode switch will be 0 ohms when closed and ∞ ohms when open. These impedances are 180° apart on the Smith Chart and therefore give $\Delta\phi = 180°$[12]. If the short circuit is considered the reference, then the open circuit is equivalent to a short circuit $\lambda/4$ away, and the effective round trip to the effective short circuit gives a $\lambda/2$ phase delay, or 180°.

A non-perfect diode will have an impedance at forward bias Z_F given by

$$Z_F = R_S + j\omega L \qquad (10\text{-}61)$$

and admittance at reverse bias Y_R given by

$$Y_R = G_P + j\omega C . \qquad (10\text{-}62)$$

As long as the normalized impedance of the diode in one bias state is half-way around the Smith Chart from the normalized impedance of the diode in the other bias state, then the phase shift will be 180°. Rotation half-way around the Smith Chart also converts normalized impedance to normalized admittance (and vice-versa); therefore, when the normalized impedance of the diode in one bias state is equal to the normalized admittance of the diode in the other bias state, then 180° phase shift will be assured. Equations (10-61) and (10-62) produce

PHASE SHIFTERS

$$\frac{Z_F}{Z_0} = \frac{Y_R}{Y_0} \tag{10-63}$$

$$\frac{R_S + j\omega L}{Z_0} = \frac{G_P + j\omega C}{Y_0} \tag{10-64}$$

$$\frac{R_S}{Z_0} = \frac{G_P}{Y_0} \tag{10-65}$$

$$\frac{\omega L}{Z_0} = \frac{\omega C}{Y_0} \tag{10-66}$$

$$Z_0 = \sqrt{\frac{L}{C}} \tag{10-67}$$

When L and C are selected to satisfy Eq. (10-67), 180° phase shift is sure to be independent of frequency. The conditions of Eq. (10-65) ensure that the magnitude of reflection coefficient is the same for both switching states. As before, resistance can be added in parallel with the diode, or Z_0 can be adjusted to equalize the reflection coefficient magnitude.

At higher frequencies, all of the parasitic elements of the diode have to be taken into account as shown in Fig. 10-15. Element aC is the junction which is shown at reverse bias. At forward bias, this capacitor is shunted by conduction current. The phase shift is

$$\Delta\phi = 2\left[\tan^{-1}\left(\frac{X_F}{Z_0}\right) - \tan^{-1}\left(\frac{X_R}{Z_0}\right)\right] \tag{10-68}$$

in which X_F and X_R are the diode reactances at forward and reverse bias. The phase shift of a MC7000 switching diode terminating 50-ohm stripline is shown in Fig. 10-16. Very wide 180° phase modulation is available up to 3.6 GHz. One degree of freedom has been unnecessarily constrained in Fig. 10-15 in setting the relationship

$$Z_0 = \sqrt{L/C} \ .$$

This degree of freedom is regained as available by defining

$$Z_0' = \sqrt{L/C} \tag{10-69}$$

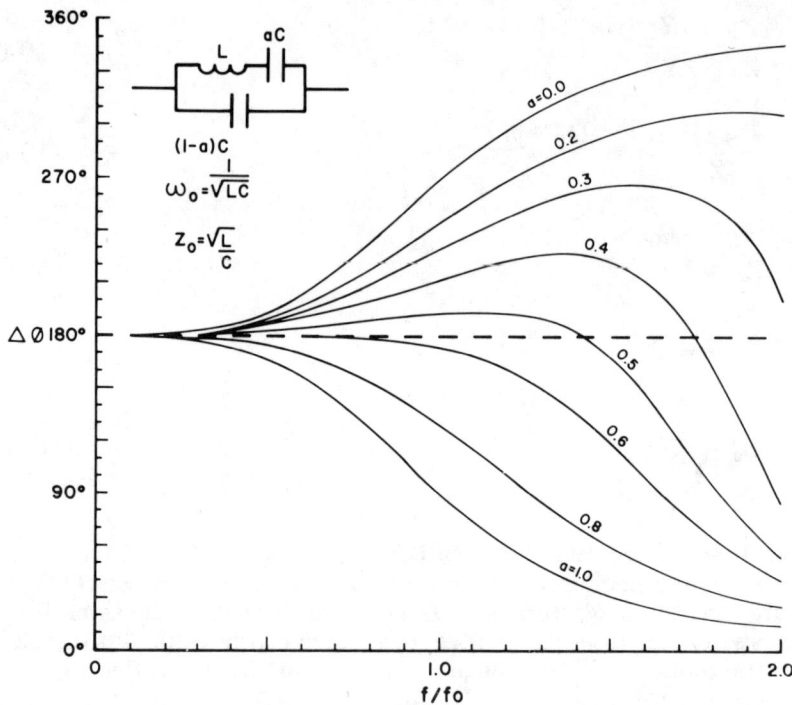

Fig. 10-15 Frequency Dependence of a Reflection Phase Modulator as Influenced by Capacitance Distribution of the Diode

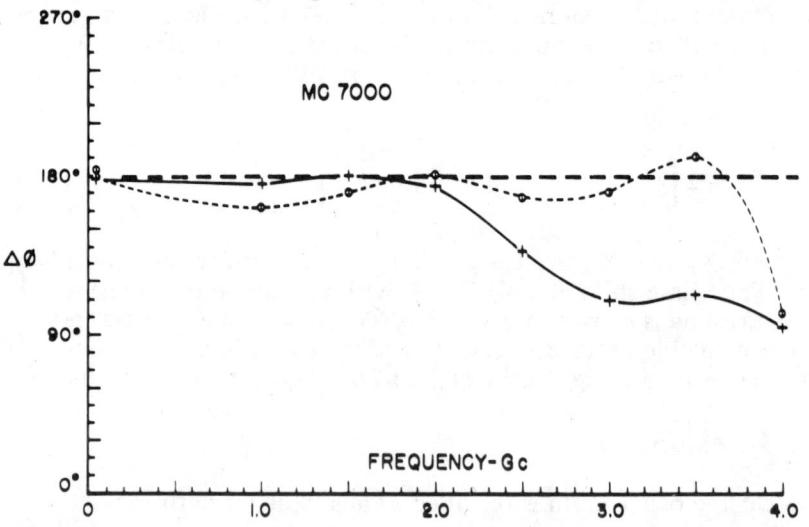

Fig. 10-16 Frequency Dependence of the Phase Modulation of the MC7000 Diode Terminating a 50-ohm Stripline Circulator

PHASE SHIFTERS

which is not necessarily equal to the characteristic impedance Z_0 of the transmission line which the diode terminates. Lowering Z_0' below Z_0 causes another wiggle to occur in the curves of Fig. 10-15. Parameters can be adjusted so that an equal ripple phase response can be achieved about any phase over a significant bandwidth. Defining

$$K = Z_0'/Z_0 \tag{10-70}$$

the equal ripple phase responses have been calculated as shown in Fig. 10-17. For $\pm 2°$ error the narrowest bandwidth (for 90°) is wider than an octave. The circuit elements for realizing these phase shifts can be calculated using Eqs. (10-71) and (10-72),

$$C = \frac{1}{Z_0' \omega_0} \tag{10-71}$$

$$L = \frac{Z_0'}{\omega_0} \tag{10-72}$$

which are derived from Eq. (10-69) and the ω_0 equation in Fig. 10-15. For example, to make a 45° $\pm 2°$ phase shift bit working from 1 GHz to 2 GHz terminating $Z_0 = 50$ ohm transmission line, the following calculations would be made.

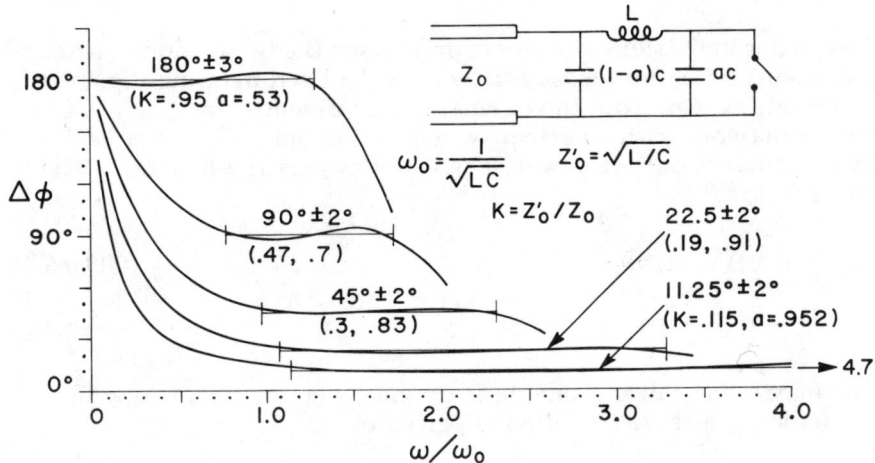

Fig. 10-17 Wideband Lumped-Element Diode Reflection Phase Shifter

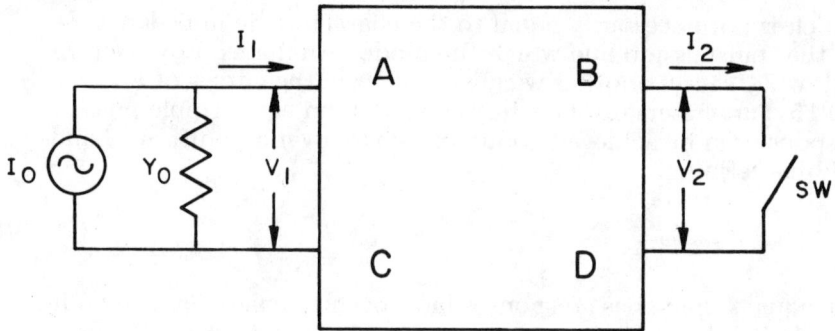

Fig. 10-18 Circuit for Calculating Power Limitations of Diode Reflection Phase Shifters

$\omega_0 = 2\pi \times 10^9$

$Z_0' = KZ_0 = (.3)(50) = 15\Omega$

$L = \dfrac{Z_0'}{\omega_0} = \dfrac{15}{2\pi \times 10^9} = 2.4 \text{ nH}$

$C = \dfrac{1}{Z_0' \omega_0} = \dfrac{1}{(15)(2\pi \times 10^9)} = 10.6 \text{ pF}$

$C_D = aC = (.83)(10.6 \text{ pF}) = 8.8 \text{ pF}$

$C_c = (1-a)C = (.17)(10.6 \text{ pF}) = 1.8 \text{ pF}$

c. Power Limitations and Insertion Loss

The power limitations and insertion loss for the broadband constant phase shift reflection phase shifter are calculated by using Fig. 10-18. The diode as seen from the circulator is represented by the ABCD matrix network with a perfect switch Sw behind it. Power reflected by the diode is output power. The equations for the ABCD matrix are as follows:

$$V_1 = AV_2 + BI_2 \qquad (10\text{-}73)$$

$$I_1 = CV_2 + DI_2 \cdot \qquad (10\text{-}74)$$

The maximum power available from the generator I_0 corresponds to incident power on the diode P_i given by

$$P_i = \tfrac{1}{2}\left(\dfrac{I_0}{2}\right)^2 \dfrac{1}{Y_0} = \dfrac{I_0^{\,2}}{8Y_0} \qquad (10\text{-}75)$$

PHASE SHIFTERS

in which I_0 is peak RF current.

When the switch Sw is closed, then $V_2 \approx 0$; Eqs. (10-73) and (10-74) then reduce to

$$V_1 = BI_2 \tag{10-76}$$

$$I_1 = DI_2 . \tag{10-77}$$

The admittance of the diode Y_1 becomes

$$Y_1 = \frac{I_1}{V_1} = \frac{D}{B} \tag{10-78}$$

Conservation of current requires that

$$I_0 = V_1 (Y_0 + Y_1) . \tag{10-79}$$

Incident power is thus given by

$$P_i = \frac{1}{8Y_0} I_0^2 = \frac{1}{8Y_0} V_1^2 (Y_0 + Y_1)^2 \tag{10-80}$$

$$= \frac{1}{8Y_0} (BI_2)^2 \left(Y_0 + \frac{D}{B}\right)^2 \tag{10-81}$$

The average power the diode can dissipate $\overline{P}_D$ is given by

$$\overline{P}_D = \tfrac{1}{2} I_2^2 R_S \tag{10-82}$$

which gives

$$I_2^2 = 2 \frac{\overline{P}_D}{R_S} \tag{10-83}$$

Combining Eqs. (10-81) and (10-83) gives the maximum average power $\overline{P}_i$ which the diode can control in the conduction state.

$$\overline{P}_i = \frac{\overline{P}_D (BY_0 + D)^2}{4 R_S Y_0} \tag{10-84}$$

Note that B is imaginary causing Eq. (10-84) to reduce to

$$\overline{P}_i = \frac{\overline{P}_D (|B|^2 Y_0^2 + D^2)}{4 R_S Y_0} \tag{10-85}$$

The ABCD matrix of Fig. 10-17 is given by

$$\begin{vmatrix} A & B \\ C & D \end{vmatrix} = \begin{vmatrix} 1 & 0 \\ j\omega(1-a)C & 1 \end{vmatrix} \cdot \begin{vmatrix} 1 & j\omega L \\ 0 & 1 \end{vmatrix} \cdot \begin{vmatrix} 1 & 0 \\ j\omega aC & 1 \end{vmatrix}$$

$$= \begin{vmatrix} 1-a\left(\frac{\omega}{\omega_0}\right)^2 & j\left(\frac{\omega}{\omega_0}\right) Z_0' \\ j\left[\frac{1}{Z_0'} - 1-a\right]\left(\frac{\omega}{\omega_0}\right)\left(\frac{\omega}{\omega_0}\right) & 1-(1-a)\left(\frac{\omega}{\omega_0}\right)^2 \end{vmatrix} \tag{10-86}$$

Substitution of B and D in Eq. (10-85) gives

$$\overline{P}_i = \frac{\overline{P}_D}{4 R_S Y_0} \left\{ K^2 \left(\frac{\omega}{\omega_0}\right)^2 + \left[1 - (1-a)\left(\frac{\omega}{\omega_0}\right)^2\right]^2 \right\} \tag{10-87}$$

The quantity in brackets can be considered as a power derating factor. For most of the 180° constant phase range the term is approximately 1.0. For mid-range of 90° it is about 2/3 and for mid-range of 45°, about 1/2.

When the absorbed power $\overline{P}_D$ is known, the insertion loss is easily calculated. The output power is $\overline{P}_i - P_D$, causing the insertion loss to be given by

$$\delta = 10 \log \left[\frac{P_i}{\overline{P}_i - \overline{P}_D} \right] = 10 \log \left[1 - \frac{\frac{1}{4 R_S Y_0}}{|B|^2 Y_0^2 + D^2} \right]$$

$$\approx 10 \log \left[1 + \frac{4 R_S Y_0}{|B|^2 Y_0^2 + D^2} \right] \tag{10-88}$$

The insertion loss given in Eq. (10-88) is for the diode conduction state. The insertion loss in the non-conduction state is usually lower and is increased to that of the conduction state by placing a resistance in parallel with the diode.

PHASE SHIFTERS

The peak power rating of the reflection, constant phase, phase shifter is calculated by considering the voltages and currents in Fig. 10-18 when the switch Sw is open. This sets I_2 such that $I_2 = 0$ and Eqs. (10-73) and (10-74) reduce to

$$V_1 = AV_2 \tag{10-89}$$

$$I_1 = CV_2 \tag{10-90}$$

The admittance of the diode Y_1 becomes

$$Y_1 = \frac{I_1}{V_1} = \frac{C}{A} \tag{10-91}$$

and Eq. (10-80) gives

$$P_i = \frac{1}{8Y_0}(AV_2)^2 \left(Y_0 + \frac{C}{A}\right)^2 \tag{10-92}$$

A diode biased to half the breakdown voltage V_B can hold off currents having peak voltage V_2 given by

$$V_2 = V_B/2 \tag{10-93}$$

Recall that $V_2 \approx V_B$ is possible when harmonics and intermodulation products will not be troublesome, ref. Chapter IV C 1. Eqs. (10-92) and (10-93) combine to give the peak incident power $\widehat{P}_i$ the diode can control

$$\widehat{P}_i = \frac{V_B^2}{32Y_0}(AY_0 + C)^2 \tag{10-94}$$

Note as before that C is imaginary causing Eq. (10-94) to reduce to

$$\widehat{P}_i = \frac{V_B^2}{32Y_0}(A^2 Y_0^2 + |C|^2) \tag{10-95}$$

3. Loaded Line Phase Shifters

a. General

The normalized ABCD matrix of two normalized susceptances B_N, shunting a transmission line and separated by $\theta = 2\pi \ell/\lambda$, as shown in

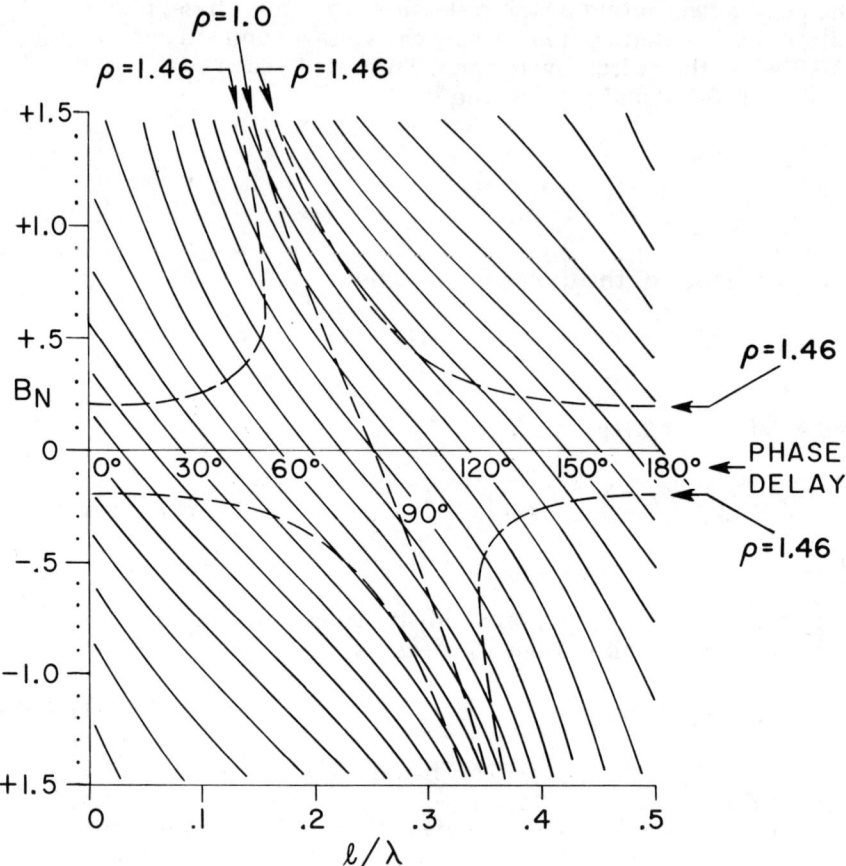

Fig. 10-19   Loaded-Line Phase Shifts and VSWR Limits

Fig. 10-3, is given by

$$\begin{vmatrix} A & B \\ C & D \end{vmatrix}_N = \begin{vmatrix} 1 & 0 \\ jB_N & 1 \end{vmatrix} \cdot \begin{vmatrix} \cos\theta & j\sin\theta \\ j\sin\theta & \cos\theta \end{vmatrix} \cdot \begin{vmatrix} 1 & 0 \\ jB_N & 1 \end{vmatrix} \quad (10\text{-}96)$$

$$\begin{vmatrix} A & B \\ C & D \end{vmatrix}_N = \begin{vmatrix} \cos\theta - B_N \sin\theta & j\sin\theta \\ j[2B_N \cos\theta + (1-B_N^2)\sin\theta] & \cos\theta - B_N \sin\theta \end{vmatrix} \quad (10\text{-}97)$$

The transmission coefficient of a normalized ABCD matrix is given by

PHASE SHIFTERS

$$S_{21} = \frac{2}{A+B+C+D} = \frac{1}{[\cos\theta - B_N \sin\theta] + j[B_N \cos\theta + (1 - \frac{B_N^2}{2})\sin\theta]}$$

(10-98)

The phase ϕ of the loaded transmission line section is given by

$$\phi = \tan^{-1}\left[-\frac{B_N \cos\theta + (1 - \frac{B_N^2}{2})\sin\theta}{\cos\theta - B_N \sin\theta}\right]$$

$$= \tan^{-1}\left[-\frac{B_N + (1 - \tfrac{1}{2} B_N^2)\tan\theta}{1 - B_N \tan\theta}\right].$$

(10-99)

Positive phase corresponds to phase advance and negative phase, to phase delay. Since the $\tan^{-1}$ function is normally taken between $\pm\pi/2$ the phase delay can be represented by

$$\phi_D = \pi + \tan^{-1}\left[\frac{B_N + (1 - \tfrac{1}{2} B_N^2)\tan\theta}{1 - B_N \tan\theta}\right]$$

(10-100)

This phase delay is shown graphically in Fig. 10-19.

Since the structure of Fig. 10-3 is considered lossless, the magnitude of the input reflection coefficient is given by

$$|S_{11}| = \sqrt{1 - |S_{21}|^2} = \sqrt{1 - \frac{1}{1 + B_N^2(\cos\theta - \tfrac{1}{2} B_N \sin\theta)^2}}$$

(10-101)

and the input VSWR ρ is given by

$$\rho = \frac{1 + |S_{11}|}{1 - |S_{11}|}.$$

(10-102)

The magnitude of VSWR that can be tolerated is determined by the amount of phase uncertainty that is caused by two interacting VSWR's. Phase uncertainty is given by[13]

$$\epsilon_\phi = \pm\sin^{-1}[|\Gamma_1| \cdot |\Gamma_2|]$$

(10-103)

in which Γ_1 and Γ_2 are the reflection coefficients of the interacting discontinuities. When both phase bits are allowed to have the same reflection coefficient, Γ, the error is given by

$$\epsilon_\phi = \pm \sin^{-1} \Gamma^2 \qquad (10\text{-}104)$$

or

$$\Gamma = \sqrt{\sin \epsilon_\phi} \qquad (10\text{-}105)$$

Allowing each bit to have $\pm 2°$ phase error gives

$$\Gamma = .187$$

or

$$\rho = 1.46$$

Using Eqs. (10-101) and (10-102), B_N and θ are calculated that give $\rho = 1.46$ which gives the dashed curves shown superimposed on Fig. 10-19.

The practical range of phase is determined by finding the range of phase available on Fig. 10-19 for a fixed ℓ/λ and bounded by the dashed curves for $\rho = 1.46$. This maximum phase shift range is shown in Fig. 10-20 as a function of ℓ/λ. $90°$ is available over two restricted ranges, while $75°$ or less is available over more than a 2:1 range in values of ℓ/λ. The limit due to $\rho = 1.2$ is also shown on Fig. 10-20 showing $45°$ or less over almost a 2:1 range in values of ℓ/λ.

b. Switching with Stubs

Some practical circuits for making loaded line phase shifters are shown in Fig. 10-21. Using the stubs, Fig. 10-21 (A) the normalized impedance of the stubs for perfect open circuit switches B_{NO} is given by

$$B_{NO} = K_S \tan K_\theta \, \theta \qquad (10\text{-}106)$$

When the switches are closed the normalized admittance of each stub B_{NS} is given by

$$B_{NS} = - \frac{K_S}{\tan K_\theta \, \theta} \qquad (10\text{-}107)$$

PHASE SHIFTERS

When Eqs. (10-106) and (10-107) are alternately put into Eq. (10-100), the difference in phases is the phase shift which is shown for various values of K_S and K_θ in Fig. 10-22. The curves stop when either phase has a VSWR of 1.46. The parameters were selected to give the $\pm 2°$ range, but the VSWR became too high with the higher phase shifts to reach the $+2°$ points. $45°$ phase shift is available over about 25% bandwidth, while $22.5°$ is available about an octave bandwidth, and $11.25°$ phase shift is available over almost two octaves bandwidth.

Both of the circuits of Fig. 10-21 tend to have the same curvature. Neither of the circuits could give more ripples than the single ripple responses shown in Fig. 10-22. An analysis permitting the diodes to be different could psosibly permit wider bandwidth than calculated from Fig. 10-22. It should be noted as a practical matter that seldom can the Z_0 in stripline be made much outside of the 25 ohm to 100 ohm range. Thus, when Z_0 is 25 ohms, $K_S = .25$ is a practical lower limit. Only at the expense of unusual characteristic impedances can wide bandwidths be realized using the simple circuits shown in Fig. 10-21.

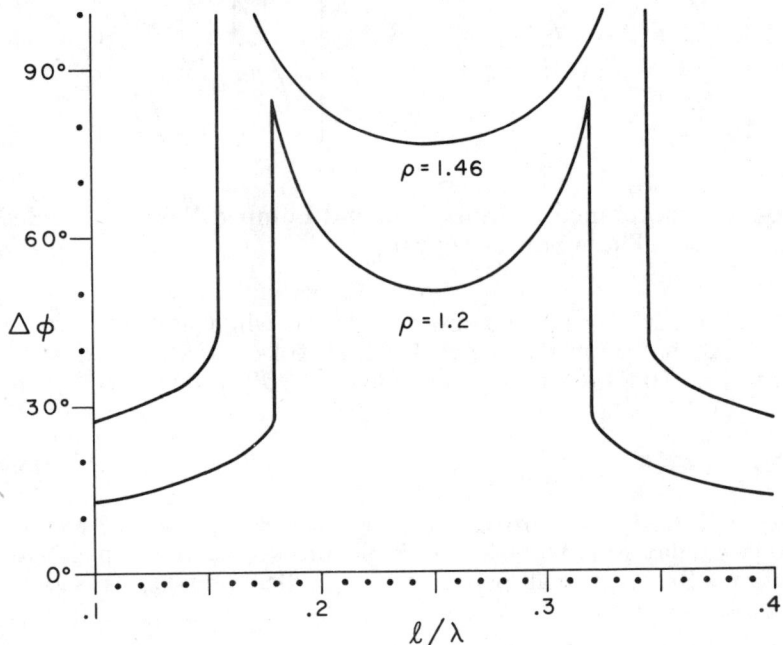

Fig. 10-20 Variable Phase Shift Ranges of the Loaded-Line Phase Shifter

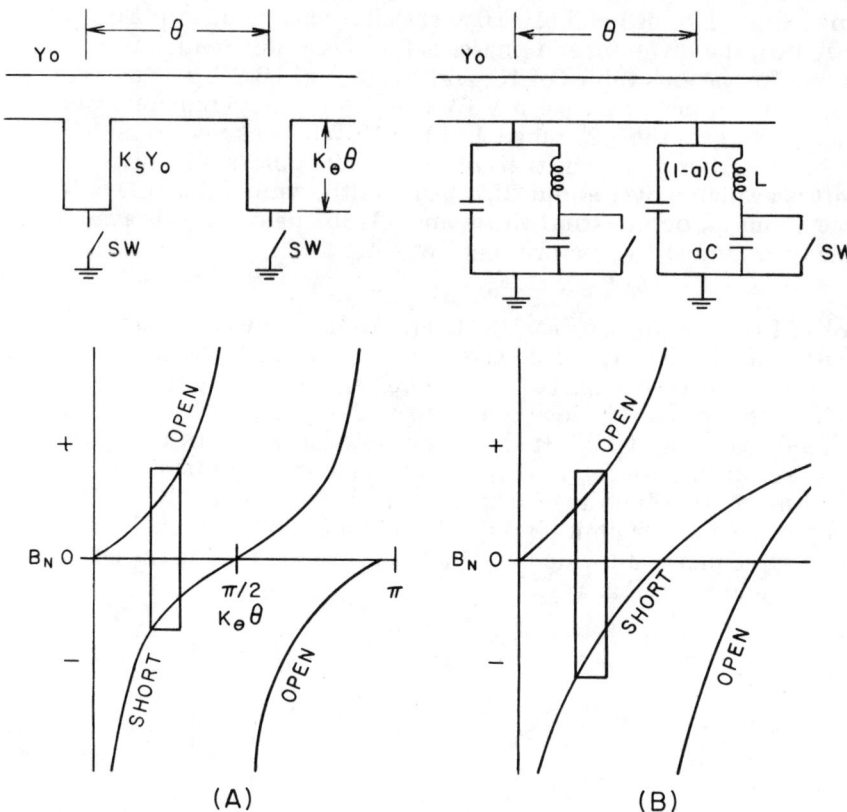

Fig. 10-21 Susceptance of Short Stub and Lumped-Element Loaded-Line Phase Shifter Diodes

Widest bandwidth is centered about $\ell/\lambda = .25$ when the stubs are $\lambda/8$ long. At that point the B_N vs. ℓ/λ lines for open and short as shown in Fig. 10-21 are parallel and Eq. (10-100) reduces to

$$\phi_D = \pi - \tan^{-1}\left[\frac{1-\tfrac{1}{2}B_N^2}{B_N}\right] \tag{10-108}$$

Open circuit diodes will introduce a positive susceptance B_N and a large phase delay ϕ_{DO} while short circuit diodes will give a negative susceptance B_N with small phase delay ϕ_{DS}. The phase shift is given by

$$\Delta\phi = \phi_{DO} - \phi_{DS} = 2\tan^{-1}\left[\frac{B_N}{1-\tfrac{1}{2}B_N^2}\right] \tag{10-109}$$

PHASE SHIFTERS

which gives the values for K_S at $\ell/\lambda = .25$ in Fig. 10-22. For example, the 22.5° curve shows 20.5° which gives $B_N = .178 = K_S$. (The susceptance of a $\lambda/8$ is $\pm j K_S Y_0$).

c. Switching with Lumped Element Diodes

Given that the susceptance curves are parallel at $\ell/\lambda = .25$ and that $B_{NO} = -B_{NS}$ the circuit element values for the diodes of Fig. 10-21(b) can be calculated. Computer calculations of a figure similar to Fig. 10-22 showed the same properties (bandwidth, centering of $\ell/\lambda = .25$).

Referring to Fig. 10-21(b), the normalized shunt susceptance of each diode when the switch Sw is closed B_{NS} is given by

$$B_{NS} = \frac{\omega(1-a)C}{Y_0} - \frac{1}{Y_0 \omega L} \cdot \qquad (10\text{-}110)$$

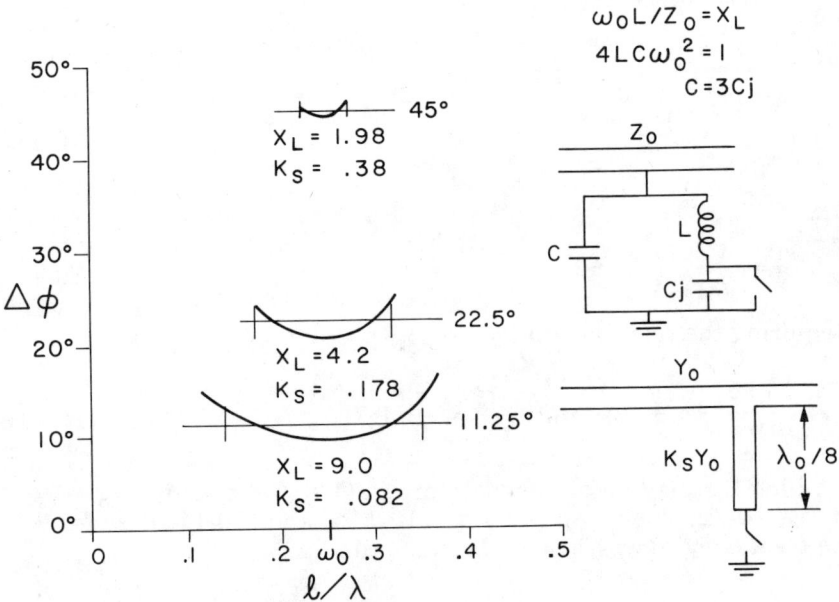

Fig. 10-22 Frequency Dependence of Loaded-Line Phase Shifters designed for Maximum Bandwidth. Curves Stop at $\rho=1.46$ Ticks Indicate end of $\pm$ 2° Range.

B_{NO}, the normalized shunt susceptance of each diode when the switch Sw is open, is given by

$$B_{NO} = \frac{\omega(1-a)C}{Y_0} - \frac{1}{Y_0\omega L - \frac{Y_0}{\omega aC}} \qquad (10\text{-}111)$$

Matching the slopes at ω,

$$\frac{\partial B_{NS}}{\partial \omega} = \frac{\partial B_{NO}}{\partial \omega}$$

gives

$$a = \frac{1}{3\omega^2 LC} = \frac{1}{3}\left(\frac{\omega_0^2}{\omega^2}\right) \qquad (10\text{-}112)$$

using the definition for $\omega_0^2 = \frac{1}{LC}$ as given in Fig. 10-17. Setting up the equation

$$B_{NO} = -B_{NS}$$

produces

$$\frac{\omega_0^2}{\omega^2} = \frac{12}{7} \qquad (10\text{-}113)$$

and

$$a = \frac{4}{7} \qquad (10\text{-}114)$$

Requiring the relationship $B_{NS} = -B_N$ produces

$$\frac{1}{\omega L Y_0} = \frac{4}{3} B_N \text{ and } \frac{\omega C}{Y_0} = \frac{7}{9} B_N \qquad (10\text{-}115) \quad (10\text{-}116)$$

in which B_N may be obtained from Eq. (10-109) or from $K_S = B_N$ in Fig. 10-22. The term ω in Eqs. (10-115) and (10-116) are for the frequency of $\lambda/4$ spacing between diodes.

d. Narrow-band, Perfect Match Phase Shifters

The above analysis has been for $\lambda/4$ spacing between diodes, because that spacing gives widest bandwidth and has been most widely used

PHASE SHIFTERS

in practice. However, phase shift greater than 45° is not possible without causing excessively high VSWR. Higher amounts of phase shift can be obtained by having $B_N = 0$ in one bias state and B_N equal to some positive number in the other bias state. The spacing between the diodes is adjusted so that $\rho = 1.0$ as on Fig. 10-19. Setting the relationship $|S_{11}| = 0$ in Eq. (10-101) gives

$$\tan \theta = \frac{2}{B_N} \tag{10-117}$$

Substituting Eq. (10-117) into Eq. (10-100) produces

$$\phi_D = \pi - \tan^{-1}\left[\frac{2}{B_N}\right]. \tag{10-118}$$

The phase delay of the line with no diodes shunting it, $B_N = 0$, ϕ_{DN} is exactly equal to its electrical length, giving

$$\phi_{DN} = \theta = \tan^{-1}\left[2/B_N\right] \tag{10-119}$$

The increased phase delay due to switching the diodes into the circuit $\Delta \phi$ is given by

$$\Delta \phi = \phi_D - \phi_{DN} = \pi - \tan^{-1}\left[2/B_N\right] - \tan^{-1}\left[2/B_N\right]$$

$$= \pi - 2 \tan^{-1}\left[2/B_N\right] \tag{10-120}$$

which reduces to

$$B_N = 2 \tan\left(\frac{\Delta \phi}{2}\right) \tag{10-121}$$

or using Eq. (10-119), Eq. (10-120) can be reduced to

$$\Delta \phi = \pi - 2\theta \tag{10-122}$$

which gives

$$\theta = \frac{\pi}{2} - \frac{\Delta \phi}{2} = \frac{2\pi \ell}{\lambda} \tag{10-123}$$

Eqs. (10-121) and (10-123) give the susceptance and spacing for a narrow band perfectly matched loaded line phase shifter.

e. Power Limitations and Insertion Loss

The derivation in this section parallels that used in Section X.A.2. for reflection phase shifters. The assumption is made that the standing wave existing between diodes $\lambda/4$ apart does not magnify or demagnify the voltages appearing across the diodes. This assumption is valid because the voltage maximum of a small shunt capacitance is $\lambda/8$ toward the generator and the voltage minimum is $3\lambda/8$ toward the generator; thus, at $\lambda/4$ the voltage is half-way between a maximum and minimum, which is the voltage of the incident wave only.

The voltages and currents on a single diode can be determined using Fig. 10-18, with the exception that the generator output impedance is $2 Y_0$ instead of Y_0. This substitution causes Eq. (10-79) to be changed to:

$$I_0 = V_1 (2Y_0 + Y_1) \tag{10-124}$$

and Eq. (10-85) becomes:

$$\overline{P}_i = \frac{\overline{P}_0 (4|B|^2 Y_0^2 + D^2)}{4 R_S Y_0} \tag{10-125}$$

The ABCD matrix for the lumped element diode is given in Eq. (10-86). For the stubs, the matrix is simply that of a transmission line having characteristic admittance $Y_0' = K_s Y_0$ and length $k_\theta \theta = 45°$.

$$\begin{vmatrix} A & B \\ C & D \end{vmatrix} = \begin{vmatrix} \cos K_\theta \theta & j Z_0' \sin K_\theta \theta \\ j Y_0' \sin K_\theta \theta & \cos K_\theta \theta \end{vmatrix}$$

$$= \begin{vmatrix} \sqrt{2} & j\frac{\sqrt{2}}{K_s Y_0} \\ j K_s Y_0 \sqrt{2} & \sqrt{2} \end{vmatrix} \tag{10-126}$$

Substitution of B and D from Eq. (10-126) into Eq. (10-125) gives

$$\overline{P}_i = \frac{\overline{P}_D}{2 R_S Y_0} \left(\frac{4}{K_S^2} + 1 \right) \tag{10-127}$$

which is the maximum incident average power that a stub loaded line phase shifter can control at center frequency.

The insertion loss of one diode stub is given by a modified Eq. (10-88).

PHASE SHIFTERS

$$\delta = 10 \log \left[1 + \frac{4 R_S Y_0}{4 |B|^2 Y_0^2 + D^2} \right] \qquad (10\text{-}128)$$

But the phase shifter has two stubs, and the insertion losses add when they are small, giving

$$\delta = 20 \log \left[1 + \frac{2 R_s Y_0}{1 + \frac{2}{K_s^2}} \right] \qquad (10\text{-}129)$$

The substitution of Eq. (10-124) for Eq. (10-79) causes Eq. (10-95) to become

$$\hat{P}_i = \frac{V_B^2}{32 Y_0} \left(4 A^2 Y_0^2 + |C|^2 \right) \qquad (10\text{-}130)$$

which is the maximum incident peak power that the loaded line phase shifter can control.

4. Hi-Lo Pass Phase Shifter

A low-pass filter comprised of series inductors and shunt capacitors provides phase delay to signals passing thru it. A high-pass filter comprised of series capacitors and shunt inductors provides phase advance. By arranging diode switches to permit switching between low-pass and high-pass, a phase shifter can be made which is smaller than the other types and has bandwidth almost as good as the lumped element reflection phase shifter.

a. General

The normalized ABCD matrix for the elements of Fig. 10-4 switched in the low-pass state is given by:

$$\begin{vmatrix} A & B \\ C & D \end{vmatrix}_N = \begin{vmatrix} 1 & jX_N \\ 0 & 1 \end{vmatrix} \cdot \begin{vmatrix} 1 & 0 \\ jB_N & 1 \end{vmatrix} \cdot \begin{vmatrix} 1 & jX_N \\ 0 & 1 \end{vmatrix}$$

$$= \begin{vmatrix} 1 - B_N X_N & j(2X_N - B_N X_N^2) \\ j B_N & 1 - B_N X_N \end{vmatrix} \qquad (10\text{-}131)$$

The transmission coefficient S_{21} of the normalized ABCD matrix is given by

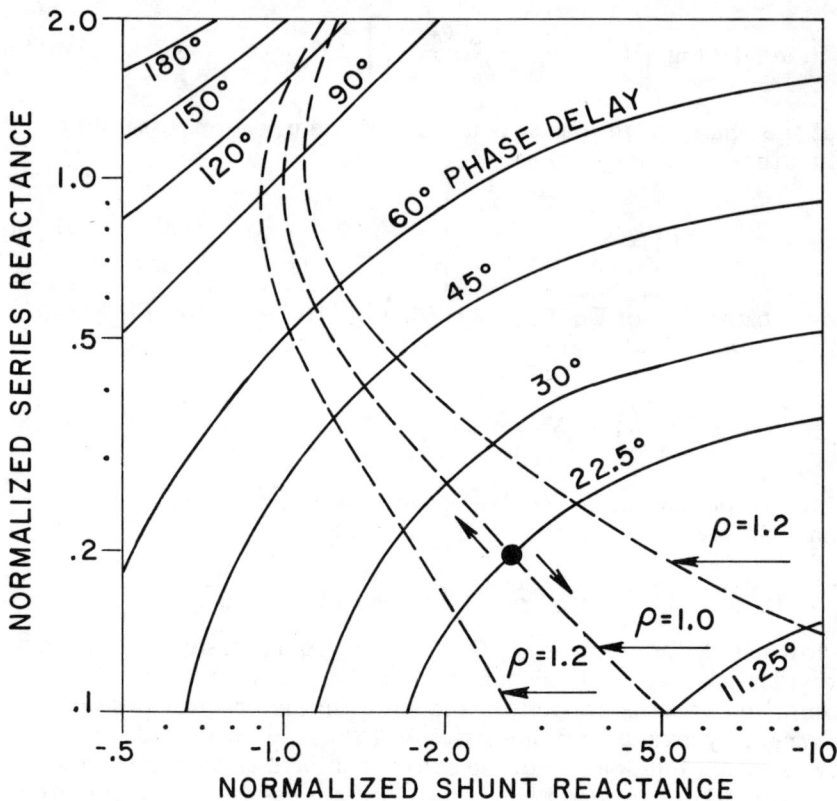

Fig. 10-23 Phase Shift of Lumped-Element Low-Pass T-Section Phase Shift Paths

$$S_{21} = \frac{2}{A+B+C+D} = \frac{2}{2(1-B_N X_N) + j(B_N + 2X_N - B_N X_N^2)} \quad (10\text{-}132)$$

The transmission phase ϕ is given by:

$$\phi = \tan^{-1}\left[-\frac{B_N + 2X_N - B_N X_N^2}{2(1-B_N X_N)}\right] \quad (10\text{-}133)$$

This phase is shown in Fig. 10-23 for a range of B_N (converted to normalized shunt reactance) and X_N. When both B_N and X_N change signs the phase remains the same but changes sign; thus, the phase shift $\Delta\phi$ caused by switching between low-pass and high-pass is twice Eq. (10-133), giving:

PHASE SHIFTERS

$$\Delta\phi = 2\tan^{-1}\left[-\frac{B_N + 2X_N - B_N X_N^2}{2(1-B_N X_N)}\right] \quad (10\text{-}134)$$

Assuming the phase shifter to be lossless, the reflection coefficient S_{11} is given by

$$|S_{11}| = \sqrt{1-|S_{21}|^2}$$

The phase shifter will be perfectly matched when the equation $|S_{21}| = 1$ is satisfied. Under conditions of match, Eq. (10-132) reduces to:

$$B_N = \frac{2X_N}{X_N^2 + 1} \quad (10\text{-}135)$$

Substitution of Eq. (10-135) in the Eq. (10-134) gives:

$$\Delta\phi = 2\tan^{-1}\left[\frac{2X_N}{X_N^2 - 1}\right] \quad (10\text{-}136)$$

which reduces to:

$$X_N = \tan\left(\frac{\Delta\phi}{4}\right) \quad (10\text{-}137)$$

Using Eq. (10-137) in Eq. (10-135) gives:

$$B_N = \sin\left(\frac{\Delta\phi}{2}\right) \quad (10\text{-}138)$$

A pi section filter instead of a T would exchange Eqs. (10-137) and (10-138) (X_N and B_N). Fig. 10-23 would also show the phase but would be for susceptances instead of reactances and the coordinates would be exchanged.

The frequency dependence of the phase shifter may be studied with the aid of Fig. 10-23. For example, a 45° phase shifter will have $X_N = .199$ and $B_N = .382$ using Eqs. (10-137) and (10-138). These two values fall at the intersection of $\rho = 1.0$ and 22.5° on Fig. 10-23. As frequency is increased in the low-pass state the series reactance increase is proportional to frequency; the shunt reactance decrease is inversely proportional to frequency. The intersection of the two reactances moves towards the upper left corner of the figure at 45° off vertical, closely following the $\rho = 1.0$ curve and increasing phase delay. On the other hand, in the high-pass state, the series reactance decreases with increasing frequency and the shunt reactance increases with increasing frequency. The intersection of the two reactances moves towards the lower right corner of the figure, again closely following the $\rho = 1.0$ curve, but now decreasing phase advance. The net effect is that the phase shifter tends to stay matched as frequency is increased, and phase delay increased in the low-pass state is compensated for by phase advance lost in the high-pass state.

The frequency dependence of the high-pass low-pass phase shifter is shown in Fig. 10-24. Low VSWR is easily obtained. $90° \pm 2°$ is obtainable over almost an octave while the smaller phase shifts are available over more than an octave. If bandwidth is limited to an octave, $90° \pm 4\%$ is possible and the smaller phase shifts have errors of $\pm 3\%$.

A more practical embodiment of the high-pass low-pass phase shifter may be as shown in Fig. 10-25. Two SPDT switches are required instead of three; and parasitics due to OFF diodes and OFF lines are more easily accounted and compensated for.

The bandwidth of the large phase shift bits may be improved by using more elements in the high-pass and low-pass circuits.

b. Power Limitations and Insertion Loss

The power limitations of the phase shifter as shown in Fig. 10-25 will be the same as for the switched line phase shifter as given in Eqs. (10-19) and (10-21). The insertion loss contributed by the diodes will be given by Eq. (10-16). Some insertion loss will also be contributed by the finite Qs of the circuit elements, and it may be necessary to "spoil" the Q in some of the elements to keep the insertion loss the same in both phase states. Another solution is to use a T circuit for high-pass and a pi circuit for low-pass.

Some caution must be exercised to avoid the regions of high insertion loss and high phase error as encountered in the switched line phase shifter and as exemplified in Fig. 10-7. Recall that the attenuation α of a series diode is given by:

$$\alpha = 10 \log \left[(1 + \tfrac{1}{2} R_N)^2 + (\tfrac{1}{2} X_N)^2 \right] \qquad (10\text{-}139)$$

PHASE SHIFTERS

in which the normalized impedance of the diode is $R_N + j X_N$. A 20 dB reactively limited isolation will have $X_N = -20$ and 10 dB, $X_N = -6$. The normalized reactance X_N of an open circuit line of length θ is given by:

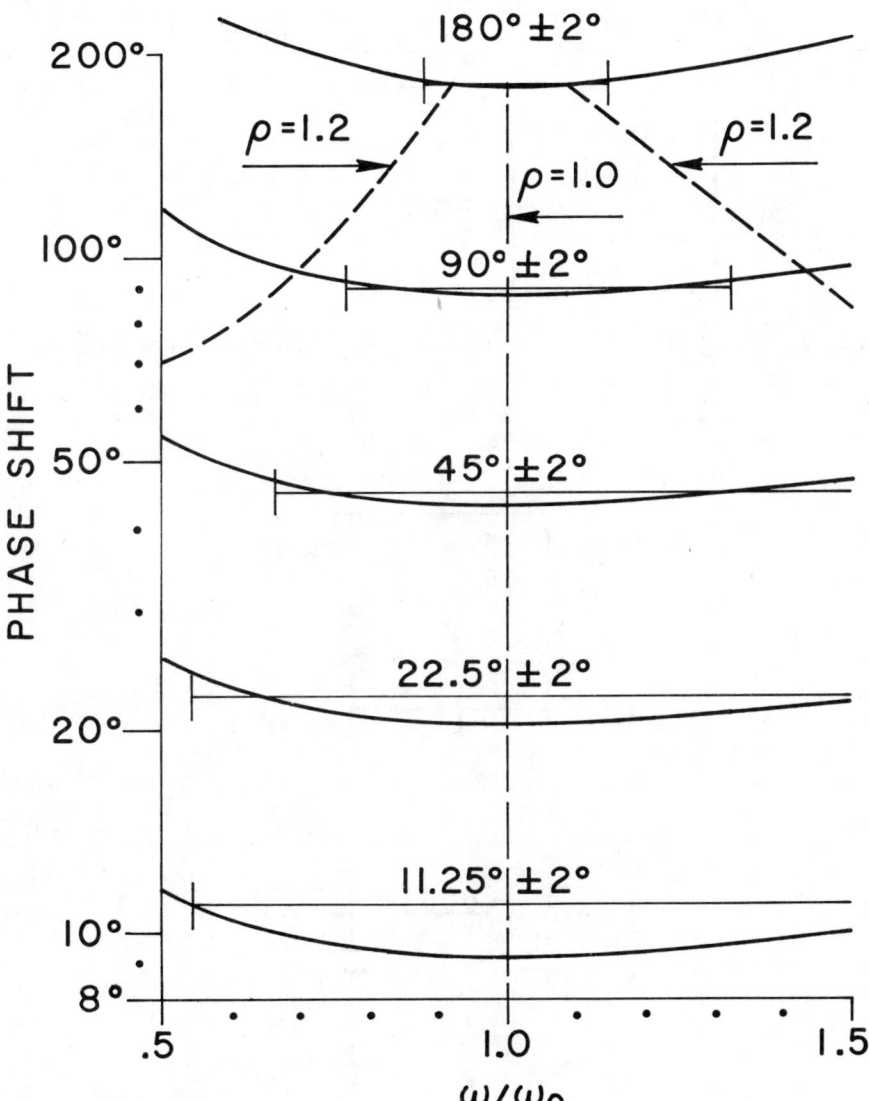

Fig. 10-24 Frequency Dependence of Lumped-Element Low-Pass High-Pass Phase Shifters

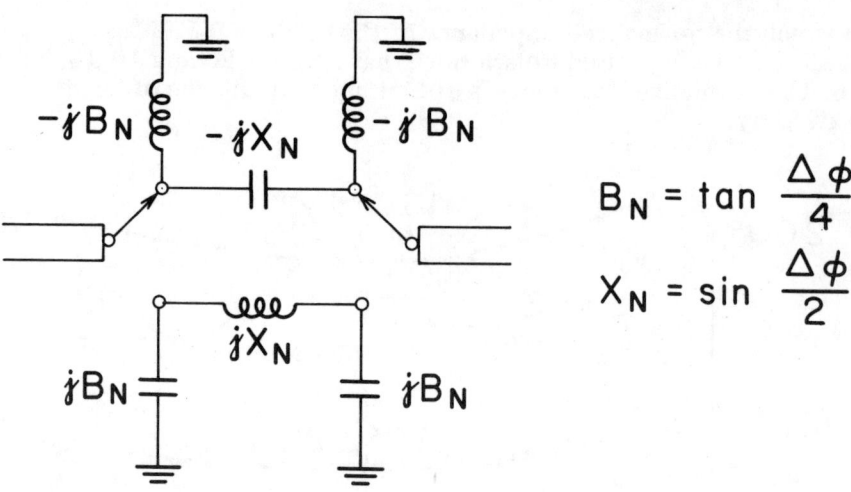

Fig. 10-25 Practical Layout of Low-Pass High-Pass Phase Shifters

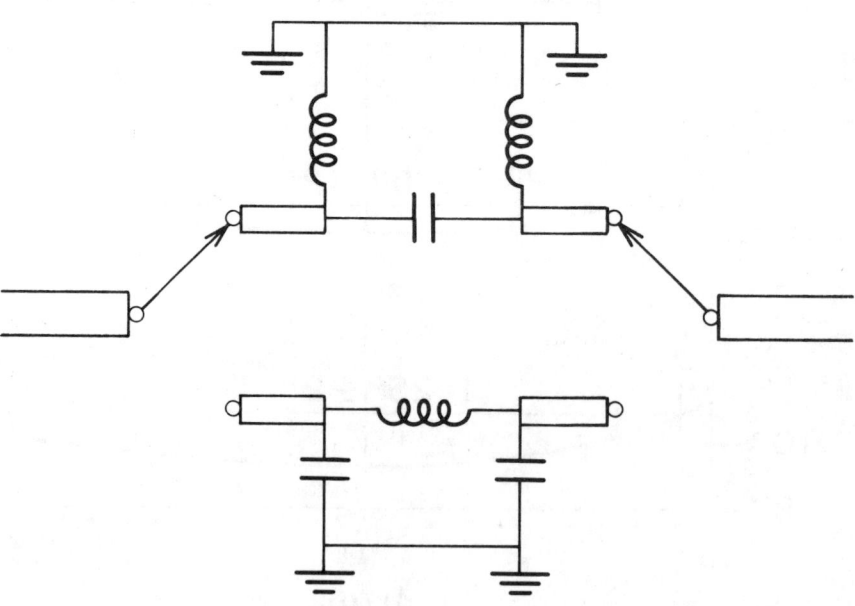

Fig. 10-26 Short Line Length Added to Fig. 10-25 to Avoid Phase Error and Insertion Loss Spikes

PHASE SHIFTERS

$$X_N = -\cot\theta \qquad (10\text{-}140)$$

The 20 dB switch appears to be 2.9° length of open circuit line and the 10 dB switch 9.5°. Recall from section X.A.1. that 180° length in the OFF line must be avoided to avoid insertion loss and phase error spikes. Since one diode is at each end of each line, the 20 dB diodes add 5.8° to the OFF line and the 10 dB diode, 19°. Therefore, phase delays of 174.2° to 101° must be avoided in the OFF lines depending on SPST diode isolation. It can be seen from Fig. 10-23 that such large phase delays are not normally encountered with simple T or pi circuits in the high-pass low-pass phase shifter. But phase advances of 5.8° to 19° must also be avoided to avoid the troublesome spikes. These advances are normally encountered with the smaller phase shift bits. The problem of too little phase delay in one path may be corrected by adding short lengths of transmission line between the T or pi sections and the diode switches, as shown in Fig. 10-26.

5. Summary of Digital Phase Shifters

The following table gives the location of information for the four types of digital phase shifters.

Type	Design Equations	Band-Width	Insertion Loss	Average Power	Peak Power
Switched Line	Fig. 10-1	Fig. 10-9	Eq. 10-16	Eq. 10-19	Eq. 10-21
Reflection-Time Delay	Fig. 10-2	Fig. 10-13	Eq. 10-40	Eq. 10-58	Eq. 10-53
Reflection-Constant Phase	Fig. 10-13	Fig. 10-17	Eq. 10-88	Eq. 10-85	Eq. 10-95
Loaded Line	Fig. 10-3	Fig. 10-22	Eq. 10-129	Eq. 10-125	Eq. 10-130
High-Low Pass	Fig. 10-4	Fig. 10-24	Eq. 10-16	Eq. 10-19	Eq. 10-21

The application often dictates which phase shift circuit is most appropriate. The switched line phase shifter is useful in circuits where it is difficult to make fine adjustments in final assembly because the phase shift is well defined by differences in line length. The reflection-time delay phase shifter also does not require fine circuit adjustments, but suffers from non-reciprocity when used in a circulator and can pick up large phase errors from the circulator.

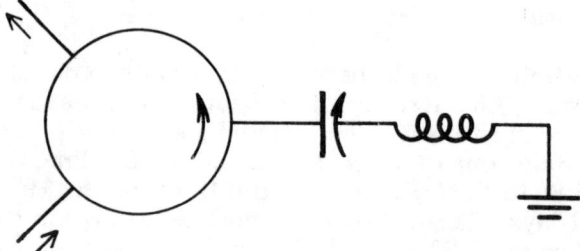

Fig. 10-27 Circuit for Continuously Variable Reflection Phase Shifters

The reflection constant-phase phase shifter easily provides octave bandwidth, but is also limited by the circulator or 3 dB coupler. The reflection constant phase bit of 180° in a 3 dB coupler has been widely adopted for integrated circuits even when other circuits are used for the other bits. The loaded line phase shifter is very useful for high-power narrow-band applications. By keeping the phase shift per bit low, high powers can be controlled. For example, a 1000V diode with R_S = 1 ohm, $\overline{P}_D$ = 2W, should be able to control 5 kW in either phase state for a 22.5° bit in 50 ohm transmission line. The high-low pass phase shifter gives wide bandwidth without needing a circulator or 3 dB coupler and is the smallest in size.

B. Continuous Phase Shifters

Continuous phase shifters have some advantages over digital phase shifters because fewer diodes are required, phase shift is smooth and continuous, and the diodes are reverse biased varactors which give very fast change of phase (no minority carriers injected by forward bias) and require very little control power. But continuous phase shifters are restricted to relatively narrow-bandwidths and relatively low power levels (~ 100 mW max). Because the diodes are reactive the circuits have to be carefully adjusted in final assembly to obtain the desired results. And because the varactor capacitance changes rapidly with voltage around zero volts bias, the phase shifters tend to generate harmonics and increase losses for higher incident power levels at low bias voltages.

1. Reflection Phase Shifter

The circuit for a reflection phase shifter is shown in Fig. 10-27. As the voltage applied to the varactor changes, the impedance of the circuit moves around the outside edge of the Smith Chart, giving relatively constant reflected power of changing phase. When the

PHASE SHIFTERS

series inductance L and minimum varactor capacitance C_{MIN} (at reverse bias) satisfy the following equations for abrupt junction varactors (capacitance proportional to the square root of applied voltage), the circuit provides 180° phase shift linearly proportional to applied voltage[14] within ±1.4% of full scale.

$$L = \frac{2.35\, Z_0}{\omega} \qquad C_{MIN} = \frac{.36}{\omega Z_0}$$

The inductance is easily realized by having a long diode lead to ground. If the capacitance cannot be obtained exactly as needed, Z_0 can be adjusted by putting a λ/4 transformer in front of the diode. At 1 GHz and Z_0 = 50 ohms, L and C are satisfied by L = 18.8 nH and C_{MIN} = 1.15 pF. It is possible to obtain 360° phase sweep with a single circulator by using two diode circuits in parallel. In one version[14] each diode circuit is designed to be linear into 100 ohms, and the two diodes are separated by a λ/4 100 ohms transmission line. In another version[15] the diodes are connected directly in parallel and the inductors behind them are different. This causes each circuit to be series resonant at a different bias, and the two circuits become parallel-resonant at an intermediate bias. Shunt tuning[14] has been used to obtain linear phase shift with pill varactors at frequencies of 5 GHz and above.

2. Loaded Line Phase Shifter

A loaded line variable phase shifter[5] is made by having each shunt susceptance as a parallel tuned circuit, as shown in Fig. 10-28. Element values are selected to go between $\pm B_N$ as prescribed in Fig. 10-3. At some middle range bias both circuits are parallel resonant and the circuit gives only 90° phase delay as prescribed by Fig. 10-19. As the bias is varied, B_N can go to ±.65 before VSWR gets too high (greater than 1.46) giving a phase sweep of about 80°. It can be seen from Fig. 10-20 that a full sweep range greater than 90° is possible by setting the condition for spacing between diodes:

$$.155 \leq \ell/\lambda \leq .187$$

No linearity will exist between phase and voltage. The diodes should be alike. The selection of diodes is not critical. The shunt inductance is selected to give the desired maximum $-B_N$ when the diode is at maximum reverse bias. The minimum reverse bias

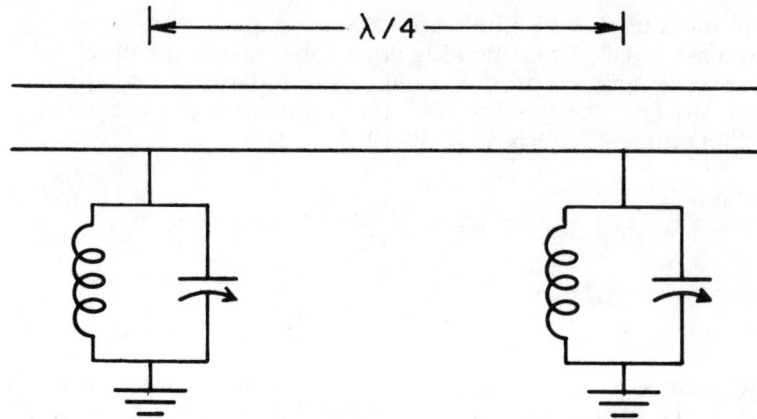

Fig. 10-28 Circuit for Continuously Variable Loaded-Line Phase Shifters

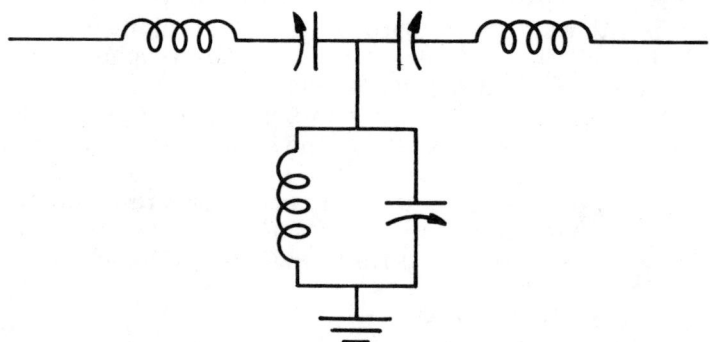

Fig. 10-29 Circuit for Continuously Variable Low-Pass High-Pass Phase Shifters

is then set by the voltage that gives the maximum desired + B_N and the variable phase is obtained by varying the bias between the maximum and minimum.

3. High-Low Pass Phase Shifter

It is possible to make the high low-pass phase shifter into a continuous device by using the circuit shown in Fig. 10-29; however, two non-linear voltages have to be generated: one for the series varactors and one for the shunt varactor. Again, the selection of diodes is not critical so long as the series diodes are the same. This phase shifter, however, can be programmed to give a full 180° of variable phase without critical VSWR problems.

REFERENCES

1. E.M. Rutz and J.E. Dye, "Frequency Translation by Phase Modulation," 1957 *IRE WESCON Conv. Rec.*, Pt. 1, pp. 201-207.

2. E.J. Wilkinson, L.I. Parad, and W.R. Connerney, "An X-Band Electronically Steerable Phased Array," *Microwave Journal*, Feb. 1964, pp. 43-48.

3. R.H. Hardin, E.J. Downey, and J. Munushian, "Electronically-Variable Phase Shifters Utilizing Variable Capacitance Diodes," *Proc. IRE*, Vol. 48, May 1960, pp. 944-5.

4. B.W. Battershall and S.P. Emmons, "Optimization of Diode Structures for Monolithic Integrated Circuits," *IEEE Trans. on Microwave Theory and Techniques*, Vol. MTT-16, No. 7, July 1968, pp. 445-450.

5. N.H. Dawirs and W.G. Swarner, "A Very Fast, Voltage-Controlled, Microwave Phase Shifter," *Microwave Journal*, June 1962, pp. 99-107.

6. J.F. White, "High Power, PIN Diode Controlled, Microwave Transmission Phase Shifters," *IEEE Trans. on Microwave Theory and Techniques*, Vol. MTT-13, March 1965, pp. 233-242.

7. P. Onno and A. Plitkins, "Miniature Multi-Kilowatt PIN Diode MIC Digital Phase Shifters," 1971 *IEEE-GMTT International Microwave Symposium Digest*, pp. 22-23.

8. R.V. Garver, "Broadband Diode Phase Shifters," *IEEE Trans. on Microwave Theory and Techniques*, Vol. MTT-20, May 1972, pp. 314-323.

9. R.W. Burns and L. Stark, "PIN Diodes Advance High-Power Phase Shifters," *Microwaves*, Nov. 1965, pp. 38-48.

10. C.H. Grauling and D.B. Geller, "A Broad-Band Frequency Translator with 30-dB Suppression of Spurious Sidebands," *IEEE Trans. on Microwave Theory and Techniques*, Vol. MTT-18, Sept. 1970, pp. 651-652.

11. R.V. Garver, D. Bergfried, S. Raff, and B.O. Weinschel, "Errors in S_{11} Measurements Due to Residual SWR of the Measurement Equipment," *IEEE Trans. on Microwave Theory and Techniques*, Vol. MTT-20, Jan. 1972, pp. 61-69.

12. R.V. Garver, "Broadband Binary 180° Phase Modulators," *IEEE Trans. on Microwave Theory and Techniques*, Vol. MTT-13, Jan. 1965, pp. 32-38.

13. E.N. Phillips, "The Uncertainties of Phase Measurement," *Microwaves*, Feb. 1965, pp. 14-21.

14. R.V. Garver, "360° Varactor Linear Phase Modulator," *IEEE Trans. on Microwave Theory and Techniques*, Vol. MTT-17, Mar. 1969, pp. 137-147.

15. B.T. Henoch and P. Tamm, "A 360° Refelction-Type Diode Phase Modulator," *IEEE Trans. on Microwave Theory and Techniques*, Vol. MTT-19, Jan. 1971, pp. 103-105.

16. G.D. Lynes, G.E. Johnson, B.E. Huckleberry and N.H. Forrest, "Design of a Broad-Band 4-Bit Loaded Switched-Line Phase Shifter," *IEEE Trans. on Microwave Theory and Techniques*, Vol. MTT-22 June 1974, pp. 693-697.

Chapter 11
Other Control Devices

A few topics do not fall within the scope of the other chapters but the devices they describe have significant usage today or have substantial potential for the future. One of these devices is the doubly balanced mixer/modulator which serves not only as a mixer and phase discriminator but also as an ON-OFF switch, variable attenuator, and 180° phase shifter (balanced modulator). The second device is a vector generator which can produce a vector of any amplitude and phase from orthogonal inputs. It lends itself nicely to single sideband modulation in that it transfers the modulation voltage directly onto the single sideband. (As modulation frequency and amplitude are changed, the frequency displacement and sideband magnitude follow exactly.) The third device is the sampling gate which was originally used for broadband spectrum generation and is presently used in the inputs of sampling oscilloscopes and the Hewlett Packard network analyzer. The final device to be discussed is bulk diode switches. These diodes raise the power limits of switching close to the power rating of the waveguide in which they are used.

A. Doubly Balanced Mixer/Modulator

The doubly balanced mixer/modulator[1,2] is shown in Fig. 11-1. Both inputs are baluns (fine wire wound on miniature ferrite cores) and Sig and LO arms can be interchanged without changing the performance of the mixer/modulator.

Mixer performance will be described first. As a mixer the LO is ideally taken to be a square wave. When point 2 is positive with respect to point 4, diodes A and B are conducting while diodes C and D are reverse biased. Point 3 is an open circuit to the Sig

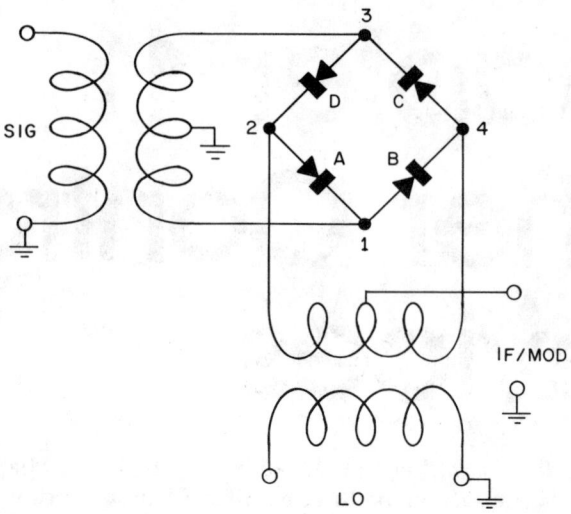

Fig. 11-1 Doubly Balanced Mixer/Modulator.

balun; therefore, the signal developed between the grounded center tap of the Sig balun and point 1 is translated to the center tap (IF output) of the LO balun. On the other half of the LO square wave cycle, diodes C and D conduct and the signal comes out the IF output shifted 180° (because it is now coming from the upper half of the Sig Balun). If Sig and LO are at the same frequency and the phases are correct (zero crossings coincide), the IF output will be a full wave rectified sine wave giving maximum DC. As the phase of the signal is shifted with respect to the LO, the DC output will fall to zero and then increase to a maximum in the opposite polarity as 180° shift is approached. When Sig and LO are at the same frequency, the mixer/modulator functions as a phase discriminator. When Sig and LO are at different frequencies, the IF output voltage goes between positive and negative sinusoidally at the difference frequency, providing the customary IF output waveform. Diode and balun symmetry provide isolation between all ports, so that very high frequency IF can be used (typically up to the lower edge of the Sig-LO band).

As a modulator, the modulation is applied at the IF/MOD terminal. Either RF terminal can serve as input and the other as output. When no voltage is applied to the IF terminal, power in the LO port causes the upper and lower paths of the diode quad to conduct alternately, and the voltage at points 3 and 1 are alternately forced to zero volts (assuming perfect balun and diode balance). No current is induced in the Sig balun, so there is no power output from the Sig balun. In conventional commercially available doubly

OTHER CONTROL DEVICES

balanced mixers, the output (isolation) is 20 dB - 30 dB below the input. Now, when a positive voltage is applied to the IF terminal, diodes A and C are in conduction while diodes B and D are reversed biased. Terminals 2 and 1 are connected together and so are terminals 3 and 4. All power in the LO balun is transferred to the Sig balun, giving a low loss (1dB - 2dB). As the IF voltage is reduced, diodes A and C draw less current, giving higher RF resistance, which produces increased insertion loss and variable attenuation. Now, as the applied voltage becomes negative, diodes D and B begin to conduct, causing points 3 and 2 to be connected together, and also points 1 and 4. LO power is again conducted to the Sig port, but $180°$ out of phase with respect to the output, with a positive voltage applied to the IF port.

Thus, the doubly balanced mixer can be used as phase discriminator, mixer, switch, variable attenuator, or $180°$ phase shifter (balanced modulator). It serves as a balanced modulator when no bias is applied to the modulation input (suppressed carrier). It serves as an amplitude modulator when the modulation input is biased, provided the modulation is smaller than the bias. Two suppressed carrier modulators can be combined with two hybrid junction couplers to provide single sideband mixing or modulation.[1]

Doubly balanced mixers are considered important as control devices because they are readily available over wide bandwidths, they are inexpensive, and are quite useful in many laboratory applications. They can control incident power up to 100 mW, can give very high speed switching, and are very flexible. If more than 20dB isolation is required in a switching application, they may be put in series observing the $\lambda/4$ spacing requirement for additive isolation.

B. Vector Generator

Vector generators find application in automatic bridge balancing (remote measurement), in cancelling LO leakage in very low noise mixers, and in doppler simulation.

If an attenuator and phase shifter were combined, a vector of any phase or amplitude (with respect to an input reference wave) could be generated. Unfortunately, the attenuator would give some phase shift and the phase shifter would give some variation in amplitude. Continuous $360°$ phase shifters also have a discontinuity between $360°$ and $0°$. This discontinuity appears as a step phase change which generates undesired sidebands in a doppler simulator or creates an instability in the servo loop of an automatic balancing

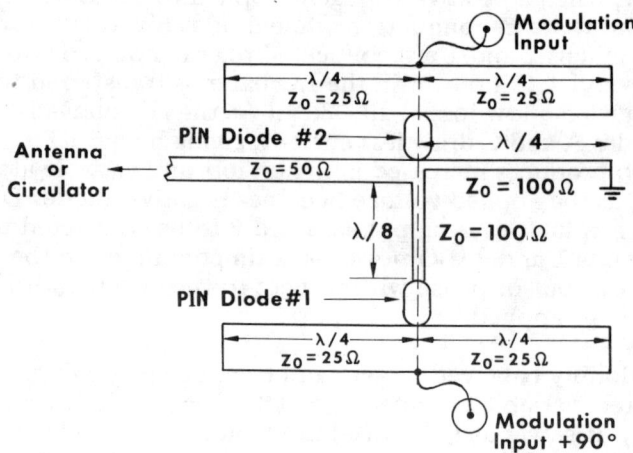

Fig. 11-2 λ/8 PIN Diode Single Sideband Modulator

bridge. These phase shifters are also narrow band and sensitive to output mismatches. A simple variable attenuator PIN diode in series with it would be a poor match. The information for controlling a vector generator is usually in the form of real and imaginary magnitudes rather than amplitude and phase, because it can be obtained from two balanced mixers having the same local oscillator but 90° difference in phase.

The vector generator most easily integrated into systems generates orthogonal vectors and combines them, using the orthogonal information directly. The balanced modulators described in Section XI-A above generate full plus-or-minus vectors. When two of these are then combined in a 90° 3-dB coupler, the full vector field is obtained.

A simpler circuit combines two PIN diodes as shown[3] in Fig. 11-2. A single PIN diode biased to 100 ohms, terminating a 100 ohm transmission line, would have an admittance at the center of the Smith Chart as shown on Fig. 11-3a. As bias is varied, the admittance would vary on the real axis. Moving toward the generator λ/8 would cause the admittance to move 90°, at which point a second 100 ohm PIN diode is mounted, giving the admittance shown in Fig. 11-3b. The λ/4 sections provide for biasing without influencing the admittance adding process. When this structure is then mounted on a 50 ohm transmission line, the admittances

OTHER CONTROL DEVICES

shown in Fig. 11-3d are presented to it. The wave reflected by the termination will be a vector of any phase or magnitude, depending on the deviation from 100 ohms bias on the PIN diodes. The reflected wave is then separated from the incident wave either by using a circulator as shown in Fig. 11-2, or by attaching two of these modulators to a 90° 3-dB coupler.

In spite of the nonlinearities of the PIN diodes and the admittance adding process this circuit gives very linear response at drives up to 10 dB loss. Numerous computer simulations and experiments have been made on this modulator which indicate that it is usually not necessary to tune out diode cartridge parasitics, that it easily gives half-octave bandwidth, that it can handle powers up to 1W, and that no dimensions are critical in fabrication. If it is desired to reduce nonlinearities even further, then a pair of modulators should be used on a 90° 3-dB coupler with the diode polarities of one modulator opposite that of the other.

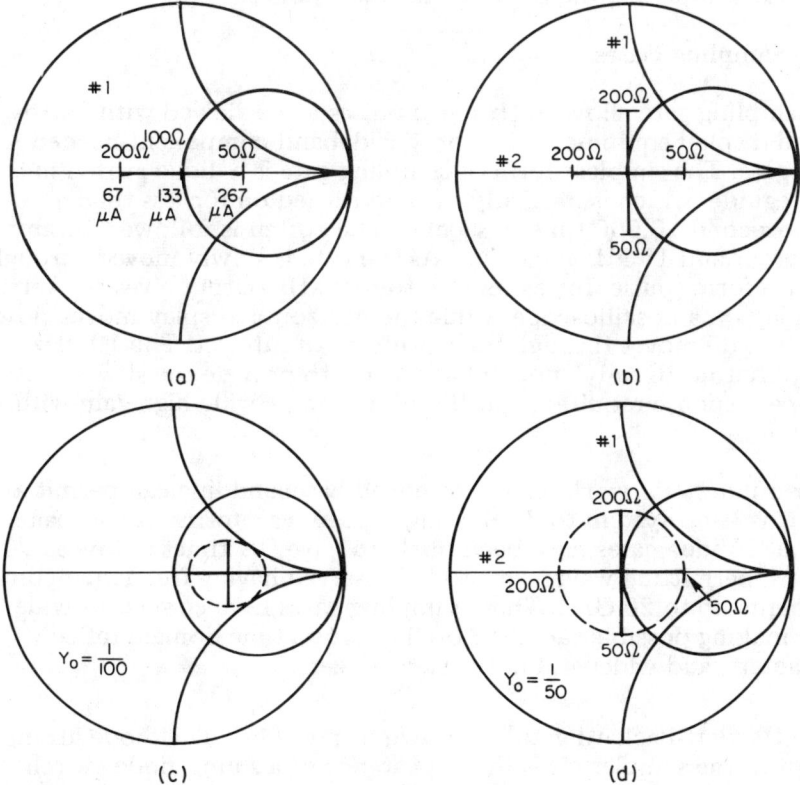

Fig. 11-3 Normalized Admittance of PIN Diodes Being Added $\lambda/8$ Apart

This vector generator is especially useful for simulating doppler. The two diodes are simply driven by a broadband video 90° phase shifter, and an antenna is connected to the reflection modulator instead of the circulator. When the two diodes are driven in quadrature, the voltage reflected from the modulator is constant in magnitude and continuously changing in phase. When the admittance of the modulator as shown in Fig. 11-3d describes a perfect circle with equally spaced reflection coefficient points occurring at equally spaced time intervals, then the output is a single frequency (single-sideband). The single sideband has amplitude proportional to the drive, and frequency displacement equal to the drive frequency. Gain compression is less than 1 dB up to the 10dB conversion loss point, as illustrated in Fig. 11-4. The most troublesome sideband in a doppler simulator is the -1 sideband (assuming the +1 sideband is the desired one). The -1 sideband is about 20 dB below the +1 sideband (20 dB suppression) up to the 10dB conversion loss point. At any given conversion loss the -1 sideband can be made to disappear completely by carefully adjusting the biases of the two diodes.

C. Sampling Gates

A sampling gate allows high speed pulses to be viewed with narrow band display equipment. The only wideband component needed is the gate. The simplest form of sampling gate is a diode switch in a waveguide, which is normally off and turned on for less than a nanosecond 10,000 times a second. The switch is followed by any detector and 10 kHz amplifier. As the gate is slowly moved through a waveform (pulse, for example), the 10 kHz output gives the vertical display on an oscilloscope, while the horizontal display moves slowly as the gate moves through the waveform of interest. The 10 kHz amplifier needs only enough bandwidth to provide the slow oscilloscope response required, and therefore can provide high gain with low noise.

A sampling gate working in a balanced baseband line can permit a narrow band system to display high speed waveforms of baseband signals. Video gates have been made to have ON times as low as 10 - 25 ps, permitting waveforms to be observed having frequency components up to 20 GHz. These sampling gates have come into wide use making possible sampling oscilloscopes, time domain reflectometers, and wideband network analyzers.

The Hewlett-Packard gate[4] is shown in Fig. 11-5. The diode arrangement in the sampler gives the appearance of a shunt diode switch, but the operation of the sampler is different. During the brief moment when the diodes are conducting, charge is transferred from the 50-ohm

OTHER CONTROL DEVICES

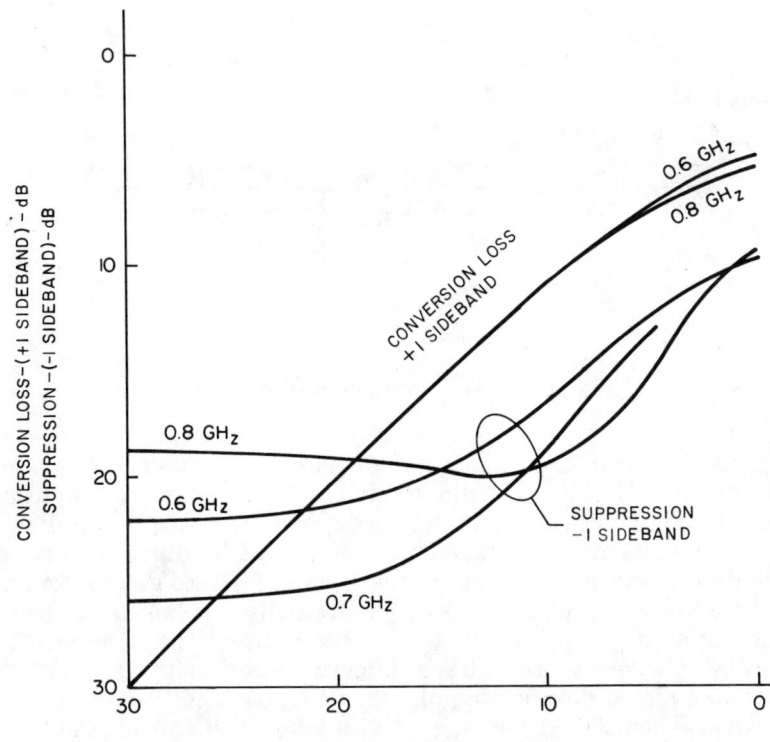

Fig. 11-4 λ/8 PIN Diode Single Sideband Modulator Conversion Loss (+1) and Suppression (-1).

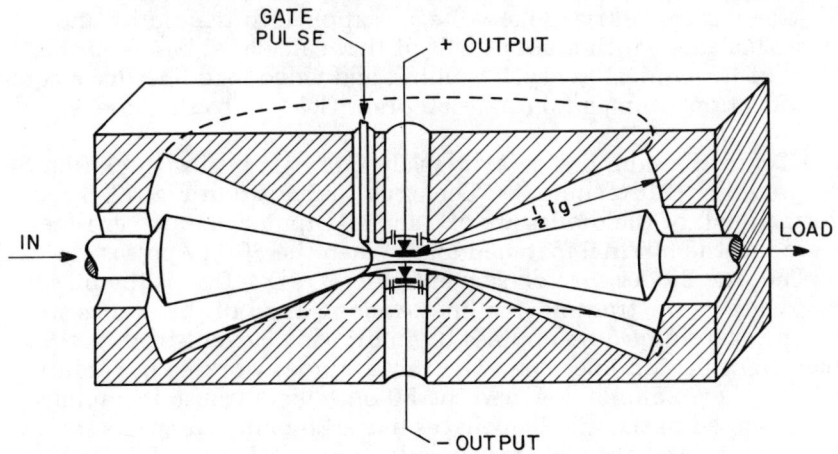

Fig. 11-5 Hewlett-Packard Sampling Gate

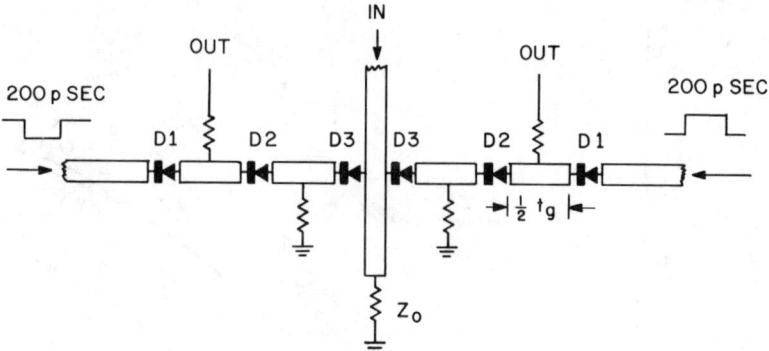

Fig. 11-6 Tektronix Sampling Gate

center conductor to the capacitors between the diodes and the sampler housing (ground). When the diodes are OFF, this charge is amplified by low frequency circuits on the output lines. The entire sampling unit is a conical cavity having a characteristic of 50 ohms with its axis on the diode output line. A step waveform applied to gate pulse line will cause the step voltage to propagate radially out on the conical transmission line until it meets the short circuit on the periphery of the cavity. The short circuit has a reflection coefficient of -1, which creates an equal and opposite pulse traveling back to the center of the cavity. When the -1 wave gets to the middle, it cancels the applied voltage bringing the diodes back to no applied voltage. The gate time, t_g, is fixed by how long it takes the pulse to propagate to the short and return. A 40 ps gate dictates a radius of 6 mm. By biasing the diodes off through the output ports and using a finite rise-time "step," a triangular voltage is applied to the diodes and the actual gate width is about half of that calculated based on the delay of the conical cavity. The input and video load lines have conical transitions to maintain Z_O = 50 ohms right up to the diodes.

The Tektronix sampling gate[5] is easily fabricated using thick film integrated circuit techniques. The circuit is shown in Fig. 11-6. Normally all of the diodes are off, and all input signal power is delivered to the internal matched load. When the 200 ps pulses are applied, the diodes turn on sequentially (D1, D2, D3) as the pulses propagate to the transmission line. As soon as all of the diodes are on, the two shunt 50-ohm lines begin drawing current from the main transmission line. The two sampling lines working in parallel behave like a 25-ohm line, and the 50-ohm main transmission line being tapped in the middle behaves like a 25-ohm source. As soon as the voltage from the main transmission line reaches the D1 diode, the trailing edge of the 200 ps pulse comes along and turns off the

OTHER CONTROL DEVICES

diodes sequentially. As D1 becomes an open circuit, the sampled wave is reflected with a reflection coefficient of +1, which doubles the voltage on the line segments between D1 and D2, By the time the +1 reflected sampled wave reaches D2, the turn-off pulse is also there and the charge of the sampled voltage is left standing on the line segments between D1 and D2. This charge is then conducted to the output circuits through the coupling resistor where it is amplified and processed for display. Diodes D3 enhance the isolation of the sampling line sections when the gates are off. The gate time is twice the propagation time of the sampling line section, which can be made quite small.

D. Bulk Diode Switches

Bulk diode switches have been the subject of study off and on since 1955.

Gunn and Hogarth[6] put a semiconductor slice across the waveguide with a current passing through it. One side of the semiconductor was shiny and had a long lifetime while the other side was rough and had a short lifetime. A magnetic field was applied perpendicular to the current so that when the current was in one direction the free carriers would be forced to the long lifetime surface, which would cause the slab to reflect incident microwave power. When the direction of current was reversed, the carriers would be forced to the short lifetime surface which would make the semiconductor slab appear as an insulator across the waveguide, and give low reflection and low insertion loss.

Arthur, et. al.,[7] devised a modulator taking advantage of the change in differential mobility when a high electric field is applied to a semiconductor.

De Ronde, et. al.,[8] developed a MM wave PIN diode modulator in which a semiconductor sheet with an extended I region was inserted through a longitudinal slit in the broad wall of reduced height waveguide. The inside of the waveguide saw only I region. With no bias, there was a low insertion loss. With modest bias, the I region became resistive, absorbing most of the incident power and providing an isolation state.

Jacobs, et al.[9] experimented with a semiconductor bar which passed through the broad waveguide walls. Its conductivity was modulated by an injection probe coming in from the middle of the narrow wall to contact the semiconductor bar.

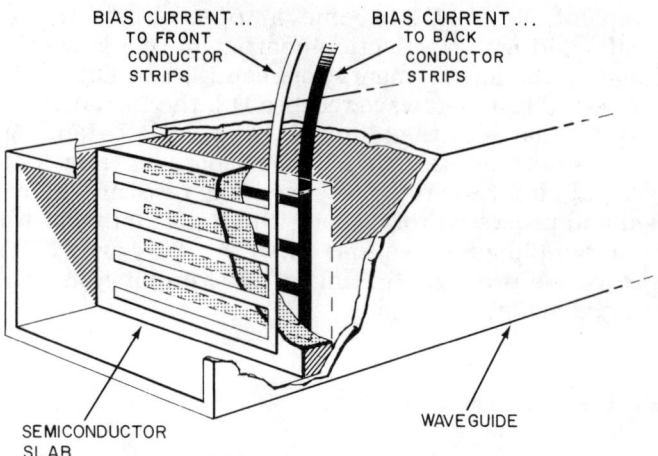

Fig. 11-7 Bulk Diode Switch

Mortenson[10] and Gutman[11] have been developing a distributed junction switch as shown in Fig. 11-7. The thin slice of semiconductor completely blocks the waveguide. When no free carriers are injected into the slice it becomes an insulator passing all incident power. One face of the semiconductor has horizontal conductors on it with long PI junctions beneath them. The other face has horizontal conductors with long IN junctions beneath them, The fingers are staggered so that each IN junction lies equidistant from two PI junctions on the other side of the slice. When the structure is biased into conduction the entire slice of semiconductor becomes very conductive and reflects incident power. This device has successfully controlled 1 MW peak, switching in 3 μsec, and giving 0.25dB insertion loss and 27 dB isolation over most of the waveguide bandwidth.

REFERENCES

1. R.B. Mouw and S.M. Fukuchi, "Broadband Double Balanced Mixer/Modulators," *Microwave Journal*, pt. 1, pp 131-134, March 1969; pt. 2, pp. 71-76, May 1969

2. R.B. Mouw, "A Broad-Band Hybrid Junction and Application to the Star Modulator," *IEEE Trans, on Microwave Theory and Techniques*, vol. MTT-16, pp. 911-918, Nov. 1968.

3. R.V. Garver, "PIN Diode Single-Sideband Modulator," *IEEE-GMTT 1970 International Microwave Symposium Digest*, IEEE Catalog No. 70C10-MTT, pp. 235-238, May 1970.

OTHER CONTROL DEVICES

4. W.M. Grove, "A DC to 12.4 GHz Feedthrough Sampler for Oscilloscopes and Other RF Systems," *Hewlett-Packard Journal*, pp. 12-15, Oct 1966.

5. G. Frye, "A New Approach to Fast Gate Design," *Tektronix Service Scope*, No. 52, pp. 8-9, Oct 1968.

6. J.B. Gunn and C.A. Hogarth, "A Novel Microwave Attenuator Using Germanium," *J. Appl. Phys.*, vol. 26, p. 353, Mar. 1955.

7. J.B. Arthur, A.F. Gibson, and J.W. Granville, "The Effect of High Electric Fields on the Absorption of Germanium at Microwave Frequencies," *J. Electronics*, vol 2, p. 145, Sept. 1956.

8. F.C. DeRonde, H.J.G. Meyer, and W.W. Memelink, "The PIN Modulator, an Electrically Controlled Attenuator for MM and Sub-MM Waves," *IRE Trans. on Microwave Theory and Techniques*, vol. MTT-8, pp. 325-327, May 1960.

9. H. Jacobs, F.A. Brand, M. Benanti, R. Benjamin, and J. Meindl, "A New Semiconductor Microwave Modulator," *IRE Trans. on Microwave Theory and Techniques*, vol. MTT-8, pp. 553-559, Sept. 1960.

10. K.E. Mortenson, J.M. Borrego, P.E. Bakeman, Jr., and R.J. Gutmann, "Microwave Silicon Windows for High-Power Broad-Band Switching Applications," *IEEE Journal of Solid State Circuits*, vol. SC-4, pp. 413-421, Dec. 1969.

11. K.E. Mortenson, A.L. Armstrong, J.M. Borrego, and J.F. White, "A Review of Bulk Semiconductor Microwave Control Components." *Proc. IEEE*, vol. 59, pp. 1191-1200, Aug. 1971.

Diode Measurement

Appendix A

1. Annotated Bibliography

Diode measurement methods have gone from measuring the junction in a point contact mixer diode to the very exact modeling needed for diodes in mm wave integrated circuits. The need for better diode measurements has kept pace with the uses of microwave diodes. Originally they were made in cartridges with a "cat's whisker" and used as mixers. The number of uses then expanded. First, their application spread to switching and control devices; then, as varactors to parametric amplifiers, phase locked oscillators, and harmonic generators;

then as tunnel diodes to low noise amplifiers and oscillators and finally as Gunn and IMPATT diodes to high level oscillators and amplifiers.

The refinement in measurement began with the work of Waltz[1] who used the point contact diode cartridge which he characterized by putting opens, shorts, and resistors in the junction and using a particular reference plane, transformer and shunt susceptance to model the cartridge. Waltz also pointed out that a tuner could be placed in front of the diode and a short behind the diode to permit the diode junction to be measured directly—the tuner and short cancelling the transformer and a shunt susceptance. Leenov[2] improved on Waltz's method in that Waltz used the slotted line in the normal manner, RF power in one end of the line, unknown on the other end, and power out the probe. At bias voltage around 0V the nonlinearity of the mixer diode generates harmonics which introduce errors in the measured impedance. Leenov put the power in the probe and put his detector on the (normally) generator port. This lowered the test signal on the diode under test by 20 dB to 30 dB without lowering measurement sensitivity. Garver and Rosado[3] cleared up a problem encountered in the Waltz technique of uncertainty in the transformer ratio due to systematic deviation of the resistor impedance from the pure resistance line. The deviation was found to be caused by the compositional structure of the resistive dice (substituted for the semiconductor dye). The resistive dye is actually high conductivity carbon particles imbedded in an insulating glue matrix. At X-band the capacitance between the particles becomes significant, which causes the resistor impedance to be partly capacitive and slightly different than the DC resistance. Garver and Rosado also allowed the cartridge to be represented by a more general network taking the form of the normal cartridge parasitic elements (in waveguide series choke inductance, cartridge capacitance, whisker inductance). Thus far the diodes were very small junction devices, point contact or gold bonded diodes.

Varactors became the subject of measurement of Houlding[4]. To avoid measuring high VSWR, he tuned the diode so that it was matched at zero bias. The effective characteristic impedance of the slotted line is then R_s. The impedance is measured as varactor bias varies from heavy forward current to maximum reverse voltage. The impedance is rotated on the Smith Chart until the reverse bias impedance circle segment lies on the $R/Z_0 = 1$ circle. The Q at the measurement frequency is the difference between the forward bias reactance and the maximum reverse bias reactance. The cutoff frequency is the Q times the measurement frequency. This same method was developed independently by Harrison[5].

As varactors became better, their R_s became lower and the matching methods became less accurate because of losses added by the matching transformer to match into a high VSWR. Bakanowski[6] developed a transmission method of measuring varactors which measures R_s very accurately. The diode is connected to the center conductor of coaxial transmission line with a sliding short circuit behind it. The diode is then biased to any voltage and the sliding short is adjusted for maximum isolation on the main line. At this setting of the sliding short, the reactance of the diode is in series resonance with it, and only R_s is shunting the transmission line. Eq. (2-9) (Chap. II) gives the value of R_s directly.

$$\eta_{\omega_0} = 20 \log \left(\frac{Z_0}{2R_s} + 1 \right) \qquad \text{(A-1)}$$

Bakanowski observed about 0.4Ω more than his calibration resistors. This was attributed to the sliding short resistance and stub attenuation. Next Bakanowski, et. al.[7,8] added capacitance information to

APPENDIX A

their measurement system by noting the length of stub required, calculating its reactance, and from it calculating the reactance and capacitance of the diode being resonated. Blake, et. al.[11] made use of Bakanowski's method in measuring cooled varactors. This method was expanded by DeLoach[10] to measuring very high cutoff frequency varactors in X-band waveguide. Rather than adjust a short for shunt resonance the isolation was measured over a range of frequencies for each setting of reverse bias. The isolation at resonance gave R_s as above. The bandwidth $\Delta\omega$ of the isolation a few dB (>6 dB) below resonance gave the tuning inductance. Eq. (2-31) (Chap. II) gives the value of L as follows:

$$\eta_{\Delta\omega} = 10 \log \left[1 + \left(\frac{Z_0}{2(\Delta\omega)L} \right)^2 \right] \quad (A-2)$$

Junction capacitance C_D is then calculated knowing the resonant frequency ω_0 and L.

$$C_D = \frac{1}{\omega_0^2 L} \quad (A-3)$$

The reverse bias diode Q, Q_D, is equal to the Q of the resonant stub as follows: A series impedance $R + jX$ has a corresponding shunt admittance Y

$$Y = \frac{R}{R^2 + X^2} - j\frac{X}{R^2 + X^2} \quad (A-4)$$

This gives isolation in the shunt attenuation equation

$$\eta = 10 \log \left[\left(\frac{G}{2Y_0} + 1 \right)^2 + \left(\frac{B}{2Y_0} \right)^2 \right] = 10 \log \left[\left(\frac{\frac{1}{2}RZ_0}{R^2 + X^2} + 1 \right)^2 + \left(\frac{\frac{1}{2}XZ_0}{R^2 + X^2} \right)^2 \right] \quad (A-5)$$

When $\eta_{\Delta\omega}$ is taken at the 3-dB points then $\eta_{\omega_0} = \eta_{\Delta\omega} + 3\text{dB}$ giving

$$\left(\frac{Z_0}{2R} + 1 \right)^2 = 2\left[\left(\frac{\frac{1}{2}RZ_0}{R^2 + X^2} + 1 \right)^2 + \left(\frac{\frac{1}{2}XZ_0}{R^2 + X^2} \right)^2 \right] \quad (A-6)$$

For high isolation (>20 dB) the "+1" terms can be neglected and equation (A-6) reduces to

$$R = X \quad (A-7)$$

R is R_s and X is ωL or over a bandwidth $(\Delta\omega)L$

$$R_s = (\Delta\omega)L \quad (A-8)$$

Using the definition of diode cutoff frequency: $\omega_c = \dfrac{1}{R_s C_D}$; and equation (A-3) in equation (A-8) give:

$$\frac{\omega_c}{\omega_0} = \frac{\omega_0}{\Delta\omega} \quad \text{or} \quad Q_D = Q \tag{A-9}$$

Another refinement was later made by Swenson and DeLoach[15] for the coaxial method which uses a sliding short. The effective R_s added by the variable length stub was accurately determined so it could be easily subtracted from that measured.

When it became possible to make varactor diodes with epitaxial processes, it became evident that part of R_s varies with reverse bias and is lower at higher reverse bias. Keywell[9] developed a method to measure cutoff frequency while low frequency AC is applied to the junction sweeping over the desired range of operation. The cutoff frequency thus measured is dynamic and matches that which it would demonstrate in a parametric circuit.

None of the above methods of measurement accurately represent the diode cartridge in any detail. This new work was started by Roberts[12], who was very careful in designing his mount and in avoiding discontinuities between the diode and the probe. His equivalent circuit, however, was still very simple. Van Iperen and Tjassens[25] later designed an even more precise bridge for measuring the diode obtaining accuracies down to ±0.1 ohm for R_s. Houlding[16] showed how more detail in the diode equivalent circuit could improve the fit between data and theory, pointing out how some of the earlier equivalent circuits could lead to 20%-40% error in R_s. Getsinger[17] went into great detail, calculating the equivalent circuit elements of a diode based on the details of its mounting and case structures. Van Iperen[21] improved on an area of Getsinger's work, in that a mode he assumed was cut off was found not to be cut off and was significant at X-band.

Another method of measuring a varactor was developed by Uhlir and advanced by Doherty[18] and Vendelin and Robinson[19]. Called the hyperbolic or power reflection technique, the diode is mounted at the end of a tapered line to a low characteristic impedance Z_0. For the reflected power technique VSWR is then measured at diode resonance (maximum reflected power). The VSWR is given by

$$\rho = \frac{Z_0}{R_s} \tag{A-10}$$

APPENDIX A

The resonance width and frequency give the other diode parameters as in the transmission method. The line loss between the diode and reflection reference short have to be taken into account for high f_c diodes. In a derivation similar to that used for Eq. (A-9), it can be shown that the resonant Q, as measured by the 3-dB bandwidth of the reflected power gives the diode Q, Q_D, as follows:

$$Q_D = \left(\frac{\rho}{2} - 1\right) Q. \tag{A-11}$$

For the hyperbolic distance method, the VSWR of the diode is measured at two different bias states, giving ρ_1 and ρ_2, and a distance between their minimum positions $\Delta \ell$ (giving phase $\psi = 2\pi \frac{\Delta \ell}{\lambda}$); The dynamic Q of the diode $\widetilde{Q}_D$ is given by:

$$\widetilde{Q} = \sqrt{\frac{(\rho_1 - \rho_2)^2 + (\rho_1^2 - 1)(\rho_2^2 - 1) \sin^2 \psi}{\rho_1 \rho_2}} \tag{A-12}$$

For a varactor diode, $\widetilde{Q}_D = Q_D$ when one bias is in the forward direction just before conduction current begins to flow. For a switching diode, $\widetilde{Q}_D$ is the quality factor that determines the switching limits as described in Chap. 4. Consider two special cases for Eq. (A-12):
Case 1: $\rho_1 = 1.0$

$$\widetilde{Q}_D = \frac{\rho_2 - 1}{\sqrt{\rho_2}} \approx \sqrt{\rho_2} \tag{A-13}$$

This agrees with Eq. (4-16) for the absorption switch with the substitution

$$\delta = 20 \log\left(\frac{\rho + 1}{\rho - 1}\right) \tag{A-14}$$

Case 2: $\psi = 90°$ (forward and reverse bias $\lambda/4$ apart) usually encountered with good switching diodes. Eq. (A-12) reduces to

$$\widetilde{Q}_D = \frac{\rho_1 \rho_2 - 1}{\sqrt{\rho_1 \rho_2}} \approx \sqrt{\rho_1 \rho_2} \tag{A-15}$$

As the operating frequencies of the diodes reached into the mm wave region the diode losses other than R_s became more prominent and had to be dealt with. Sard[20] derived equivalent circuits and methods to determine the elements having losses.

A very useful standard[24] has been published, clearly defining the three methods for measuring varactors (junctions only). The standard compares the methods and shows where care needs to be exercised with each method. The *first method* measures the diode as it is at the end of a transmission line. Exercises are carried out to model the black box between the junction and measurement probe. This is the method used in references 1, 2, 3, 9, 12, 16, 17, 18, 19, 20, 21, 25. The *second method* places a tuner in front of the diode so that the junction impedance is measured directly. The black box is tuned out by the transformer. This method was used in references 1, 4, 5. The *third method* measures the transmission loss through a shunt resonant diode as used in references 6, 7, 8, 10, 11, 15.

Several measurement methods have been derived which are only of use for switching diodes. Johnston, et al[13] and Galvin and Uhlir[14] developed methods for measuring the impedance of PIN diodes during the switching transient. Johnston, et al measured the voltage and current on the diode with resistive probes at 250 MHz. The point-by-point phases and amplitudes give the diode impedance. Galvin and Uhlir used a conventional bridge at 1000 MHz. The bridge output was viewed on an oscilloscope synchronized to the switching pulse. As the bridge was adjusted a null could be made to appear on the scope for only one short segment of the pulse at a time. The impedance of the diode at that time was the setting of the bridge. As the null was moved point-by-point through the pulse, the diode impedance was measured during the pulse.

Another measurement method unique to switching diodes was derived by Garver[22]. This method is disclosed in detail in Chapter V, Control Element Design, A. TEM Transmission Line, 4. Special Model for Shunt Diode Stubs, Measurement. The basic problem is that when a diode is measured at the end of a transmission line using VSWR methods, the residual VSWR of the measurement instrument introduces large phase errors[23]. A 1.2 residual VSWR will introduce a phase error as high as 20°. This method puts the diode on the end of a shunt stub and measures the transmission loss past the stub. The stub resonant frequencies are measured very accurately with and without the diode to give the diode reactance without intervening error mechanisms. Errors with the VSWR method were as high as 25% while with the resonant stub method they are as low as 0.2%.

Switching diode quality can also be measured from the tuned insertion loss δ and isolation η assuming the tuning elements are very low loss (non-absorption switch). The $\tilde{Q}$ is given by Eq. (4-8) and in Garver[26].

APPENDIX A

$$\widetilde{Q}_D = \left[\frac{10^{\eta/20} - 1}{10^{\delta/20} - 1} - 1 \right]^{1/2} \qquad (A-16)$$

A study of the measurement methods used on diodes in the past leads to certain basic principles to apply for the measurements to be most useful. 1. The diode should be mounted as it will be mounted in the intended application. Any contribution of the mounting structure to the diode impedance will then be included in the measurement. 2. Where phase is important the diode should not be measured using a VSWR-Smith Chart method, but rather using a transmission or reflection resonance method, unless precise corrections are made for intervening mismatches. 3. The diode should be measured as closely as possible to performing its intended function--only far enough away to give the generality needed for overall efficiency. In other words, for Principle 1 it makes no sense to measure the diode junction in great detail and the cartridge in great detail if they are always going to be used together in an intended application. They should be measured and modeled as a unit in the most practical mount that will be used. For Principle 2, some diode parameters (resistance vs. bias current) are not phase critical, so they can be measured using network analysers or slotted lines. BUT the reactance of the diode at forward and reverse bias is phase critical, so VSWR-Smith Chart methods should not be used for them without precise corrections. For Principle 3 a diode to be used over a wide bandwidth should be measured over a wide bandwidth.

2. Diode Measurement (Diode and Junction together)

For this requirement and those that follow, adequate diode measurements can be made with a computer corrected measurement system by putting the diode in a well designed mount and requiring the computer to make corrections. Care should be exercised in measuring varactor diodes and detector diodes so that the measurement AC voltage is significantly lower than the bias voltage. Switching diodes are normally poor rectifiers and not as sensitive to incident power level as varactors and detectors. Therefore, switching diodes may be measured with 1-10 mW incident power, while the others may be restricted to 1 μW (10mV AC). The remaining measurement methods presented here will give accurate measurements of switching diodes without necessarily using computer corrected measurement systems.

When a diode will be used in series (or shunt) in a transmission line, it is recommended that the diode be mounted as it will be used, and

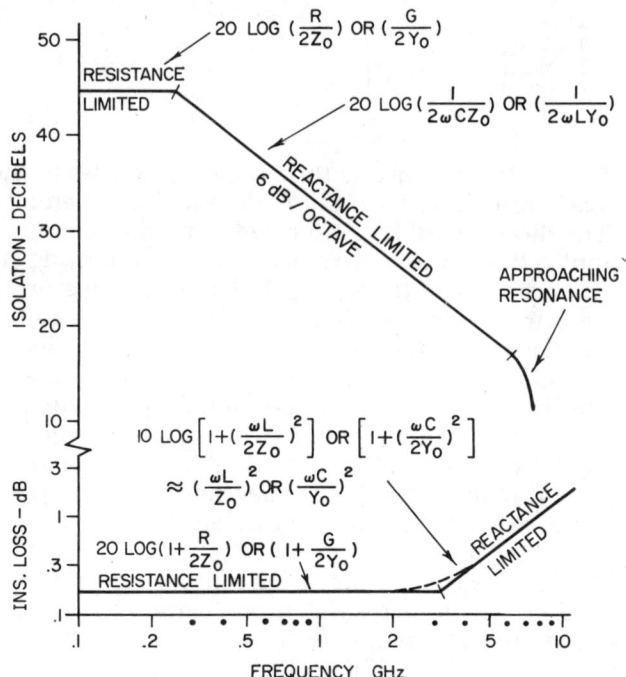

A-1 Influence of Diode Quivalent Circuit Elements on Baseband Switching

that the isolation and insertion loss versus frequency be measured and plotted as in Fig. 2-12. It will be obvious from these measurements if the characteristics are being resistively limited (insertion loss or isolation constant with frequency) or if a resonance occurs near the desired operating frequency as shown in Fig. A-1. Diode parameter equations were found using equations (2-8), (2-9), (2-31), (2-35), and (2-36).

For shunt stub applications the diode should be measured using the equivalent line length model and the measurement method described in V,A,4,b. This method can also give very complete impedance information on the diode by using all of the measured data to plot the diode impedance on the Smith Chart (or calculate it) as explained in Section V, A, 4. For example, consider the stub data at 1GHz in Fig. 5-12. The shunt 30-cm stub resonated at 1.000 GHz, giving 28dB isolation at resonance. With the forward biased diode on the stub the resonance occurred at 0.953 GHz and had 19 dB isolation at resonance. A 0.953 GHz resonance corresponds to a 31.5 cm long line, so the equivalent length of the forward biased diode is 1.5 cm. Using

APPENDIX A 305

Eq. (5-11) or Fig. 5-13, the 28 dB stub isolation indicates .18 dB stub attenuation, while the 19 dB with the diode indicates a total equivalent stub attenuation of .60 dB. The diode thus has .42 dB attributed to it which, according to Eq. (5-8) or Fig. 5-13, corresponds to a VSWR of 24. The 1.5 cm at 1 GHz (λ = 30 cm) gives ℓ/λ = .05. These combine to give a normalized impedance of .05 + j .33, which corresponds to 2.5 ohm in series with 2.6 nH.

3. Cartridge (and Mount) Measurement

Frequently the performance of a circuit will be limited by the diode cartridge or mount rather than the junction properties. To measure the cartridge (and mount), either cartridges with an open and short or a cartridge with a well characterized junction should be used. If possible, a cartridge with a resistor in place of the junction would also be helpful (necessary if the evaluation is to be done at a single frequency). Any of the above measurement techniques can then be applied to the calibration cartridges. At any frequency, the cartridge may be represented by a pi or tee equivalent circuit or any other arrangement of three independent elements. In broad-band applications, however, the cartridge will have to be represented by elements associated with the cartridge and mount structures as disclosed by Getsinger[17] and van Iperen[21].

4. Junction Measurement

The cost and performance of a diode are dependent on the quality of the junction. Therefore the diode junctions have to be measured with precision. The diode cartridge and mount are always between the measuring instrument and the junction. The cartridge and mount have to be either tuned out or evaluated. The following procedure will give a substantial degree of accuracy for most of the diodes currently available.

When the diode is carefully mounted, C_c, as shown in Fig. 2-3, is virtually zero. This is accomplished in a series mounted glass diode by having metal around the glass, which is well connected to ground. Having the center conductor of the coax or stripline run up to the glass also prevents extraneous parasitic elements from becoming significant. The pill diode is mounted by removing slightly more dielectric than required for the diode diameter. This extra dielectric removed creates a void of capacitance for the distributed capacitance of the transmission line. The void is then filled by the insertion of the diode with its small C_c. The capacitance removed from a

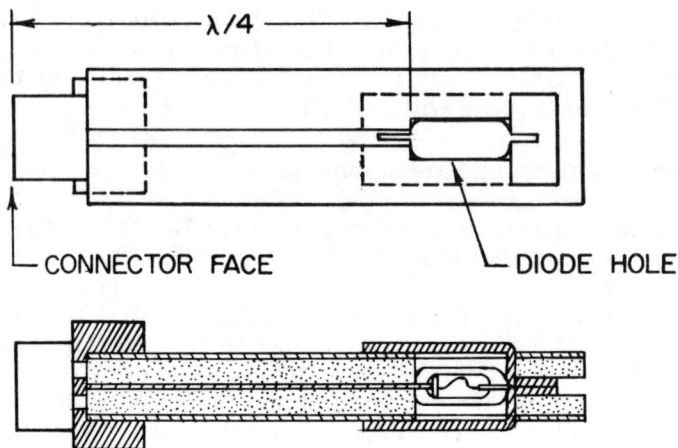

A-2 Stripline Measurement Mount for Glass Packaged Diode

transmission line by removing dielectric is $Z_0 \sqrt{\epsilon} \, (\epsilon-1)/c$ in which ϵ is the dielectric constant and $c(3 \times 10^{10}$ cm/sec) is the speed of light. 50-ohm stripline gives .45 pF/cm. In order to prevent the inductance from becoming unnecessarily large, the pill is buried in metal as much as possible to keep the center of the ceramic insulator centered between the center conductor and ground plane. Ideally, the distrance from center conductor to ground plane should be equal to the height of the exposed ceramic. With this careful mounting of the diode, the only parasitic element will be the series inductance (L_w as shown in Fig. 2-3).

The most common instruments for measuring impedance are the slotted line and network analyzer. A measurement mount for junctions in glass diodes is shown in Fig. A-2. The diode is placed $\lambda/4$ from the connector face so that the capacitive discontinuities associated with the connector will not introduce S_{11} phase error when the diode is a high impedance. (The admittance of an open circuit diode will have infinite admittance at the connector face. More or less shunt susceptance at the connector will not change this admittance[22].) An open circuit (no diode) is used for establishing a reference plane, because it will also have no connector induced phase error. The effective length of the open circuit is not corrected for

APPENDIX A 307

because it is still present during measurements. Any error in line length because of this very small error is included in the L_w that is measured and subtracted out later. Next, the diode is placed in the mount, crushing the dielectric where the input must now include the volume of the diode lead. The diode lead should be cut to one or two strip widths to permit crushing. Note in Fig. A-2 how the diode is surrounded by metal (soldered at its periphery to the ground plane), and how stout the ground return is behind the diode. A rule of thumb is that the width of the grounding sheet should be three times as wide as the diode hole. Another device to minimize C_c is to mount the diode so that the junction is adjacent to the input, and the whisker goes to the ground return. Any distributed capacitance associated with the whisker is thus kept shunted to ground by the whisker inductance.

The mount is connected to the measuring instrument directly. Any biasing must be done from the generator port of the slotted line, or with the terminals provided in the network analyser. Generator mismatch on a slotted line introduces no errors, so the quality of the bias T is not critical. But if the network analyzer does not have built-in bias provisions, considerable care should be exercised, because generator mismatch does introduce errors when impedance is measured with a directional coupler. When a slotted line is used for diode measurements as compared to an uncorrected network analyzer, there are fewer uncertainties.

When the diode is put into forward bias, it will be inductive. This inductance is transformed by the $\lambda/4$ to be a shunt capacitance at the connector face. More or less capacitance will be added by the imperfections of the connector. When converted back to the impedance plane of the diode input, the connector capacitance will be transformed to more or less inductance of L_w. Typical impedance measurements are shown in Fig. A-3. The figure shows $L_w = 2$ nH. Since the junction is being measured and not the cartridge, the fact that L_w contains connector imperfections is of no consequence. Connector imperfections giving a residual VSWR of 1.05 will cause an uncertainty of about $\pm 5°$ (i.e. $100°$ $(\rho_m - 1)$ in $\omega L_w/Z_0$). This inductive reactance could be from 0.2 to 0.3, or $.25 \pm 20\%$. No matter what $\omega L_w/Z_0$ is, what is measured is subtracted from the measured impedances to give the junction impedance very accurately. All of the significant errors have been combined with the measured L_w which is subtracted directly from the diode measured impedance.

In the event very high VSWR's are expected, line attenuation should be taken into account. This attenuation can be calculated from the

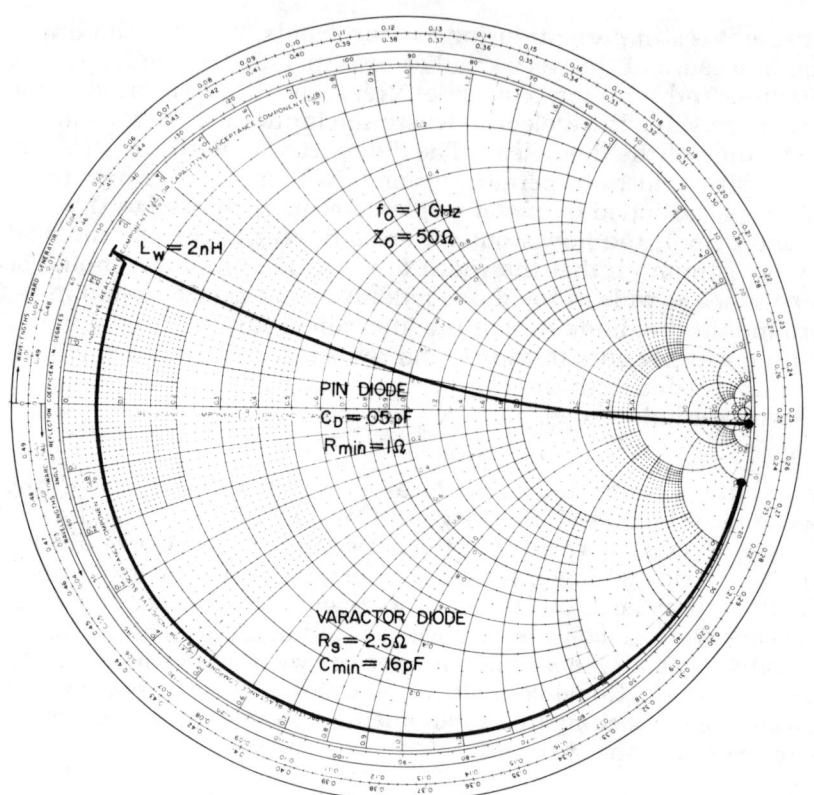

A-3 Typical Diode Impedances

VSWR of the open circuit. Since it is not radiating into free space, it should be infinite except for the attenuation between the open and the probe. This line attenuation could be compensated for by a gain adjustment on a network analyzer, but other error mechanisms are likely to be greater than line losses (on the analyzer). For high VSWR the actual VSWR, ρ_A, of the device being measured is related to the measured VSWR, ρ_M, and the calibrating VSWR, ρ_C, (of the open circuit) by the following relationship:

$$\frac{1}{\rho_A} = \frac{1}{\rho_M} - \frac{1}{\rho_C} \qquad \text{(A-17)}$$

or

$$\rho_A = \left(\frac{\rho_C}{\rho_C - \rho_M}\right) \rho_M \qquad \text{(A-18)}$$

APPENDIX A 309

For very high VSWR it will also be necessary to put a low-pass filter between the probe on the slotted line and the detector. Harmonics generated by the diode will be travelling from the diode to the generator. If the generator is a fair match at the harmonic, then any deep nulls at the VSWR minimums will be partially filled. The detector will never sense a power lower than the detected level of the harmonics.

When varactor or detector junctions are being measured, the test signal input should be the slotted line input port (to keep the test signal power low) as in Fig. A-4.

The measurement example above was for a diode in a glass cartridge. This example was selected because the glass package has the largest parasitics of the diodes in wide use. Diode junctions in other cartridges can be measured with fewer problems using the principles above to avoid unnecessary errors.

5. PIN Diode Resistance Measurement

PIN diode resistance for $R/Z_0 < 2$ is easily taken directly off of the Smith Chart. For $R/Z_0 > 5$ the resistance is more accurately calculated directly from the VSWR.

$$R = \rho Z_0 \tag{A-19}$$

When this is done the distortion of C_D is of no consequence. The current vs. resistance will have curves similar to those shown in Fig. A-5. The coefficient of RF conductivity α_{RF} can be calculated directly from any resistance point.

$$R = \frac{1}{\alpha_{RF} I} \tag{A-20}$$

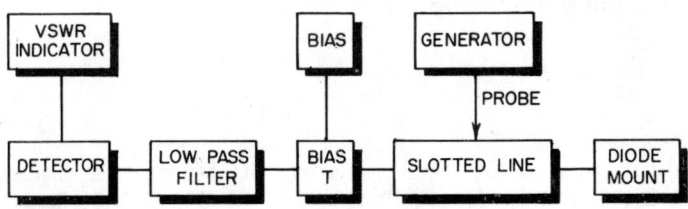

A-4 Microwave Circuit for Measuring Diode Impedance

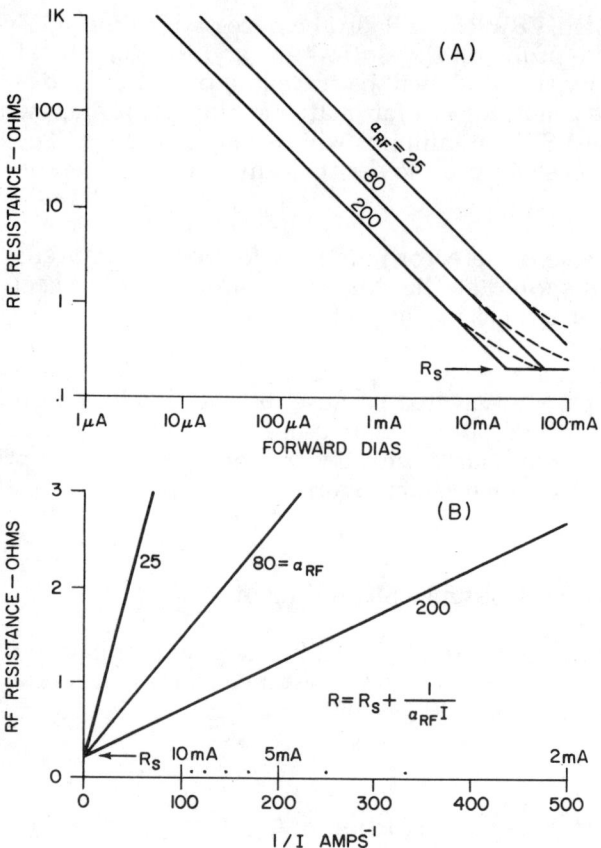

A-5 Current Dependence of Diode RF Resistance

It should be noted in Fig. A-5 (a) that the current conductivity law is not exactly

$$\frac{1}{R} = \alpha_{RF} I^{1.0} \qquad (A\text{-}21)$$

but closer to the relationship

$$\frac{1}{R} = \alpha_{RF} I^{.87} \qquad (A\text{-}22)$$

The approximation of Eq. (A-21) is adequate for most applications when α_{RF} is selected to fit the data to Eq. (A-21) at the center of the resistance range of interest.

APPENDIX A

6. Diode Lifetime Measurement

The storage properties of the diode junction influence the switching time and limiting level of the diodes. The measurement circuit for diode lifetime is shown in Fig. A-6. The diode is mounted in a 50-ohm structure. The bias T is not of the ordinary type because it must pass the high frequency components of the fast rise pulse (0.5 nsec→700 mHz) and not cause droop for long pulses.

$$C \geqslant t/d \text{ Farads} \tag{A-23}$$

in which t is pulse length in seconds and d is droop in percent. For example, a 100 nsec pulse with 5% droop would require a capacitance $C \geqslant 100 \times 10^{-9}/5 = .02\ \mu F$. The inductor is similarly constrained

$$L \geqslant 10^4\ t/d \text{ Henrys} \tag{A-24}$$

The same pulse requires an inductor given by $L \geqslant 200\ \mu H$.

To measure lifetime the diode is forward biased, I_F (~10mA) causing charge Q to be stored in the depletion region (I layer)

$$Q = I_F\ \tau \tag{A-25}$$

in which τ is the carrier lifetime. When the reverse bias pulse reaches the diode, the junction appears as a short circuit until all of the free carriers are swept from the depletion region. The reverse current I_R is determined by the pulse voltage and the 100-ohm loop resistance. When the free charge is depleted, the current drops abruptly (time T_R). The charge pulled out of the diode is $I_R T_R$ which is equal to the charge stored, giving

$$\tau = \frac{I_R}{I_F} T_R. \tag{A-26}$$

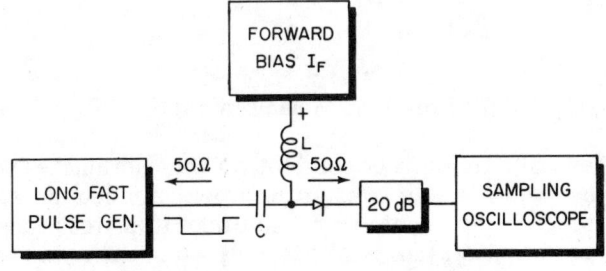

A-6 Diode Recovery Measurement Circuit

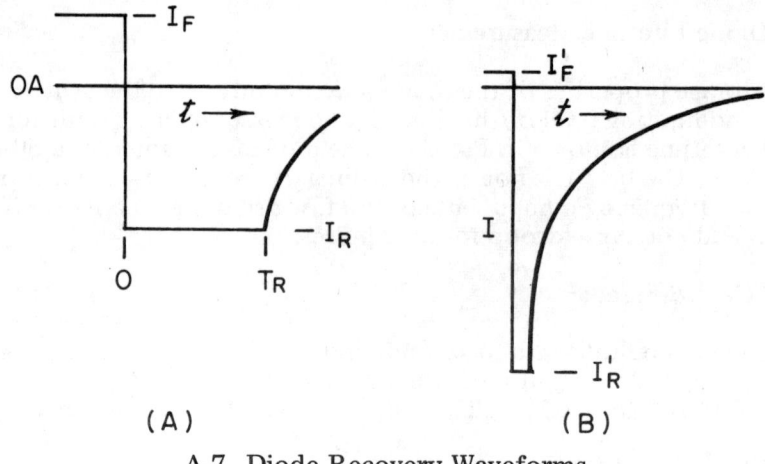

A-7 Diode Recovery Waveforms

I_F and I_R may be varied over a substantial range and still give the same τ. The reverse recovery pulse should be short enough to have a sharp break at T_R.

For small forward current (~1mA) and large reverse current (~50mA), another property of the recovery time can be observed. These are the trapped carriers which cause the exponential decay following the constant current part of the recovery pulse waveform as shown in Fig. A-7. These carriers are not free and come out of their traps when the trapping time constants release them.

They are not pulled out more quickly by higher reverse voltages. They are filled by forward current, and therefore the area under the curve is proportional to forward current as in Eq. (A-25) (the trapping lifetime will be different than the free carrier lifetime). This time constant may be determined from the plot of the reverse current as in Fig. A-8.

7. Power Dissipation Limit, P_D, Measurement

The power limit of a diode is dependent on the thermal resistance from the diode junction to a constant temperature heat sink. The thermal resistance is represented by θ in units of degrees Celsius per watt. The diode junction temperature is T_J while the heat sink temperature is T_s; then P_D is given by

APPENDIX A

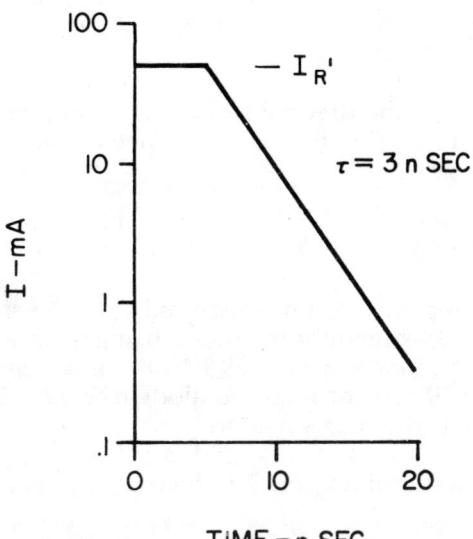

A-8 Plot of Slow Recovery Tail Showing Trapped Carrier Lifetime

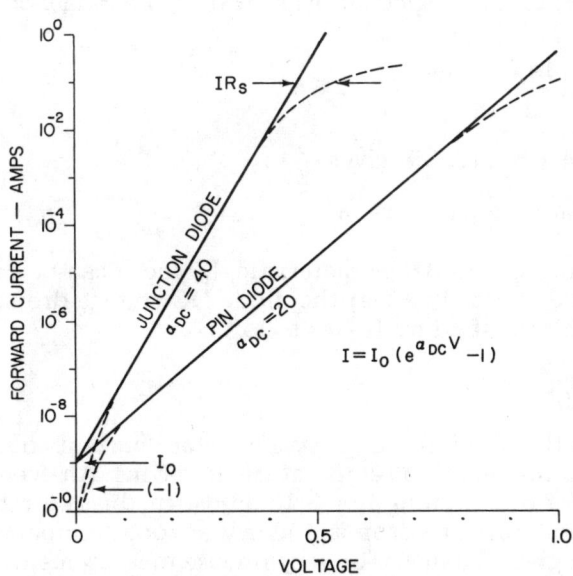

A-9 Diode Forward Characteristic

$$P_D = \frac{T_J - T_s}{\theta} \tag{A-27}$$

θ is comprised of θ_D, the thermal resistance from diode junction to diode exterior and θ_M, the thermal resistance of the diode mount.

$$\theta = \theta_M + \theta_D \tag{A-28}$$

Measurement of θ

Overall thermal resistance can be measured with the aid of a heater and thermometer. Fortunately the diode being measured can perform both functions. The diode has about 1 Volt voltage drop at forward bias. By putting 100 mA through the diode it will be dissipating 0.1 watt. Therefore, the diode is a heater.

The diode is also a thermometer. The forward characteristic is given by

$$I_F = I_0 \left(\epsilon^{\frac{qV_f}{nKt}} - 1 \right) \tag{A-29}$$

in which I_0 is leakage current ($\sim 10^{-8}$ Amps), T is temperature in degrees kelvin, q/k = 12,000° K/V, and n = 1 for rectifiers and varactors and 2 for PIN diodes. This characteristic is shown in Fig. A-9. The -1 in Eq. (A-29) can be neglected over most of the range. Solving for V_F gives

$$V_F = n \frac{\ln (I_F / I_0)}{12,000} T \tag{A-30}$$

For $I_F \approx 1mA$ Eq. (A-30) gives

$$V_F \text{ (in mV)} = nT \tag{A-31}$$

By measureing the diode characteristic, I_0 and n can be determined and I_F can be selected so that the diode DC voltage drops reads $1mV/°K$. This satisfied by I_{FT} as follows:

$$I_{FT} = I_0 \epsilon^{12/n} \tag{A-32}$$

Thus, when the diode is biased by a constant current source at I_{FT}, V_F in mV is the junction temperature in kelvins. An even simpler procedure for determining I_{FT} is to adjust the bias current on the diode until the forward drop is 300 mV at room temperature. 100°C should then give 370 mV. For very precise measurements a calibration curve should be run with the diode in an oven because I_0 changes slightly with temperature.

APPENDIX A

The procedure for measuring θ is to put the diode in the mount with the heat sink attached. Forward bias the diode to a power level that is less than its anticipated rated power level getting

$$P_D = I_F V_F \qquad (A\text{-}33)$$

After the temperature has stabilized, the bias to the diode is quickly reduced to I_{FT} (no more quickly than 10 lifetimes, about 20 μsec). The diode voltage will give the junction temperature T_J and the heat sink temperature T_S can be measured with a conventional thermometer. The overall thermal resistance is then given by

$$\theta = \frac{T_J - T_s}{I_F V_F} \qquad (A\text{-}34)$$

The diode junction as a thermometer can also be used to measure the thermal time constants of the diode. I_{FT} as described so far has been selected to give convenient temperature readings and to be low to prevent the temperature bias current from generating heat. To a good first approximation the diode can be subjected to a large current pulse, and the voltage waveform will give the diode temperature response. Unfortunately, the diode thermal time constant is on the order of the diode lifetime, so another method is more accurate. The bias current is adjusted so the diode junction is 50 ohms, and the diode mounted at the end of a 50 ohm transmission line with tuners to make sure the parasitic reactances are tuned out. The tuners will also match into the diode resistance in case exactly 50 ohms is not obtained. The diode is then subjected to short modest-power RF pulses at a low duty cycle while the voltage waveform on the diode is observed as shown in Fig. A-10. As the RF pulse length is increased, the peak temperature will increase linearly at first but then begin to reach a saturation temperature, T_H, which is the same as it would have for a constant CW power at the pulse peak as shown in Fig. A-11. The temperature should follow a curve described by

$$T = T_o + (T_H - T_o)(1 - \epsilon^{-t/\tau_{TH}}) \qquad (A\text{-}35)$$

in which T_o is the temperature just before the pulse and t is the pulse length. When T_H is known τ_{TH} can be measured from the slope of Fig. A-11 around $t = 0$.

$$\tau_{TH} = \frac{T_H - T_o}{T - T_o} t \qquad (A\text{-}36)$$

τ_{TH} is the intersection of the CW and erg limited power curves as given in Fig. 4-16.

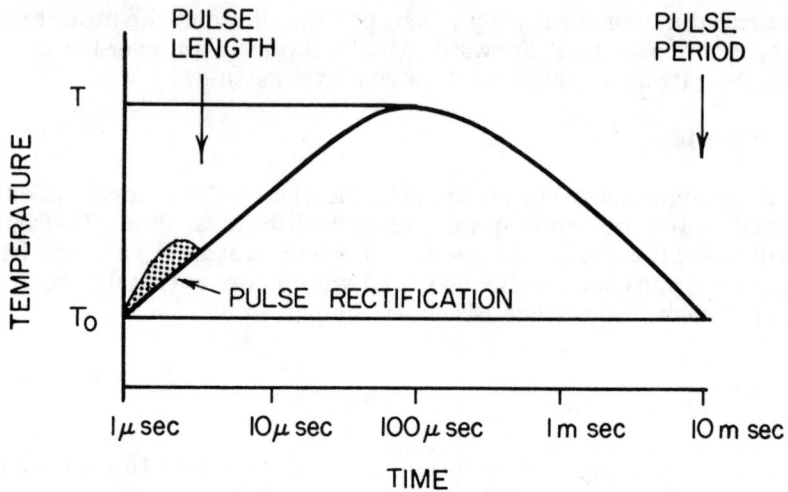

A-10 Diode Temperature Response to an RF Pulse

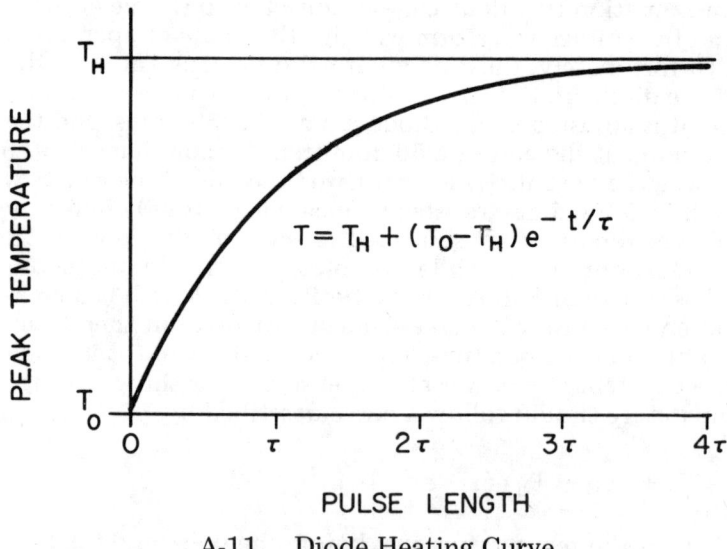

A-11 Diode Heating Curve

The delay of peak temperature shown in Fig. A-10 is caused by the heat being generated in the I-region and the temperature being measured at the PI and IN junctions at the ends of the I region. The heat must propogate the short distance to the junction, causing the slight

APPENDIX A

delay. The peak temperature is also slightly lower than the peak in the I region (estimate 10% lower). Significant damage from heat would not necessarily occur in the I region in any case, but would be more likely to occur at the PI and IN junctions where the doping gradients are steep.

The diode thermal resistance θ_D can be measured by mounting the diode on a good heat sink with a thermometer attached at the base of the diode. θ_D is measured as with Eq. (A-34).

The mount thermal resistance θ_M is measured using a bimetallic thermocouple, as shown in Fig. A-12. The diode is run at some constant power $I_F V_F$. The thermocouple measures the temperature difference between the two junctions (twisted wire pairs). One junction is put at the base of the diode T_M, while the other is put on the heat sink T_s, which is maintained at a constant temperature. The thermal resistance is given by

$$\theta_M = \frac{T_M - T_s}{I_F V_F} \tag{A-37}$$

Measurement of Maximum T_J

There are a number of factors that determine the maximum temperature a diode should be allowed to reach. The very worst case of diode burnout would be instant destruction due to high temperature. But a subtle factor that must be considered is the level of reliability required. The higher the temperature at which the diode is operated continuously, the more atoms will move around inside over a long period of time. If a diode is being built into a complex device that must operate continuously for 20 years it may be necessary for the diode have a 1000 year life expectancy. Only when these effects are taken into account will "solid-state reliability" be a reality.

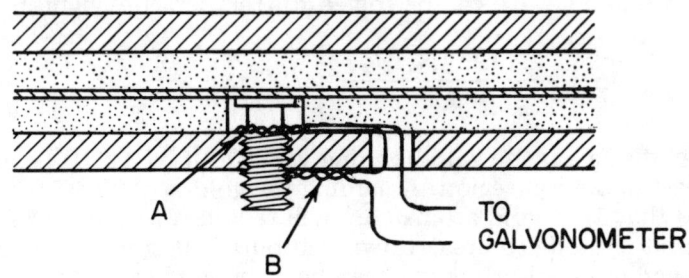

A-12 Structure for Measuring Diode Mount Thermal Resistance

The diffusion of atoms in a semiconductor is given by Lin[27]

$$\frac{N_x - N_0}{N_s - N_0} = \epsilon^{-\frac{x^2}{4Dt}} \qquad (A\text{-}38)$$

in which N_x is the impurity density after diffusion, N_0 is the impurity density before diffusion, N_s is the impurity density at the source, x is the distance from the source to a point in the medium, D is the diffusion constant, and t is the time for which diffusion has been allowed to take place. The equation is for a thin wall of N_s surrounded on both sides by N_0 at t = 0, but it is approximately correct (and easy to use) for two semi-infinite regions of N_s and N_0 in contact. For $N_0 \ll N_x < N_s$ Eq. (A-38) can be written

$$N_x = N_s \, \epsilon^{-\frac{x^2}{4Dt}} \qquad (A\text{-}39)$$

The diffusion coefficient is given by

$$D = A\epsilon^{-B/T} \qquad (A\text{-}40)$$

taken from the graphs of Sze[28], D is in units of cm^2/sec, A and B are given in the table below, and T is temperature in kelvin (K).

Diffusion Constants in S_i (Silicon)		
Element	A	B
Au (gold)	10^5	3.9×10^4
B (boron-P type dopant)	100	4.6×10^4
P (phosphorus-N type dopant)	50	4.6×10^4
A_S (arsenic-N type dopant)	10	4.6×10^4

Eqs. (A-39) and (A-40) can be solved for temperature giving

$$\frac{1}{T} = \frac{1}{B} \ln \left[\frac{4A\, t\, \ln(N_s/N_x)}{x^2} \right] \qquad (A\text{-}41)$$

A typical diode has the gold bonding wire .0005" (2×10^{-4} cm) from the closest depletion region. Assuming the gold is at $N_s = 10^{23}$ atoms/cm^3 and that 10^{16} gold atoms/cm^3 will reduce V_B to 50 V (which will cause the diode to break down and burn out completely in some high power application), then T can be calculated as a function of time as shown in Fig. A-13. To meet the 1000 year requirement, the

APPENDIX A

diode junction would have to operate at less than 415°C. Other factors may cause failure at a lower temperature. For example, the gold-silicon eutetic melts at 375°C. If the diode is experiencing any physical stress while above 375°C, the bond will break. If the diode is to operate for a shorter life, higher temperatures may be possible, but soon the intrinsic doping concentration will reduce V_B. For example, at 600°C the intrinsic doping concentration is 6×10^{15}, which will lower V_B to about 80 V. Bakanowski[29] did some burnout experiments of the type diode described here, observing 40 minutes burnout at an estimated 650°C (within 50°C of the 700°C predicted in Fig. A-13).

The present method to prevent burnout is to keep diode junctions in the 100°C — 200°C maximum temperature range, having characterized them at room temperature. More power could be handled by diodes by studying their burn-out properties in more detail and by characterizing them at higher temperatures.

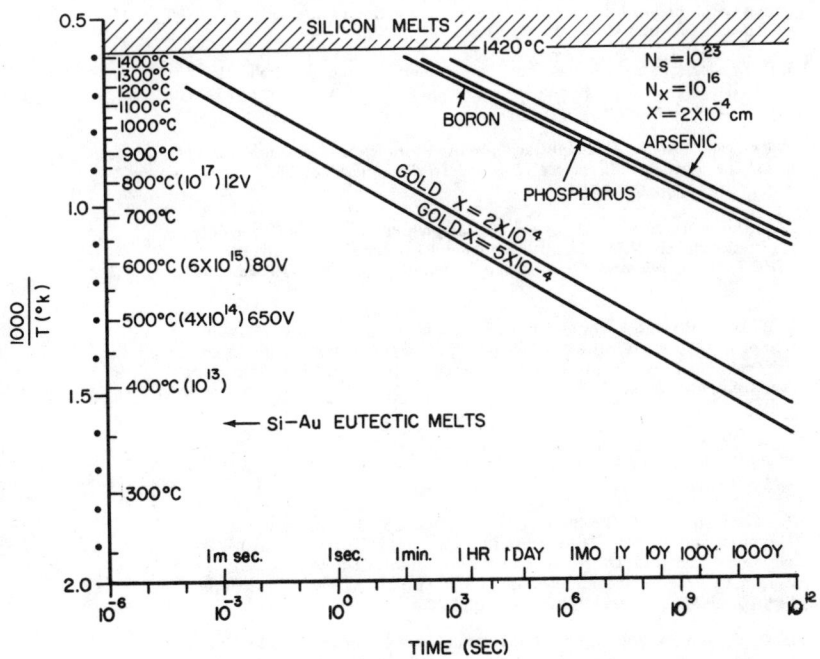

A-13 Junction Life Dependence on Temperature

REFERENCES

1. M.C. Waltz, "A Microwave Resistor for Calibration Purposes," Bell Telephone Labs., *Third Interim Report on Task 8 (Crystal Rectifiers)*, Contract DA-36-039-SC-5589; April 1955. Also *Microwave Journal*, May 1959, pp. 23-27.

2. D. Leenov, "Small-Signal Admittance of a Gold-Bonded Diode," Bell Telephone Labs., *Tenth Interim Report on Task 8 (Crystal Rectifiers)*, Contract DA-36-039-SC-5589; January 1957.

3. R. V. Garver and J. A. Rosado, "Microwave Diode Cartridge Impedance," *IRE Trans. on Microwave Theory and Techniques*, Vol. MTT-8, January 1960, pp. 104-107.

4. N. Houlding, "Measurement of Varactor Quality," *Microwave Journal*, January 1960, pp. 40-45.

5. R. I. Harrison, "Parametric Diode Q Measurements," *Microwave Journal*, May 1960., pp. 43-46.

6. A. E. Bakanowski, "Discussion of the Transmission Method for Small Signal Characteristics of Varactor Diodes," Bell Telephone Labs., *Fifth Interim Report on Microwave Solid-State Devices*, Contract DA-36-039-SC-85325, August 1961.

7. A. E. Bakanowski, J. H. Forster, and J. C. Irvin, "Determination of Varactor Diode Capacitance at 1 GC Using Transmission Techniques," Bell Telephone Labs., *Eighth Interim Report on Microwave Solid-State Devices*, Contract DA-36-039-SC-85325, May 1962.

8. J. H. Forster, "Varactor Impedance Measurement Techniques," Bell Telephone Labs., *Final Report on Microwave Solid-State Devices*, Contract DA-36-039-SC-85325, September 1962.

9. F. Keywell, "On the Resonant-Cavity Method for Measurement of Varactors," *IRE Trans. on Microwave Theory and Techniques*, Vol. MTT-10, Nov. 1 1962, pp. 567-572.

10. B. C. DeLoach, "A New Microwave Measurement Technique to Characterize Diodes and an 800-Gc Cutoff Frequency Varactor at Zero Volts Bias," *IEEE Trans. on Microwave Theory and Techniques*, Vol. MTT-12, Jan. 1964, pp. 15-20.

11. C. Blake, L. Bowles, F. Dominick, W. Getsinger, and E. McCurley, "Evaluation of Loss Characteristics of High-Quality Varactor Diodes," *1964 International Solid-State Circuits Conference Digest*, pp. 42-43, Also *Microwaves*, Jan. 1965, pp. 18-23.

12. D. A. E. Roberts, "Measurement of Varactor Diode Impedance," *IEEE Trans. on Microwave Theory and Techniques*, Vol. MTT-12, July 1964, pp. 471-475.

13. R.L. Johnston, D. F. Ciccoletta, and B. C. DeLoach, Jr., "Techniques for Measuring the Transient Microwave Impedance of PIN Diodes." Bell Telephone Labs., *Eighth Quarterly Report on Microwave Diode Research*, Contract DA-36-039-SC-89205, Sept. 1964; Also *1965 International Solid-State Circuits Conference Digest*, pp. 102-103.

14. R. Galvin and A. Uhlir, Jr., "Transient Microwave Impedance of PIN Switching Diode," *IEEE Trans. On Electron Devices*, Vol. ED-11, Sept. 1964, p. 441.

APPENDIX A

15. R. C. Swenson and B. C. DeLoach, "Microwave Diode Characterization-Correction Considerations for a Fixed-Frequency Transmission Measurement," Bell Telephone Labs., *Eleventh Quarterly Report on Microwave Diode Research*, Contract DA-36-039-SC-89205, June 1965.

16. N. Houlding, "Varactor Measurements and Equivalent Circuits," *IEEE Trans. on Microwave Theory and Techniques*, Vol. MTT-13, Nov. 1965, pp. 872-873.

17. W. J. Getsinger, "The Packaged and Mounted Diode as a Microwave Circuit," *IEEE Trans. on Microwave Theory and Techniques*, Vol. MTT-14, Feb. 1966, pp. 58-69.

18. W. E. Doherty, Jr., "Precise Cutoff Frequency Characterization of Varactor Diodes," *1966 International Solid-State Circuits Conference*, pp.74-75.

19. G. D. Vendelin and S. A. Robinson, "A Power Reflection Technique for Characterization of High Quality Varactor Diodes," *IEEE Trans. on Microwave Theory and Techniques*, Vol. MTT-14, Dec. 1966, pp. 603-608.

20. E. W. Sard, "A New Procedure for Calculating Varactor Q from Impedance Versus Bias Measurements," *IEEE Trans. on Microwave Theory and Techniques*, Vol. MTT-16, Oct. 1968, pp. 849-860.

21. B. B. van Iperen, "Impedance Relations in a Diode Waveguide Mount," *IEEE Trans. on Microwave Theory and Techniques*, Vol. MTT-16, Nov. 1968, pp. 961-963.

22. R. V. Garver, "New Model for Microwave Diodes," *1969 European Microwave Conference*; also *Harry Diamond Laboratories Tech. Report 1460*, Jan. 1970.

23. R. V. Garver, D. E. Bergfried, S. J. Raff, and B. O Weinschel, "Errors in S_{11} Measurements due to the Residual SWR of the Measuring Equipment," *IEEE Trans. on Microwave Theory and Techniques*, Vol. MTT-20, Jan 1972, pp. 61-69.

24. IEEE Standard No. 318, "Varactor Measurements," *IEEE Trans. on Electron Devices*, Vol. ED-18, Sept. 1971, pp. 816-830.

25. B. B. van Iperen and H. Tjassens, "An Accurate Bridge Method for Impedance Measurements of Impatt Diodes," *Microwave Journal*, Nov. 1972, pp. 29-33.

26. R. V. Garver and M. E. Hines, "Fundamental Limitations in RF Switching Using Semiconductor Diodes," *Proc. IEEE*, Vol. 52, Nov. 1964. pp. 1382-1384.

27. H. C. Lin, *Integrated Electronics*, Holden-Day, San Francisco, 1967, p. 128.

28. S. M. Sze, *Physics of Semiconductor Devices*, Wiley Interscience, New York, N. Y., 1969, p. 31.

29. A. E. Bakanowski, "Some Thermal Considerations in the Design of Graded Junction Silicon Varactor Diodes of the Mesa Type," Bell Telephone Labs., *Ninth Report on Microwave Solid-State Devices*, Contract DA-36-039-SC-73224, May 1959.

Diode Junction Theory

Appendix B

1. Basic Relationships

In a high concentration of thermally excited particles there is a tendency for those particles to diffuse until a uniform distribution exists over the larger confining volume. The diffusion particle motion is described by the diffusion coefficient D.

$$\frac{\partial N}{\partial t} = D \frac{\partial^2 N}{\partial x^2} \tag{B-1}$$

For silicon, D varies about 10 cm^2/sec depending on doping density, temperature, and whether the doping is N- or P-type.

When a piece of N-type semiconductor is joined to a piece of P-type semiconductor, the high concentration of electrons at the junction attempts to diffuse into the P-type material. The electrons, being loosely held to the donors of the N-type semiconductor, don't get very far. When they escape, they leave behind a positive charge. The field from the positive charge serves to counteract the diffusion so that some equilibrium is reached with a finite number of electrons just over the junction into the P-type material. The penetration of the electrons is found by equalizing the current due to diffusion and the current induced by the field[1] giving:

$$I_n = q \left[D \frac{\partial n}{\partial x} - \mu n \frac{\partial^2 \psi}{\partial x^2} \right] = 0 \tag{B-2}$$

where q is electronic charge, n is electron density, μ is mobility (~400 cm^2/V-sec) and ψ is electrostatic potential. A similar argument applies to the holes drifting into the N-type material.

The number of free electrons in the N-type semiconductor is given approximately by the Boltzmann energy distribution,

$$n = n_i e^{\frac{q}{kT}(\psi - \varphi_n)} \qquad (B\text{-}3)$$

where ψ is the potential mid-way between the valence band and conduction band, and φ_n the Fermi level of the electrons.

Although the electrons (holes) are free in that they have not recombined with any donors (acceptors), they are tightly bound near the junction by the electrostatic field and leave on the other side of the junction a layer of N-type (P-type) material depleted of free carriers. Having no carriers the depletion region is an insulator.

The width of the depleted region is calculated using Poisson's equation,[2]

$$\frac{d^2\psi}{dx^2} = -q \frac{n}{\epsilon}, \qquad (B\text{-}4)$$

where ϵ is the permittivity of the silicon semiconductor crystal
eq. (B-4) is solved by integrating two times with respect to x. The first constant of integration is found by requiring the electric field to be zero at the edges of the depletion region. The second constant of integration is found by requiring the potentials to be equal at the junction. The potentials ψ_p and ψ_n in the acceptor and donor materials must also be matched. The width for an abrupt junction is given by

$$h = \left[\frac{2\epsilon\phi}{q}\left(\frac{1}{N_a} + \frac{1}{N_d}\right)\right]^{1/2} \cdot \left[1 - \frac{V}{\phi}\right]^{1/2} = h_o \left(1 - \frac{V}{\phi}\right)^{1/2} \qquad (B\text{-}5)$$

in which V is applied voltage, ϕ is contact potential ($\varphi_n - \varphi_p$), and N_a and N_d are the acceptor and donor doping densities.

The depletion region becomes wider with increasing reverse bias because the applied voltage is added to the potential induced by the diffusion current. This added voltage pulls more carriers out of each side, the new carriers coming from the outer edges.

The junction width h is used to calculate the capacitance and breakdown voltage of the diode. Junction capacitance is given by

APPENDIX B

$$C_J = \frac{\epsilon A}{h} \quad \text{Farads} \tag{B-6}$$

in which A is the area of the junction; and breakdown voltage V_B is given by

$$V_B = h \,(5 \times 10^5 \text{ V/cm}) \tag{B-7}$$

where h is calculated at V_B.

The voltage current characteristic of the junction is dependent on the lifetime of the carriers τ which have been injected across the junction. The lifetime varies from greater than one microsecond to less than 0.1 nanosecond. The lifetime gives rise to a diffusion length L as follows:

$$L = \sqrt{D\tau} \tag{B-8}$$

A 400 nanosecond lifetime in silicon will give rise to a diffusion length of 20 microns. This is to be compared to a junction width of 8 microns (V_B = 50 V). A lifetime of 0.1 nanosecond, on the other hand, will give L = .3 microns. When h > L, diode theory applies. The carriers recombine before they get across the depletion region. The major voltage drop is at the junction. The current is determined by the density of carriers that surmount the barrier in accordance with their Boltzmann distribution, Eq. (B-3). The current is given by

$$I = I_B \left(e^{\frac{qV}{kT}} - 1 \right) \tag{B-9}$$

These short diffusion lengths are typical of metal-semiconductor junctions as with point contact diodes and Schottky barrier (hot carrier diodes). Because of their very short lifetime they are good rectifiers and very fast switches.

When h<L diffusion theory applies. The major voltage drop is beyond the depletion region. Eq. (B-2) (not equal to zero) applies and the current is given by

$$I = I_o \left(e^{\frac{qV}{kT}} - 1 \right) \tag{B-10}$$

in which I_o is quite different from I_B. I_o = f(D,n,L) while I_B = f(m_e,n) in which m_e is the effective mass of the carriers.

2. PIN Diode

A PIN diode is made by putting high conductivity P-type silicon on one side of a piece of intrinsic (undoped) silicon and high conductivity N-type silicon on the other side. At zero and reverse bias the diffusion and electrostatic potential combine to hold the electrons from the N-type silicon just inside the I region adjacent to the N side and the holes just inside the I region adjacent to the P side. P-I and I-N junctions are formed by the free carriers pulled out of the respective P and N materials. The net effect is two series diodes with the I region between them as shown in Fig. B-1. At forward bias the properties of the PIN junction will depend on carrier lifetimes in the different regions. When the lifetimes are low (L<d) the PIN diode behaves like two diodes in series at forward bias, which causes a DC characteristic as shown in Fig. B-2. For a given current, the voltage suffers two forward biased junction voltage drops. When the lifetimes are long (L>h), the forward biased diode behaves like a single diode. Modeling the junction becomes complicated when the lifetime in the I region is much higher than in the N and P regions. To a fair approximation, the PIN diode behaves somewhere in between: (L ≈ h).

a. Capacitance (non-conduction)

The impedance of the PIN diode at zero bias can be calculated from the equivalent circuit shown in Fig. B-3. The P-I and I-N junctions have high capacitance and high resistance because they are thin junctions and their depletion regions are free of carriers giving low leakage current. The I region, on the other hand, is relatively thick, giving low capacitance, and has some residual carriers, giving a lower shunt resistance. As reverse bias is applied and increased, the depletion regions of the PI and IN junctions increase in width until the entire I region is swept free of carriers. At maximum reverse bias, the entire PIN junction blends into one PI-IN junction which has capacitance approaching C_I and resistance approaching R_{PI} (R_{IN}).

At low frequencies the impedance of the I region is lower than that of the junctions, therefore, the low frequency non-conduction impedance will be dominated by the two junctions in series. At high frequencies the junctions are low impedances by virtue of their large capacitance, therefore, the high frequency non-conduction impedance is dominated by the I region.

The diode as shown in Fig. B-3 would have 5 pF in parallel with 20MΩ at low frequencies and 0.1 pF in parallel with 10 kΩ at high frequencies.

APPENDIX B

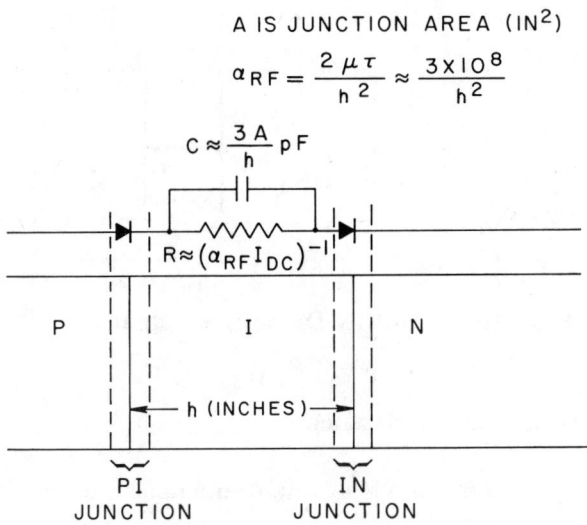

B-1 PIN Diode Equivalent Circuit

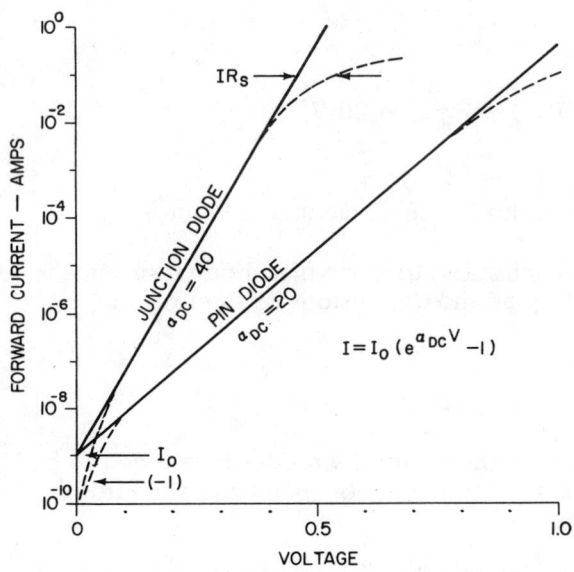

B-2 Diode DC Forward Characteristic

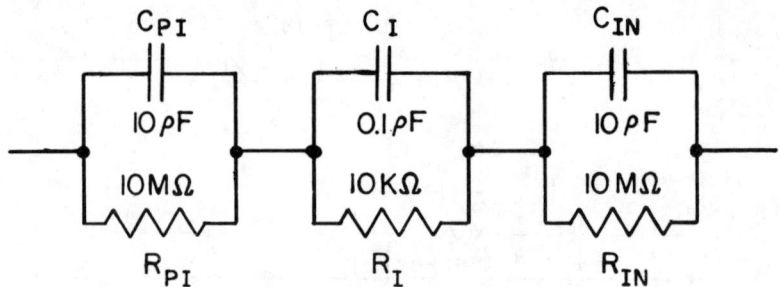

B-3 Zero Bias PIN Diode Equivalent Circuit

b. Forward Bias Resistance

The forward bias resistance R at low frequencies can be obtained by taking $\left(\frac{\partial I}{\partial V}\right)^{-1}$ from Fig. B-2. Most PIN diodes are close to the PIN curves, giving

$$R = \left(\frac{\partial I}{\partial V}\right)^{-1} = \frac{1}{\frac{1}{2}\frac{q}{kT} I} = \frac{1}{\alpha_{DC} I} \tag{B-11}$$

$$\alpha_{DC} = \frac{1}{2} \frac{q}{kT} = 20 \text{ V}^{-1} \tag{B-12}$$

in which α_{DC} is the DC coefficient of conductivity.

At high frequencies, the forward bias diode resistance is determined by the resistivity ρ and dimensions of the I region[3]

$$R = \rho \frac{h}{A} \tag{B-13}$$

The impedance of the PI and IN diodes is assumed to be lower than that of the I region at high frequencies and forward bias just as it is at non-conduction.

Resistivity ρ is the reciprocal of conductivity σ in which conductivity is given by

APPENDIX B

$$\sigma = nq\mu + pq\mu \approx 2nq\mu \tag{B-14}$$

where n and p are the electron and hole densities in the I region. Each electron injected into the I region stimulates a hole to be injected from the other side of the I region in order to maintain charge neutrality. Therefore, the injection rate is twice the bias current. When a diffusion limited diode is forward biased, the injected carriers recombine exponentially with lifetime τ. The total unrecombined charge Q is given by the integral of the current 2I, (injection rate) over the time the injected carriers are free.

$$Q = 2I\tau = (n + p) q\, Ah = 2nq\, Ah \tag{B-15}$$

The free charge density will be

$$n = \frac{I\tau}{qAh} \tag{B-16}$$

Combining Eq. (B-13) (B-14) and (B-16) gives

$$R = \frac{h^2}{2\mu\tau} \frac{1}{I} = \frac{1}{\alpha_{RF}\, I} \tag{B-17}$$

$$\alpha_{RF} = \frac{2\mu\tau}{h^2} \tag{B-18}$$

It should be pointed out that α_{RF} is a constant only to the first approximation. α_{RF} is dependent on μ, which is dependent on the carrier concentration in the I region. The mobility of holes in silicon varies from 600 cm²/V-sec for 10^{14} to 200 cm²/V-sec for 10^{18} at room temperature. The net effect is that $R \propto I^{-.87}$ instead of $R \propto I^{-1}$.

3. Diode Transient Properties

For a PIN diode to be put into the conduction state, a certain amount of charge must be injected into it in accordance with Eq. (B-15). A low impedance source can inject this charge without significant delays. Extracting the charge, however, is another problem.

The lifetime in the P and N regions is small compared to the lifetime in the I region. Referring to Fig. B-4, as the diode is pulsed into conduction, the carriers build up uniformly across the I region. Nothing resists them, and there is practically no diffusion tail beyond the I

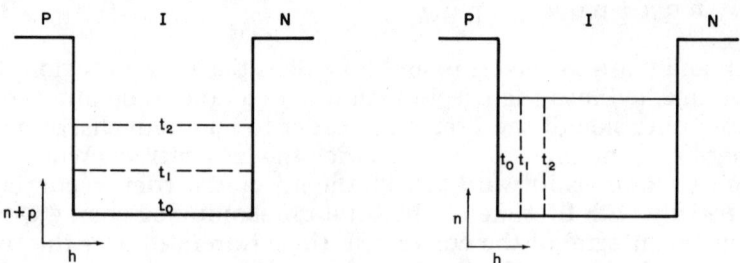

B-4 Stored Charge Distribution in the I-Region During Forward and Reverse Switching Transients

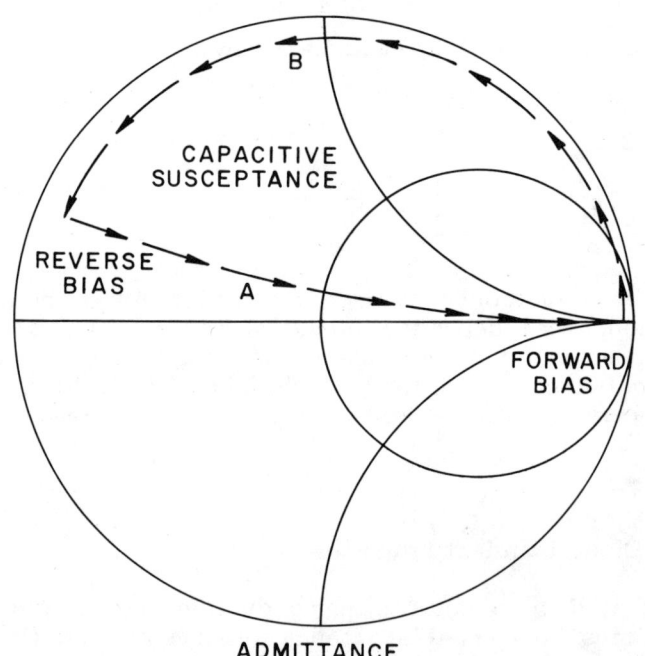

B-5 RF Impedance of a PIN Diode During Switching Transients

region into the P or N material. The RF admittance of the I region remains resistive throughout the turn-on transient,[4] as shown in Fig. B-5. When the diode is reverse biased, however, practically no elec-

APPENDIX B

trons come out of the P side into the I region (because of the short lifetime in the P material), so a depleted junction begins to form immediately at the PI junction. As time progresses, the electrons leave the I region as a wall of carriers heading back to the N side. There are many electrons between the moving wall and the N side, so that the structure is an insulator (the depleted I region) whose width is expanding, with conductors at either side. The holes are similarly moving back toward the P side with a widening insulator on the right. When the two walls meet in the middle the conductive volumes no longer overlap and the structure becomes a capacitor with seperating plates. The RF admittance is that of a capacitor going from infinite capacitance to some minimum, determined by the I region geometry (the steady state I region capacitance). This variable capacitance admittance is shown in Fig. B-5. When these carriers return to their respective junctions, the current stops, abruptly giving the property used in step recovery diodes.

REFERENCES

1. S. M. Sze, *Physics of Semiconductor Devices*, Wiley, N.Y., N.Y., 1969.

2. H. A. Watson, *Microwave Semiconductor Devices and Their Circuit Applications*, McGraw-Hill, N.Y., N.Y., 1969, p.102.

3. D. Leenov. "The Silicon PIN Diode as a Microwave Radar Protector at Megawatt Levels," *IEEE Trans. On Electron Devices*, Vol. ED-11, February 1964, pp. 53-61.

4. R. Galvin and A. Uhlir, Jr., "Transient Microwave Impedance of PIN Switching Diode, "*IEEE Trans. On Electron Devices*, Vol. ED-11, Sept. 1964, p. 441.

Filter Theory

Appendix C

Filter theory is useful for diode switching in two respects: (a) It is useful for analysing circuits containing the parasitic capacitance and inductance of diodes which limit the band-width of insertion loss and isolation. When these elements can be made part of a filter structure, these limitations can be significantly raised. (b) Filters are required to separate the modulation and carrier currents in high-speed switching. The modulation currents are at video frequencies while the carrier currents are centered about the microwave carrier frequency. Filters permit these currents to be sorted in the frequency domain.

Filter theory makes use of (1) a basic low-pass lumped-element filter prototype; (2) equations and tables for determining prototype element values based on characteristic impedance, maximum VSWR, or insertion loss and number of elements; and (3) rules for changing the prototype into high-pass or band-pass, for using transmission line elements in place of lumped elements (Richards equation), and for changing the microwave structure to be more realizable (Kuroda's Identity). Additional comments will be made in the appendix about the Bode matching integral limit.

Filter theory is covered very thoroughly in the "Filter Bible."[1]

1. Prototype

The basic prototype circuit is shown in Fig. C-1. The five-element filter (n = 5) has a maximum insertion loss δ_m (in dB) ripple with 2½ = $(\frac{n}{2})$ ripples, and rises above that maximum insertion loss at frequency ω_1'. As long as generator and load impedances are equal, the insertion loss will be zero at zero frequency, and the structure will re-

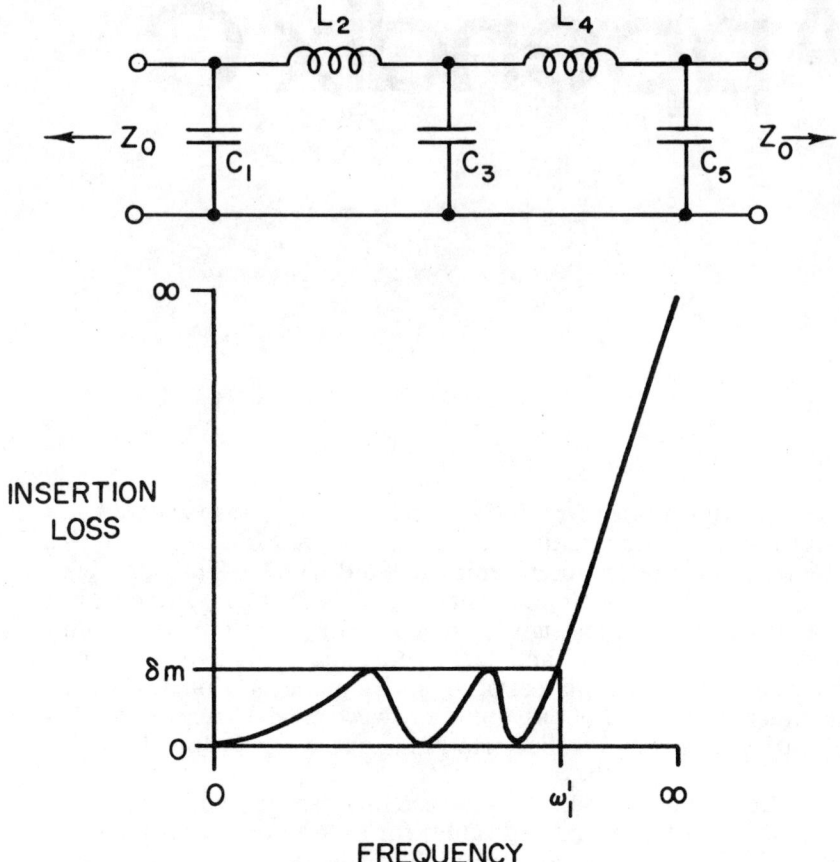

C-1 Low-Pass Filter Prototype

quire an odd number of elements.

The elements for a low-pass prototype must alternate between shunt capacitor and series inductor. The structure can begin with either inductor or capacitor. The index associated with each element is given by k ($C_1 \Rightarrow k = 1$). Then the following equations define the element values:

$$\frac{\omega_1' C_k}{Y_0} = g_k \tag{C-1}$$

$$\frac{\omega_1' L_k}{Z_0} = g_k \tag{C-2}$$

APPENDIX C

$$g_1 = \frac{2a_1}{\gamma} \tag{C-3}$$

$$g_k = \frac{4a_{k-1} a_k}{b_{k-1} g_{k-1}} \tag{C-4}$$

$$a_k = \sin\left(\frac{2k-1}{2n}\pi\right) \tag{C-5}$$

$$b_k = \gamma^2 + \sin^2\left(\frac{k}{n}\pi\right) \tag{C-6}$$

$$\gamma = \sinh\left[\frac{1}{2n} \ln\left(\coth\frac{\delta_m}{17.37}\right)\right] \tag{C-7}$$

$$\delta_m = 20 \log\left(\frac{\rho_m + 1}{2\sqrt{\rho_m}}\right) \approx (\rho_m - 1)^2 \tag{C-8}$$

in which ρ_m is the maximum VSWR. ω' is used for the prototype while ω is reserved for later use with transmission line elements.

The values of g_k are given in Table C-1 and Fig. C-2. Note that $g_k = g_{n-k+1}$ and that the g_k values are the normalized shunt susceptance or series reactance of the various elements at cutoff frequency.

Below ω_1' insertion loss is given by δ while above ω_1' it is considered isolation η which is given by the following:

$$\delta = 10 \log\left\{1 + \epsilon \cos^2\left[n \cos^{-1}\left(\frac{\omega'}{\omega_1'}\right)\right]\right\} \tag{C-9}$$

$$\eta_{ER} = 10 \log\left\{1 + \epsilon \cosh^2\left[n \cosh^{-1}\left(\frac{\omega'}{\omega_1'}\right)\right]\right\} \tag{C-10}$$

$$\epsilon = 10^{\delta_m/10} - 1 = \frac{(\rho_m - 1)^2}{4\rho_m} \tag{C-11}$$

$$\eta_{ER} \approx 6(n-2) - 10 \log\left(\frac{1}{\delta_m}\right) + 20n \log\left(\frac{\omega'}{\omega_1'}\right) \tag{C-12}$$

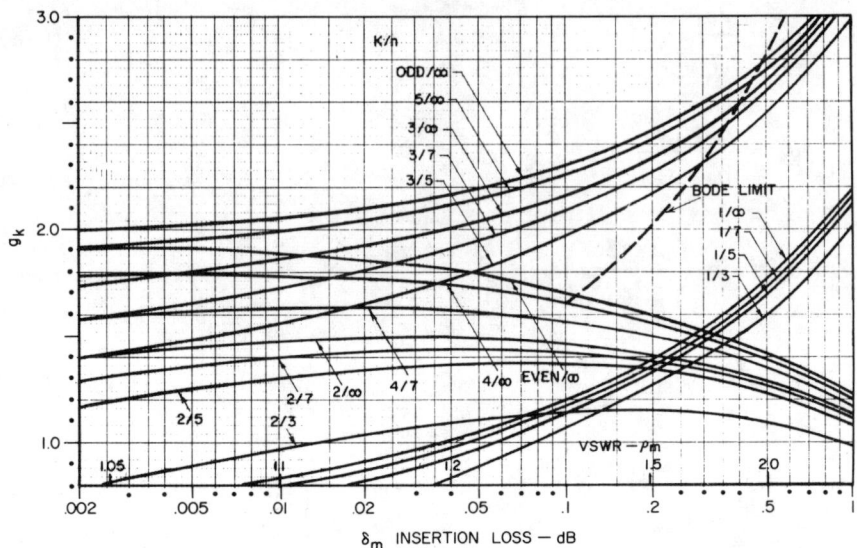

C-2 Tchebyscheff (Equal Ripple) Coefficients

For the maximally flat filter:

$$\delta = \eta = 10 \log \left[1 + \left(\frac{\omega'}{\omega'_1} \right)^{2n} \right] \tag{C-13}$$

$$\eta_{MF} \approx 20 n \log \left(\frac{\omega'}{\omega'_1} \right) \tag{C-14}$$

By defining the frequency at which the maximally flat filter gives insertion loss δ as ω'_δ, the isolation of Tchebysheff (equal ripple) and maximally flat filters can be compared. Eq. (C-14) becomes

$$\eta_{MF} \approx -6 - 10 \log \left(\frac{1}{\delta_m} \right) + 20n \log \left(\frac{\omega'}{\omega'_\delta} \right) \tag{C-15}$$

The equal ripple filter has more isolation than the maximally flat filter given by

$$\eta_{ER} - \eta_{MF} = 6(n-1) \tag{C-16}$$

It is interesting to note in Fig. C-2 that for filters with large n the middle elements are nearly the same and the only difference in elements is the matching elements on each end.

APPENDIX C

2. Transformations

a. High-Pass Prototype

The low-pass prototype is converted to a high-pass prototype by using series capacitors and shunt inductors as shown in Fig. C-3. The element values are to be found using the following equations:

$$\frac{\omega_1' C_k}{Y_0} = \frac{1}{g_k} \tag{C-17}$$

$$\frac{\omega_1' L_k}{Z_0} = \frac{1}{g_k} \tag{C-18}$$

Eqs. (C-3) - (C-12) are the same except of course that δ is above ω_1' and η is below ω_1'.

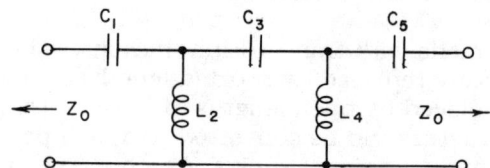

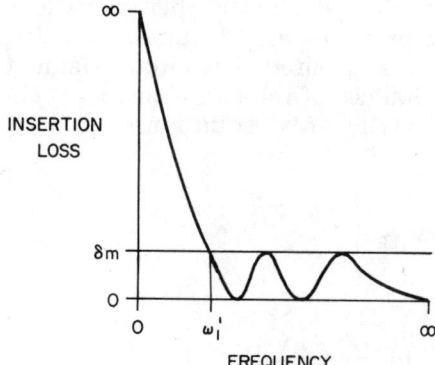

C-3 High-Pass Filter Prototype

b. Band-Pass Prototype

The low-pass prototype filter is converted to a band-pass filter simply by resonating all of the elements at the center (geometric) frequency of the pass-band. First the ripple, bandwidth (in MHz) and number of elements are selected to give the maximum insertion loss and bandwidth desired in the pass-band. This produces a low-pass prototype as in Fig. C-1. Then given upper frequency ω_U and lower frequency ω_L the center frequency ω_0 is calculated.

$$\omega_0 = \sqrt{\omega_U \omega_L} \tag{C-19}$$

Then inductors are selected to parallel resonate with the shunt capacitors and capacitors are picked to series resonate with the series inductors as follows:

$$L_1 = \frac{1}{\omega_0^2 C_1} \quad \text{and} \quad C_2 = \frac{1}{\omega_0^2 L_2} \quad \text{etc.} \tag{C-20}$$

The resonating elements simply shift the response up to a higher frequency.

c. Transmission Line Elements

When filters are made with transmission lines instead of lumped elements, inductors are replaced by shorted lengths of line ($<\frac{\lambda}{4}$) and capacitors are replaced by open lengths of line ($<\frac{\lambda}{4}$) as shown in Fig. C-4. This structure can be converted to a high-pass filter by making shorts opens and opens shorts. As a high-pass filter it is actually a band-pass filter because of the cyclic repetition of the attenuation profile. The stubs are a quarter wavelength long at ω_0 which produces infinite attenuation. The characteristic impedance of the stubs are selected to give the desired ripple. ω_0 is selected to put the isolation where most desired. The circuit elements are selected by setting equal the reactances of the two elements at the insertion loss band edge as follows (Richards[2] equations):

$$\frac{\omega_1' C_k}{Y_0} = g_k = \frac{Y_{0CK}}{Y_0} \tan\left(\frac{\pi}{2} \frac{\omega_1}{\omega_0}\right) \tag{C-21}$$

$$\frac{\omega_1' L_k}{Z_0} = g_k = \frac{Z_{0LK}}{Z_0} \tan\left(\frac{\pi}{2} \frac{\omega_1}{\omega_0}\right) \tag{C-22}$$

APPENDIX C

Eqs. (C-9)–(C-15) may be used with the frequency transformation derived as follows:

in general
$$\omega' L_k = Z_{0Lk} \tan\left(\frac{\pi}{2} \frac{\omega}{\omega_0}\right) \quad \text{(C-23)}$$

at ω_1 (ω_1')
$$\omega_1' L_k = Z_{0Lk} \tan\left(\frac{\pi}{2} \frac{\omega_1}{\omega_0}\right) \quad \text{(C-24)}$$

therefore in general

$$\frac{\omega'}{\omega_1'} = \frac{\tan\left(\frac{\pi}{2} \frac{\omega}{\omega_0}\right)}{\tan\left(\frac{\pi}{2} \frac{\omega_1}{\omega_0}\right)} \quad \text{(C-25)}$$

d. Kuroda's Identity[3]

The realization of filter structures with TEM transmission lines is often difficult without the aid of Kuroda's Identity. For example in order to realize the transmission line circuit shown in Fig. C-4

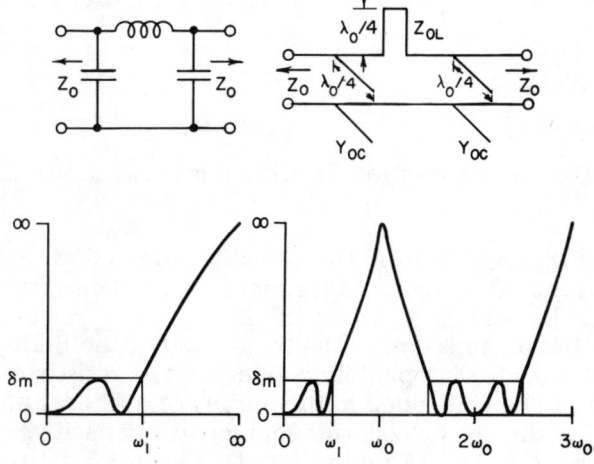

C-4 Transformation of Circuit Elements from Lumped Elements to Transmission Line Elements (Richards' Equation)[2]

it would be necessary to make a series shorted quarter wavelength stub. This may be done in coax by putting the stub back inside the center conductor of one of the lines in series with the stub. It is more difficult in stripline. Kuroda's identity allows the structure of Fig. C-4 to be transformed to all shunt stubs with the identical response. Fig. C-5 gives these identities. An example of the application of Kuroda's identity is given in Chapter III, "Biasing Circuits" section B, "Designing for Low VSWR."

3. Bode Matching Integral[4]

When a resistive element R has shunt capacitance C associated with it there is a limit of the bandwidth that it can be matched into. The limit is given by

$$\int_{\omega_a}^{\omega_b} \ln\left|\frac{1}{\Gamma}\right| d\omega = \frac{\pi}{RC} \tag{C-26}$$

in which the reflection coefficient is constant from ω_a to ω_b and unity elsewhere.

$$\ln\left|\frac{1}{\Gamma}\right| = \frac{\pi}{\Delta\omega\, CR} = \frac{\pi X_c}{Z_0} = \pi X_{CN} \tag{C-27}$$

in which $|\Gamma|$ is magnitude of reflection coefficient ($\rho = \frac{1 + |\Gamma|}{1 - |\Gamma|}$). This limit is plotted in Fig. C-6. Converting Γ to insertion loss

$$\delta = 10 \log \frac{1}{1 - |\Gamma|^2} \tag{C-28}$$

and using 50 ohm transmission line, the limit is as shown in Fig. C-7.

The point of this calculation is to show that the limitations on the bandwidth imposed by diode capacitance C in a transmission line of characteristic impedance Z_0 is not the Bode limit as might be generally applied but a higher limit. The reason is that the Bode limit applies to a capacitor inseparably connected to a resistor while a diode capacitance can be embedded in the middle of a filter. Referring to Fig. C-2, g_k is the normalized susceptance of a capacitive element at the upper end of the insertion loss band, refer to Eq. (C-1). The Bode limit applies to the "1/∞" curve while the limit for an embedded capacitor is the highest "odd/∞" curve. The limits are compared in Fig.

APPENDIX C

ALL LINES ARE $\lambda_0/4$

$$Y_1' = Y_0 + Y_0^2/Y_1$$

$$Y_0' = Y_0 + Y_1$$

$$Z_0' = Z_0 + Z_1$$

$$Z_1' = Z_0 + Z_0^2/Z_1$$

C-5 Kuroda's Identity[3]

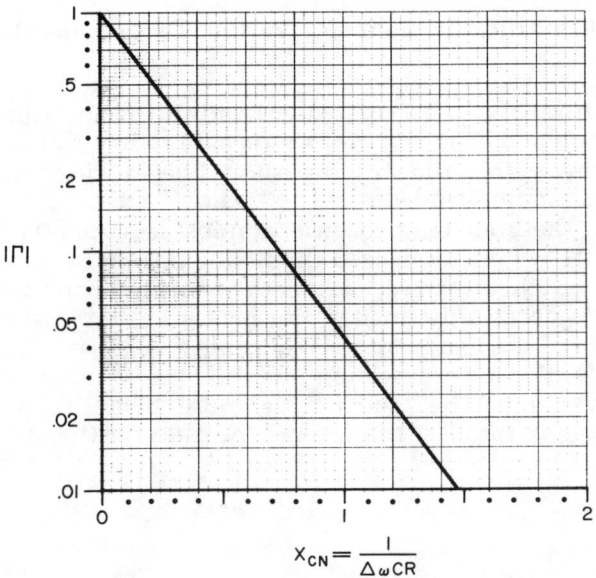

C-6 Bode Matching Integral Limit[4]

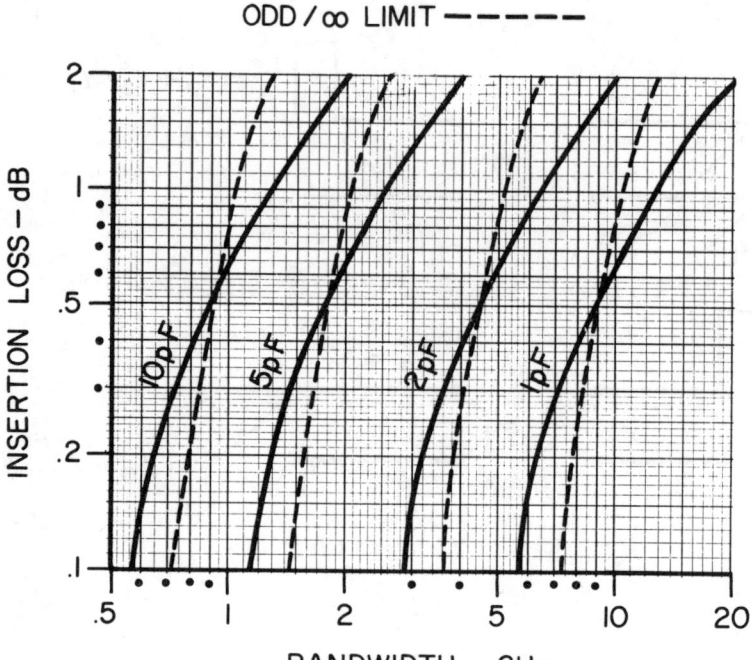

C-7 Bode and Realizable Limits for Various Diode Capacitances

C-7. Note also from Fig. C-2 that greater bandwidth is possible when the diodes are the odd elements in the structure rather than the even elements. In other words, the structure should begin with a shunt capacitor for maximum bandwidth. The "odd/∞" curves give better bandwidth than the Bode limit for insertion loss less than 0.5 dB. In no case do the "1/∞" curves exceed the Bode limit. Because the ripple is better matched than ρ_m part of the time, the bandwidth with the ripple is necessarily smaller. For high insertion loss (greater than 1 dB), 50% of the bandwidth is realizable (for "1/∞"). At 0.1 dB, 73% is realizable. For lower insertion loss the filter structure comes even closer to the limit. The "odd/∞" is about 1 greater than the "1/∞" in all cases so that it is the favored location for the diode.

APPENDIX C

TABLE C-1
TCHEBYSHEFF COEFFICIENTS

VSWR ρ_m	δ_m(dB)	n	g_1	g_2	g_3	g_4	g_5	(g_{12}) g_6	(g_{13}) g_7
1.05	.0026	(25)	(.7088)	(1.4098)	(1.7621)	(1.7894)	(1.9356)	(1.9177)	(2.0141)
		7	.6624	1.3097	1.6143	1.5924	1.6143	1.3097	.6624
		5	.6181	1.2026	1.4130	1.2026	.6181		
		3	.4861	.8259	.4861				
1.1	.01	(25)	(.8370)	(1.4712)	(1.8663)	(1.7845)	(2.0086)	(1.8818)	(2.0704)
		7	.7953	1.3917	1.7466	1.6329	1.7466	1.3917	.7953
		5	.7547	1.3040	1.5754	1.3040	.7547		
		3	.6274	.9688	.6274				
1.2	.04	(25)	(1.0138)	(1.4950)	(2.0071)	(1.7381)	(2.1219)	(1.8081)	(2.1701)
		7	.9765	1.4353	1.9116	1.6277	1.9116	1.4353	.9765
		5	.9359	1.3677	1.7693	1.3677	.9395		
		3	.8187	1.0896	.8187				
1.36	.1	(25)	(1.2153)	(1.4676)	(2.1761)	(1.6544)	(2.2722)	(1.7053)	(2.3116)
		7	1.1811	1.4228	2.0966	1.5733	2.0966	1.4228	1.1811
		5	1.1468	1.3712	1.9750	1.3712	1.1468		
		3	1.0315	1.1474	1.0315				
1.54	.2	(25)	(1.4047)	(1.4136)	(2.3460)	(1.5632)	(2.4316)	(1.6027)	(2.4661)
		7	1.3722	1.3781	2.2756	1.5001	2.2756	1.3781	1.3722
		5	1.3394	1.3370	2.1660	1.3370	1.3394		
		3	1.2275	1.1525	1.2275				
2.0	.5	(25)	(1.7682)	(1.2826)	(2.6997)	(1.3869)	(2.7748)	(1.4134)	(2.8046)
		7	1.7372	1.2583	2.6381	1.3444	2.6381	1.2583	1.7372
		5	1.7058	1.2296	2.5408	1.2296	1.7058		
		3	1.5963	1.0967	1.5963				
2.66	1	(25)	(2.1975)	(1.1287)	(3.1515)	(1.2033)	(3.2223)	(1.2218)	(3.2501)
		7	2.1664	1.1116	3.0934	1.1736	3.0934	1.1116	2.1664
		5	2.1349	1.0911	3.0009	1.0911	2.1349		
		3	2.0236	.9941	2.0236				
4.1	2	(25)	(2.8980)	(.9234)	(3.9360)	(.9732)	(4.0079)	(.9853)	(4.0358)
		7	2.8655	.9119	3.8780	.9535	3.8780	.9119	2.8655
		5	2.8310	.8985	3.7827	.8985	2.8310		
		3	2.7107	.8327	2.7107				
Maximally Flat = 3dB @ ω_1'		7	0.4450	1.247	1.802	2.000	1.802	1.247	0.4450
		5	0.6180	1.618	2.000	1.618	0.6180		
		3	1.000	2.000	1.000				

REFERENCES

1. G.L. Matthaei, L. Young, and E.M.T. Jones, *Microwave Filters, Impedance-Matching Networks, and Coupling Structures*, McGraw-Hill Book Co., New York, N.Y., 1964.

2. P.I. Richards, "Resistor-Transmission-Line Circuits," *Proc. IRE*, vol. 36, pp. 217-220, February, 1948.

3. N. Ozaki and J. Ishii, "Synthesis of a Class of Strip-line Filters," *IRE Trans. on Circuit Theory*, vol. CT-5, pp. 104-109, June, 1958.

4. H.W. Bode, *Network Analysis and Feedback Amplifier Design*, Princeton, N.J.: Van Nostrand, 1945.

Appendix D
Matrix Relationships

ABCD Matrices

Conventional loop and nodal analyses are convenient for circuits comprised of all lumped elements. However, when distributed elements (transmission line elements) become mixed with lumped elements, methods of calculation using matrices are more convenient. The well-known matrix methods are impedance matrix (Z), admittance matrix (Y), wave cascading coefficients matrix (r), generalized circuit constants matrix (ABCD), and scattering matrix (S)[1]. Of these types of matrices, two stand out as most useful for microwave circuits. Measurements are most often made in terms of S parameters. S_{11} is identical to Γ, the voltage reflection coefficient, and the basic parameter of the Smith Chart. Calculations are most easily made using ABCD matrices, because: (a) lumped elements and transmission line elements are related to the matrix elements quite simply; and (b) elements are cascaded simply by multiplying their matrices.

Beatty and Kerns[1] give the definitions of and interrelationships between the various matrices even when input and output impedances are unequal. The more restrictive case given here is adequate for most problems and is all that is required for the derivations presented in this book. The overall input and output impedances are assumed to be equal. The terms of the ABCD matrix are defined in Fig. D-1. A is a voltage transforming term; D, a current transformer; B, a series impedance term; and C, shunt admittance.

The representation of various elements by ABCD matrices is given in the upper half of Fig. D-2. Z and Y are complex numbers. A lossey transmission line may be represented either by using hyperbolic functions or by putting the appropriate attenuator next to each lossless line. This latter is often preferred, because hyperbolic functions with complex arguments are not always available in Fortran IV.

ABCD MATRIX DEFINITIONS

$$u_1 = Au_2 + Bi_2 \quad (D-1)$$
$$i_1 = Cu_2 + Di_2 \quad (D-2)$$

$$\begin{bmatrix} u_1 \\ i_1 \end{bmatrix} = \begin{bmatrix} A & B \\ C & D \end{bmatrix} \begin{bmatrix} u_2 \\ i_2 \end{bmatrix} \quad (D-3)$$

SYMMETRICAL: $\quad A = D \quad$ (D-4)
RECIPROCAL: $\quad AD - BC = 1 \quad$ (D-5)
LOSSLESS RECIPROCAL: A&D−REAL; B&C−IMAGINARY

D-1 ABCD Matrix Definition

The relationships permitting ABCD to be converted to S parameters are given in the lower portion of Fig. D-2. The main terms of interest are S_{21} and S_{11}. Attenuation is given by

$$\alpha = 20 \log \left| \frac{1}{S_{21}} \right| = 20 \log \left| \frac{A+B+C+D}{2} \right| \quad (D-10)$$

using normalized ABCD. Transmission phase shift is given by

$$\Phi = -\tan^{-1}\left[\frac{\text{Im}(A+B+C+D)}{\text{Re}(A+B+C+D)} \right] \quad (D-11)$$

The reflection coefficient is given by S_{11}

$$\Gamma = S_{11}. \quad (D-12)$$

From this can be calculated the VSWR,

$$\rho = \frac{1 + |\Gamma|}{1 - |\Gamma|} = \frac{1 + |S_{11}|}{1 - |S_{11}|}. \quad (D-13)$$

APPENDIX D

The operations for combining ABCD matrices in cascade and in parallel are given in Fig. D-3. Cascade combination is straight matrix multiplication. The relationships for parallel combination were derived by converting to Y matrices, adding, and converting back to ABCD matrices.

The relationships of Figs. D-2 and D-3 and Eqs. (D-10) - (D-13) are easily set up as Fortran IV subroutines to make the simulation of microwave circuits in the computer a relatively simple matter. The relationships are also useful for deriving analytic expressions for the performance of microwave circuits as is done many times in this book.

ABCD MATRIX RELATIONSHIPS

		GENERAL ABCD	NORMALIZED ABCD	
SERIES IMPEDANCE:	—⟋⟋⟋— Z	$\begin{bmatrix} 1 & Z \\ 0 & 1 \end{bmatrix}$	$\begin{bmatrix} 1 & Z/Z_0 \\ 0 & 1 \end{bmatrix}$	
SHUNT ADMITTANCE:	Y	$\begin{bmatrix} 1 & 0 \\ Y & 1 \end{bmatrix}$	$\begin{bmatrix} 1 & 0 \\ Y/Y_0 & 1 \end{bmatrix}$	
TRANSMISSION LINE LENGTH:	θ, Z_0	$\begin{bmatrix} \cos\theta & jZ_0\sin\theta \\ jY_0\sin\theta & \cos\theta \end{bmatrix}$	$\begin{bmatrix} \cos\theta & j\sin\theta \\ j\sin\theta & \cos\theta \end{bmatrix}$	
ATTENUATOR α IN dB	α	$\begin{bmatrix} \frac{1}{2}(10^{\alpha/20}+10^{-\alpha/20}) & \frac{Z_0}{2}(10^{\alpha/20}-10^{-\alpha/20}) \\ \frac{Y_0}{2}(10^{\alpha/20}-10^{-\alpha/20}) & \frac{1}{2}(10^{\alpha/20}+10^{-\alpha/20}) \end{bmatrix}$	DROP Z_0 AND Y_0	
TRANSFORMER:	$n:1$	$\begin{bmatrix} n & 0 \\ 0 & 1/n \end{bmatrix}$	$\begin{bmatrix} n & 0 \\ 0 & 1/n \end{bmatrix}$	
INPUT REFLECTION COEFFICIENT:	$S_{11} = \dfrac{A+BY_0-CZ_0-D}{A+BY_0+CZ_0+D}$		$\dfrac{A+B-C-D}{A+B+C+D}$	(D-6)
REVERSE VOLTAGE TRANSFER COEFFICIENT:	$S_{12} = \dfrac{2(AD-BC)}{A+BY_0+CZ_0+D}$		$\dfrac{2(AD-BC)}{A+B+C+D}$	(D-7)
FORWARD VOLTAGE TRANSFER COEFFICIENT:	$S_{21} = \dfrac{2}{A+BY_0+CZ_0+D}$		$\dfrac{2}{A+B+C+D}$	(D-8)
OUTPUT REFLECTION COEFFICIENT:	$S_{22} = \dfrac{-A+BY_0-CZ_0+D}{A+BY_0+CZ_0+D}$		$\dfrac{-A+B-C+D}{A+B+C+D}$	(D-9)

D-2 ABCD Matrix Relationships

MICROWAVE DIODE CONTROL DEVICES

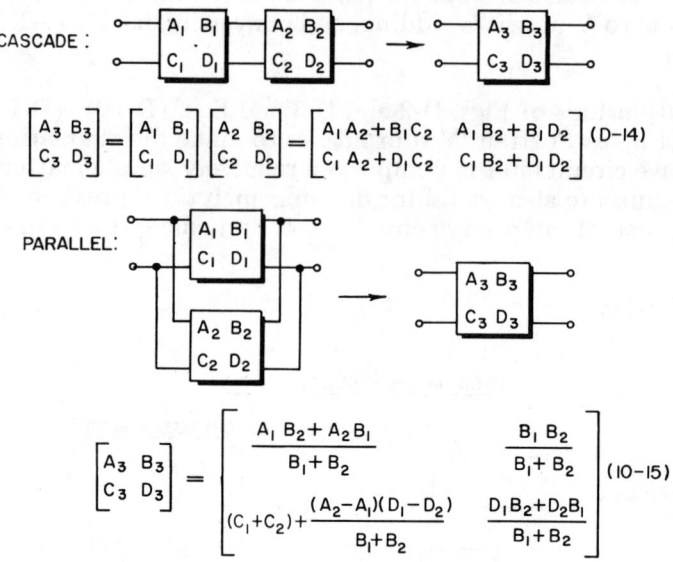

D-3 ABCD Matrix Operations

REFERENCE

1. R.W. Beatty and D.M. Kerns, "Relationship between Different Kinds of Network Parameters, Not Assuming Reciprocity or Equality of the Waveguide or Transmission Line Characteristic Impedances," *Proc IEEE*, Vol. 52, Jan. 1964, p. 84.

Appendix E
Transmission Line Topics

The basic transmission line equations are derived by considering the voltages and currents on a two-wire (Lecher) transmission line as shown in Fig. E-1. The voltages and currents are related as follows:

$$\Delta V = Z(\Delta x)(-I) \qquad \frac{dV}{dx} = -ZI \qquad \frac{d^2 V}{dx^2} = ZYV = \gamma^2 V \qquad \text{(E-1)}$$

$$\Delta I = Y(\Delta x)(-V) \qquad \frac{dI}{dx} = -YV \qquad \frac{d^2 I}{dx^2} = ZYI = \gamma^2 I \qquad \text{(E-2)}$$

in which Z is impedance per unit length and Y, admittance per unit length.

$$Z = R + j\omega L \qquad\qquad Y = G + j\omega C \qquad \text{(E-3)}$$

$$\gamma = \sqrt{ZY} = \sqrt{(R + j\omega L)(G + j\omega C)} \qquad \text{(E-4)}$$

$$\gamma = \alpha + j\beta \qquad \text{(E-5)}$$

For low loss lines

$$\alpha = 0 \qquad \beta = \omega \sqrt{LC} = \frac{\omega}{v_\epsilon} = 2\pi \frac{f}{v_\epsilon} = 2\pi/\lambda_\epsilon \qquad \text{(E-6)}$$

in which v_ϵ is the velocity of propogation in the transmission line and λ_ϵ is the wavelength in the transmission line (usually dielectric ϵ).

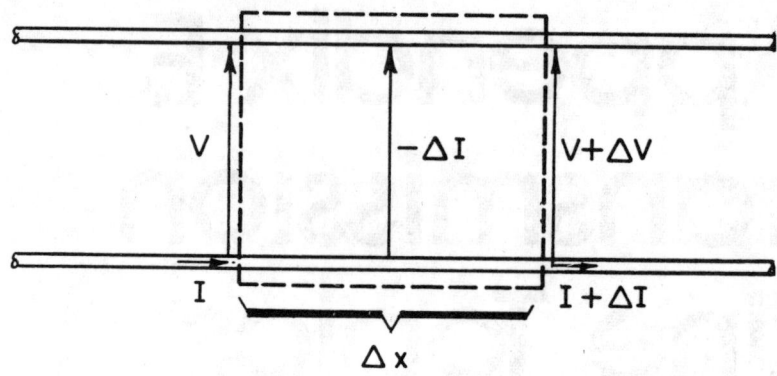

$$V = V_0 \cos \omega t = V_0 \operatorname{Re}\{e^{j\omega t}\}$$
$$I = I_0 \cos \omega t = I_0 \operatorname{Re}\{e^{j\omega t}\}$$

E-1 Incremental Voltages and Currents on a Two-Wire Transmission Line

1. Basic Solution

Eqs. (E-1) and (E-2) are second order differential equations having solutions of the form:

$$V(x) = V_i e^{-\gamma x} + V_r e^{\gamma x} \tag{E-7}$$

or

$$I(x) = I_i e^{-\gamma x} + I_r e^{\gamma x} \tag{E-8}$$

in which the subscript i indicates a forward traveling incident wave (from left to right), and r indicates reflected wave (right to left). Eq. (E-7) is arbitrarily selected as the solution to use for further development.

2. Reflection Coefficient, Γ

The reflection coefficient is defined as

$$\Gamma = \frac{V_r}{V_i} \tag{E-9}$$

causing Eq. (E-7) to read

$$V(x) = V_i(e^{-\gamma x} + \Gamma e^{\gamma x}) \tag{E-10}$$

APPENDIX E

3. VSWR, ρ

The voltage Standing Wave Ratio VSWR is the ratio of the maximum voltage V_{max} on the transmission line to the minimum voltage V_{min}.

$$\rho = \frac{V_{max}}{V_{min}} \tag{E-11}$$

Values of x can be selected which give

$$V_{max} = V_i(1 + |\Gamma|) \tag{E-12}$$

$$V_{min} = V_i(1 - |\Gamma|) \tag{E-13}$$

giving

$$\rho = \frac{1 + |\Gamma|}{1 - |\Gamma|} \tag{E-14}$$

4. Basic Solution Further Developed

Beginning with Eq. (E-10) the current is found using Eq. (E-1)

$$V(x) = V_i(e^{-\gamma x} + \Gamma e^{\gamma x}) \tag{E-10}$$

$$\frac{dV}{dx} = -ZI \qquad I(x) = -\frac{1}{Z}\frac{dV}{dx} = -\frac{V_i}{Z}(-\gamma e^{-\gamma x} + \Gamma \gamma e^{\gamma x}) \tag{E-15}$$

$$\frac{V(x)}{I(x)} = \frac{Z\, V_i(e^{-\gamma x} + \Gamma e^{\gamma x})}{\gamma\, V_i(e^{-\gamma x} - \Gamma e^{\gamma x})} = \sqrt{\frac{Z}{Y}}\, \frac{(e^{-\gamma x} + \Gamma e^{\gamma x})}{(e^{-\gamma x} - \Gamma e^{\gamma x})} \tag{E-16}$$

5. Characteristic Impedance, Z_0

When there is only a forward traveling wave ($\Gamma = 0$) the ratio of voltage to current is the characteristic impedance of the transmission line.

$$Z_0 = \frac{V_i(x)}{I_i(x)} = \sqrt{\frac{Z}{Y}}\, \frac{(e^{-\gamma x} + 0)}{(e^{-\gamma x} - 0)} = \sqrt{\frac{R + j\omega L}{G + j\omega C}} \approx \sqrt{\frac{L}{C}} \tag{E-17}$$

the last approximation holding for low loss lines. Thus Eq. (E-16) reduces to

$$\frac{V(x)}{I(x)} = Z_0\, \frac{e^{-\gamma x} + \Gamma e^{\gamma x}}{e^{-\gamma x} - \Gamma e^{\gamma x}} \tag{E-18}$$

6. Impedance at a Point, Z_T

For a transmission line terminated by an impedance Z_T at $x = 0$

$$Z_T = \frac{V(0)}{I(0)} = Z_0 \frac{1 + \Gamma_T}{1 - \Gamma_T} \tag{E-19}$$

in which Γ_T is the reflection coefficient of Z_T in its electrical plane ($x = 0$). Solving for Γ_T

$$\Gamma_T = \frac{Z_T - Z_0}{Z_T + Z_0} \tag{E-20}$$

7. General Transmission Line Equation

The input impedance of the transmission line Z_{IN} a distance x toward the generator will be given by

$$Z_{IN} = Z(-x) = \frac{V(-x)}{I(-x)} = Z_0 \frac{e^{\gamma x} + \Gamma_T e^{-\gamma x}}{e^{\gamma x} - \Gamma_T e^{-\gamma x}} = Z_0 \frac{(Z_T + Z_0)e^{\gamma x} + (Z_T - Z_0)e^{-\gamma x}}{(Z_T + Z_0)e^{\gamma x} - (Z_T - Z_0)e^{-\gamma x}} \tag{E-21}$$

$$= Z_0 \frac{Z_T \dfrac{e^{\gamma x} + e^{-\gamma x}}{2} + Z_0 \dfrac{e^{\gamma x} - e^{-\gamma x}}{2}}{Z_0 \dfrac{e^{\gamma x} + e^{-\gamma x}}{2} + Z_T \dfrac{e^{\gamma x} - e^{-\gamma x}}{2}} = Z_0 \frac{Z_T \cosh \gamma x + Z_0 \sinh \gamma x}{Z_0 \cosh \gamma x + Z_T \sinh \gamma x} \tag{E-22}$$

For a low loss line

$$Z_{IN} = Z_0 \frac{Z_T \cos\left(2\pi \dfrac{x}{\lambda_\epsilon}\right) + jZ_0 \sin\left(2\pi \dfrac{x}{\lambda_\epsilon}\right)}{Z_0 \cos\left(2\pi \dfrac{x}{\lambda_\epsilon}\right) + jZ_T \sin\left(2\pi \dfrac{x}{\lambda_\epsilon}\right)} \tag{E-23}$$

All of the above equations may be equally well expressed in admittance by substituting $1/Y$ for Z in all places.

8. Special Cases of Transmission Line Equation

1. Line short circuit terminated: $Z_T = 0$

$$Z_{IN} = jZ_0 \tan\left(2\pi \frac{x}{\lambda_\epsilon}\right) \tag{E-24}$$

APPENDIX E

The transmission line is pure reactance, positive at short length (inductive).

$$Z_{IN} = jZ_0\left(2\pi \frac{x}{\lambda_\epsilon}\right) = jZ_0(\beta x) = j\sqrt{\frac{L}{C}}\,\omega\sqrt{LC}\quad x = j\omega Lx \quad \text{(E-25)}$$

The reactance of a short circuit terminated line is shown in Fig. E-2. At $x/\lambda_\epsilon = 1/8$, $X_{IN} = Z_0$. At $x/\lambda_\epsilon = 1/4$ $X_{in} = \infty$. A quarter wavelength shorted stub will be an open circuit. When placed in shunt with a transmission line, it will provide a DC path to ground and will have no effect on propagation in the main transmission line. When the stub is $\lambda/2$ long it will be the same as zero length.

2. Line open circuit terminated: $Z_T = \infty$

$$Z_{IN} = -jZ_0 \cot\left(2\pi \frac{x}{\lambda_\epsilon}\right) \quad \text{(E-26)}$$

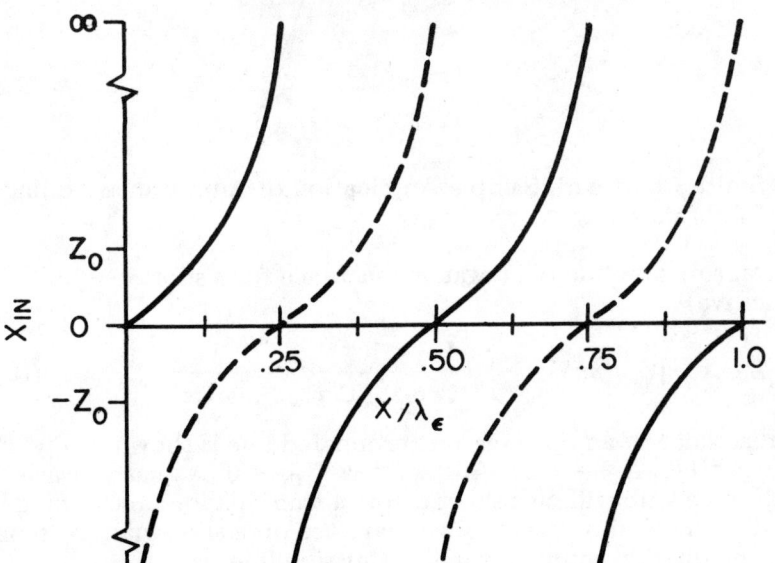

E-2 Reactances of Short Circuit and Open Circuit Terminated Transmission Lines

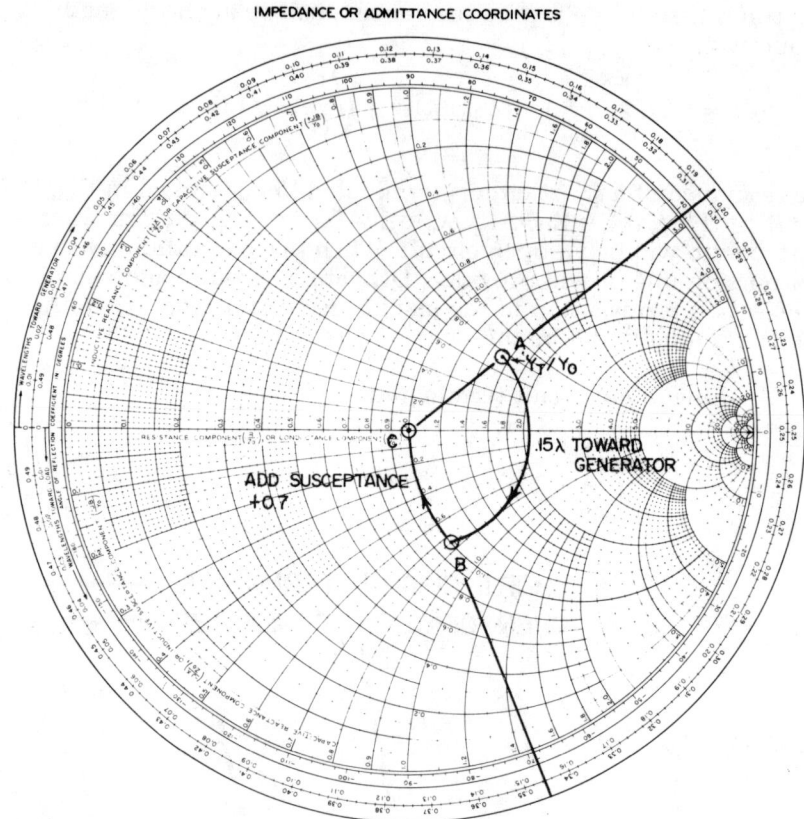

E-3 Smith Chart with Sample Application of Admittance Adding to Input Matching

The transmission line is a negative reactance for a short length (capacitive)

$$Z_{IN} = -jZ_0/(\beta x) = -j\sqrt{\frac{L}{C}}\frac{1}{\omega\sqrt{LC}\,x} = -\frac{j}{\omega C x} \quad \text{(E-27)}$$

The reactance of an open circuit terminated line is shown in Fig. E-2. At $x/\lambda_\epsilon = 1/8$, $X_{IN} = -Z_0$. At $x/\lambda_\epsilon = \frac{1}{4}$, $X_{IN} = 0$. A quarter wavelength open stub will be a short across a transmission line. A diode placed on the end of the stub will have the opposite switching sense from one directly shunting the transmission line.

3. Line a quarter wavelength long: $2\pi \dfrac{x}{\lambda_\epsilon} = \dfrac{\pi}{2}$

APPENDIX E

$$Z_{IN} = Z_0^2/Z_T, \quad Z_{IN}/Z_0 = Z_0/Z_T = Y_T/Y_0 \tag{E-28}$$

The transmission line is an inverting transformer. Z_T can be matched to Z_{IN} by selecting Z_0

$$Z_0 = \sqrt{Z_{IN} Z_T} \tag{E-29}$$

The characteristic impedance of a quarter wavelength stub terminated by a switching diode will transform the impedance of the switching diode.

9. The Smith Chart

Although Eq. (E-23) is quite complex, it can easily be used with the aid of a Smith Chart as shown in Fig. E-3. The Smith Chart is a plot in the Γ plane.

Given

$$\Gamma = |\Gamma| e^{j\theta} \tag{E-30}$$

the radius is $|\Gamma|$ and θ is the phase angle. The angle θ is marked around the outside edge. The radius $|\Gamma|$ can be calculated from ρ using Eq. (E-14) solved for $|\Gamma|$.

$$|\Gamma| = \frac{\rho - 1}{\rho + 1} \tag{E-31}$$

The normalized impedance Z_T/Z_0 is also shown as a grid on the Smith Chart which can be calculated using Eqs. (E-19) and (E-20). Note that for a short circuit $Z_T = 0, \Gamma_T = -1, |\Gamma| = 1, \theta = 180°$ ($V_r = -V_i$). For an open circuit $Z_T = \infty, \Gamma_T = +1, |\Gamma| = 1, \theta = 0°$, ($V_r = +V_i$). For a matched load $\Gamma_T = 0, Z_T = Z_0, (Z_T/Z_0 = 1), \rho = 1.0$.

The Smith Chart is also marked along the outer edge in wavelengths toward generator and wavelengths toward load (away from the generator). This sense of movement pertains to the point of observation (to the user's eye looking towards the termination). As the observer moves toward the generator (like walking backwards on a railroad track), the radius, $|\Gamma|$, remains constant while θ decreases in accordance with the distance moved. Note that zero wavelength is taken at the $\theta = 180°$ point. This is done because measurements are usually made with respect to a short circuit ($Z_T = 0, \theta = 180°$) as a reference. Consider the impedance of a short circuit as the observer backs away from it, moving towards the generator. It is inductive in accordance with Eqs. (E-24) and (E-25). Similarly as the observer backs away from an open circuit ($Z_T = \infty$), he sees a capacitive reactance in accordance with Eqs. (E-26) and (E-27).

Note that half-way around the Smith Chart is a quarter wavelength. This indicates that point symmetry which uses the Smith Chart center converts normalized impedance to normalized admittance in accordance with Eq. (E-28). Note that the normalized impedance curves on the Smith Chart are also labeled normalized admittance. Most of the operations stated above for normalized impedance apply to normalized admittance. If measurements are made with reference to an open circuit ($Y_T = 0$), the markings in wavelengths toward generator are correct. The only marking that is not correct is the reflection coefficient angle which is off by 180°. ($Y_T = 0$ obviously gives $\theta = 0°$.) This subtle point should be carefully taken into account when writing computer programs to do transmission line calculations. In all other respects, the Smith Chart represents normalized impedances and normalized admittances equally well.

The Smith Chart is an invaluable aid for calculating impedance or admittance from slotted line measurements. The VSWR gives the radius $|\Gamma|$ in accordance with Eq. (E-31). The location of V_{min} gives θ. V_{min} occurs where the impedance of Z_T/Z_0 is transformed to its lowest value.

$$\frac{Z_{IN}}{Z_0} = \frac{1}{\rho_T} \tag{E-32}$$

Consider an inductive termination—it is equivalent to a short length of short circuit terminated line. Its minimum will be to the right (away from the generator) of that of a short circuit reference. Thus the Smith Chart markings "wavelengths toward generator (load)" have the opposite sense of the shift in minimum position when the beginning point is taken as the short circuit reference. For making slotted line measurements, this difference in sense should be taken into account to avoid confusion.

It is also useful to observe that the VSWR is equal to R_T/Z_0 when $R_T/Z_0 \geq 1$. ($X_T/Z_0 = 0$).

$$\rho = \frac{R_T}{Z_0} \tag{E-33}$$

This identity is easily seen by noting the similarity between Eqs. (E-14) and (E-19) when Γ_T is real and positive ($\theta = 0°$). The radius $|\Gamma|$ can be found directly from the VSWR by measuring the distance from the Smith Chart center to the point on the R/Z_0 line equal to ρ. The nodal shift, when the unknown is replaced by a short circuit, gives the angular location of Γ. The radial location of Γ in the complex plane is the distance $|\Gamma|$ from the Smith Chart center.

APPENDIX E

The Smith Chart is frequently used as an aid in improving the match of devices. The matching elements are normally shunt elements, and they are normally put on the input port (generator side) of the device. Using shunt elements makes it advantageous to work in normalized admittance. Adding a shunt capacitance adds a positive susceptance which corresponds to clockwise movement on the constant conductance circles. The matching element has to be added on the generator side of the device, so the observer must move toward the generator (back up) to determine where to add the capacitive stub. Moving the point of observation toward the generator also corresponds to clockwise rotation on the Smith Chart (with the chart center as center). Therefore the normal matching procedure is to plot the admittance of the device (at its input plane) on the Smith Chart and move clockwise in combinations of constant radius and constant conductance to bring the admittance to the chart center. For example, to match the point A ($Y_T/Y_0 = 1.5 + j\,.75$), the observer moves .15 wavelengths towards the generator, taking the admittance to point B ($Y/Y_0 = 1.0 - j\,.75$), at which point he adds a normalized susceptance of $+.75$ to take the admittance to the matched condition, point C ($Y/Y_0 = 1.0 + j\,0$). The only circumstances that would produce counter-clockwise rotation are the use of an inductive stub or the possibility of adding something on the load side of the device.

Driver Circuits

Appendix F

For best performance a diode switch should be biased in both states. Under most conditions PIN diodes are reverse biased only enough to hold off rectified current (10V—100V). But when harmonic generation, intermodulation products, or insertion loss must be low then the PIN diode must be biased to half the breakdown voltage V_B. The reverse bias voltage V_R would be

$$V_R = \frac{V_B}{2} \tag{F-1}$$

For lower power the bias need not be as high, but is a function of the incident peak power $\hat{P}_i$ as follows:

$$V_R = 2\sqrt{2\hat{P}_i Z_o} \quad \text{for series diodes} \tag{F-2}$$

$$V_R = \sqrt{2\hat{P}_i Z_o} \quad \text{for shunt diodes} \tag{F-3}$$

The forward bias determines how low the insertion loss of a series diode will be or how high the isolation of a shunt diode will be as discussed in Chapter II, Basic Concepts, section D, Variable Attenuation. The forward voltage drop is approximately 1 V and the forward bias current is I_F.

The simplest bias circuit is shown in Fig. F-1. The series resistor is determined by

$$R = \frac{V_R - 1}{I_F} \tag{F-4}$$

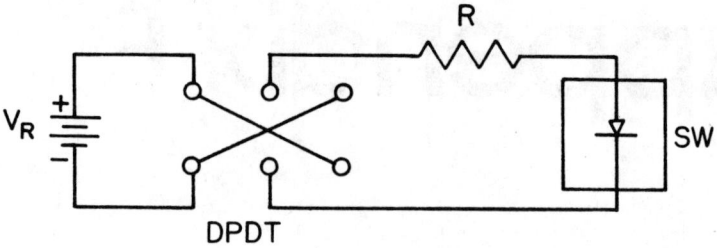

F-1 Manual Switching Driver Circuit

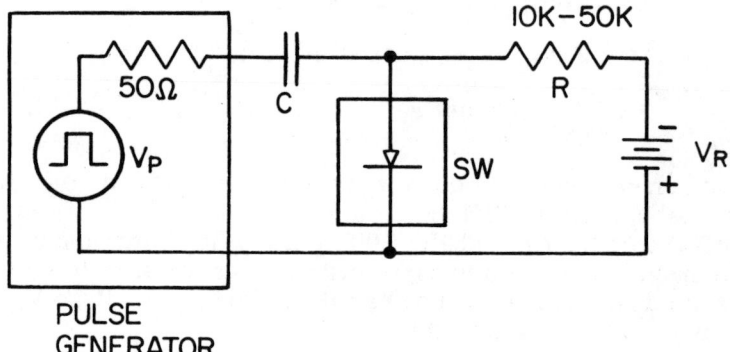

F-2 High-Speed Pulsing Driver Circuit

Nominal values of V_R and I_F are 20V and 100 mA respectively. Under these conditions, the battery has to be able to supply 2 W power and R has to be able to dissipate it.

In many applications, the diode is normally in one state and put into the other state only for short periods of time (pulsed). It is most efficient to arrange the switch so that it is reverse biased normally and pulsed into conduction. Fig. F-2 shows the circuit for a pulse driver. The pulse voltage must be large enough to overcome V_R. It also must be high enough to overcome the 1V forward drop of the diode. Therefore, the pulse voltage must satisfy

$$V_P \geq V_R + 1 \tag{F-5}$$

The series capacitor should be part of an RC time constant longer than the pulse length t_p to prevent pulse sag (integration). The resistive part of the time constant is made up of the 50 ohm pulse generator impedance and the 10 ohm diode resistance. The series capacitor should satisfy

APPENDIX F

$$C > \frac{t_p}{60} \qquad (F\text{-}6)$$

The mismatch of the conducting diode causes the pulse to be reflected back into the pulse generator. This pulse can be prevented from having undesirable transients either by making sure that the pulse generator has 50 ohm output impedance, or by using a long cable so that reflections appear after the switch has gone into the non-conducting state. The second-time-around echo will normally be too low to overcome V_R. A resistor may have to be put across the capacitor if the pulse duty factor is high because the rectification of the switching diode will add to V_R and require V_P to be even larger.

If the diode is to be biased into conduction and pulsed into non-conduction, the circuit of Fig. F-2 will be adequate provided V_R and R are adjusted to give I_F. If R is greater than 500 ohms, the pulse voltage on the diode will be 2 V_P. The requirement on C is also relaxed as follows:

$$C > \frac{t_p}{50 + R} \qquad (F\text{-}7)$$

For this polarity of switching, C can also be replaced by a diode (or multiple diodes in series to assure that their forward drop is greater than that of the switching diode). With a diode there is no limit to the pulse length.

When switching is done with a PIN diode the diode tends to slow the switching. The carriers take time after they are injected into the I region before they begin to recombine and change the I region conductivity. Time is also necessary to pull the carriers out of the I region when going to the non-conduction state. The charge stored in the I region Q_I is proportional to the forward current I_F and minority lifetime τ.

$$Q_I = I_F \tau \qquad (F\text{-}8)$$

To have rapid switching of a PIN diode, this charge must be rapidly injected and removed. The slow switching causes the waveform to be partially integrated. Partial differentiation of the drive pulse will compensate for the slow switching. A circuit to accomplish this is shown in Fig. F-3. The charge Q_c stored in C_τ is given by

$$Q_c = C_\tau E \qquad (F\text{-}9)$$

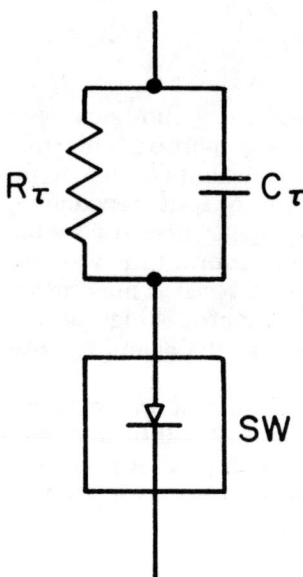

F-3 Switching Pulse Shaping Circuit

in which E is the voltage across C_τ. This is the same voltage as appears across R_τ.

$$E = I_F R_\tau \qquad (F\text{-}10)$$

Setting $Q_I = Q_c$ in Eqs. (F-8)–(F-10) gives

$$R_\tau C_\tau = \tau \qquad (F\text{-}11)$$

For Eq. (F-11) all of the charge in C_τ was assumed to be transferred to the I region. C_τ discharges through R_τ directly and has the resistance of the external pulse circuit in series with the switching diode. Thus R_τ should be large and the external pulse circuit resistance should be low. Normally, it is reasonable to allow these resistances to be moderate, and to adjust $R_\tau C_\tau$ such that $Q_C > Q_I$. The extra current causes no problem in diode turn-on. Only if the reverse bias is near V_B does it cause problems in that breakdown current may flow after the charge is cleared out of the I region.

When R_τ and C_τ are used, diode turn-on is limited by the RC time constant of C_τ and the R composed of R_τ and the external pulse circuit. The external pulse circuit resistance should be low. For fast switching time diode turn-off is sped up by the circuit, but there is a

APPENDIX F

limit to the speed with which it can be turned off. The circuit pulls most of the free carriers out of the I region quickly, but to have high resistance the I region must be free of all carriers. The high injection current does not remove carriers caught in traps in the I region. Carriers are also injected beyond the P-I and I-N junctions where their population falls off. When the diode is pulsed off, these carriers continue to emerge from the I region for a while after the stored charge is depleted depending on the abruptness of the junctions and the purity of the I region.

A transistor circuit for maintaining low pulse circuit driving impedance[1] is shown in Fig. F-4. It must satisfy $C_- \gg \dfrac{I_F}{V_-} \tau$. Another circuit using a toroid inductor to store charging and discharging energy[2] is shown in Fig. F-5. Normally, the transistor is off, and both diodes are back biased. When the transistor is first turned on, most of the current flows through the switching diode limited by R_F. Eventually, the current builds up in L and D_1, leaving the switching diode with only as much current as it needs to stay on. When the transistor is turned off, the current in L wants to continue, so it draws the charge out of the switching diode.

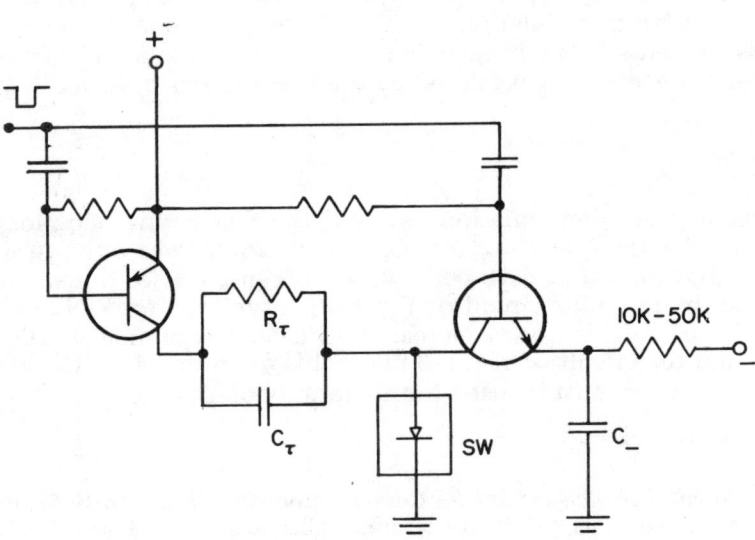

F-4   Low Impedance Driver Circuit

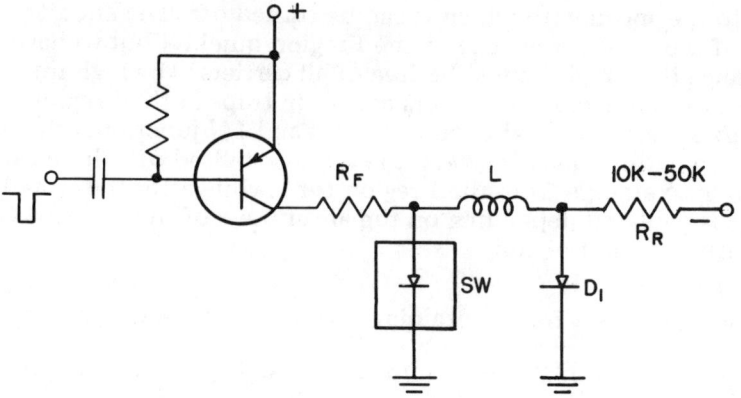

F-5 Inductive Pulse Shaping Circuit

For all of the circuits given, it might seem that when more than one diode is to be switched the diodes could be driven by connecting them in parallel. But variation from diode to diode could cause a severe imbalance in current distribution. This imbalance is easily counteracted by placing a few ohms in series with each diode.[2]

The slow processes in PIN diodes also reduce modulation index when sinusoidal or video modulation is applied at too high a frequency. When the modulation is narrow band, the decrease in modulation efficiency is easily accommodated by increasing the drive power. But when wideband modulation is being applied, a compensation circuit needs to be used. The frequency at which the modulation efficiency begins to decrease ω_c is related by the lifetime τ as follows:

$$\omega_c = \frac{1}{\tau} \tag{F-12}$$

Below ω_c a given modulation power gives a fixed conversion loss to both sidebands. Above ω_c, a fixed modulation power will give sidebands that roll off at 6 dB per octave. ω_c can be raised one or two decades by using the circuit of Fig. F-3. A typical τ for PIN diodes is 1.0–0.1 μsec. This causes decreased efficiency beginning at .16–1.6 MHz and the circuit of Fig. F-3 raises this to around 10 MHz. Shorter lifetime diodes must be used for higher modulation rates.

REFERENCES

1. Anon. *Biasing and Driving Considerations for PIN Diode RF Switches and Moaulators,* Hewlett-Packard Application Note 914, Jan. 1967.

2. C.J. Georgopoulos, "PIN-Driver Design Saves Time," *Microwaves*, pp. 50-55, Aug. 1972.

Author Index

Adams, D.K., et. al., 71
Armistead, M.A., et. al., 1
Arthur, J.B., et. al., 293
Assaly, R.N., 156

Bakanowski, A.E., 298, 319
Bakanowski, A.E., et. al., 298
Baskin, P., 233
Battershall, B.W., et. al., 13, 283
Beatty, R.W., et. al., 345
Blake, C., et. al., 299
Bloom, M., 12
Bode, H.W., 158, 191, 340
Bourdreau, C.A., et. al., 4
Brown, N.J., 225
Brule, J.J., 233
Burns, R.W., et. al., 283

Ciccoletta, D.F., et. al., 120
Clar, P., 13
Cole, F.S., 12

Dawirs, H.N., et. al., 12, 283
DeLoach, B.C., 299
DeRonde, F.C., et. al., 293
Doherty, W.E., Jr., 300

Ekinge, R., et. al., 213

Fisher, R.E., 182
Forster, J.H., 320
Franco, A.G., et. al., 156
Frye, G., 295

Galvin, R., et. al., 302, 331
Garver, R.V., 12, 13, 43, 182, 283, 284, 294, 302, 305
Garver, R.V., et. al., 12, 13, 43, 83, 156, 233, 297, 302
Georgopoulos, C.U., 366
Getsinger, W.J., 300, 305
Grace, D.J., 12
Grauling, C.H., et. al., 283
Grove, W.M., 295
Gunn, J.B., et. al., 293
Gutman, R.J., 294

Hardin, R.H., et. al., 283
Harrison, R.I., 298

Henderson, K.W., et. al., 74
Henoch, B.T., et. al., 284
Higgins, V.J., 156
Hines, M.E., 94, 96, 114, 120
Hoover, J.C., 213
Houlding, N., 298, 300
Hunton, J.K., et. al., 13, 121, 213

Jacobs, H., et. al., 293
Jakes, W.C., Jr., 156
Johnston, R.L., et. al., 302

Kawakami, S., 88
Keywell, F., 300
Krupke, W.F., et. al., 233
Kuroda, 58, 60, 71, 339
Kurokawa, K., et. al., 93, 96

Leenov, D., 223, 233, 297, 331
Leenov, D., et. al., 12
Lin, J.C., 318
Lynes, G.D., et. al., 247, 284

Matthaei, G.L., et. al., 13, 65, 342
McDermott, M.M., et. al., 57
Mortenson, K.E., et. al., 294
Mott, H., et. al., 121
Mouw, R.B., 57, 294
Mouw, R.B., et. al., 294
Muehe, C.E., 54, 193
Mumford, W.W., 80, 176

Nelson, W.W., 195

Oelke, H. 233
Onno, P., et. al., 13, 283
Ozaki, N., et. al., 343

Peppiatt, H.J., et. al., 156
Phillips, E.N., 284
Podell, A., 247
Pound, R.V., 82
Prufer, C.A., 213

Rebsch, D.L., 156
Richards, P.I., 338
Riebman, L., 12
Roberts, D.A.E., 300
Roberts, D.A.E., et. al., 43, 213
Rutz, E.M., et. al., 12, 43, 283
Ryder, R.M., et. al., 4

Sard, E.W., 301
Sicotte, R.L., et. al., 121
Stern, E., 12
Swenson, R.C., et. al., 300
Sze, S.M., 318, 331

Tenenholtz, R., 233
Tevelow, F.L., 13
Thornton, C.G., et. al., 12

Tolkien, J.R.R., 12
Torrey, H.C., et. al., 43

Uhlir, A., 12

Van Iperen, B.B., 305
Van Iperen, B.B., et. al., 300
Vandelin, G.D., et. al., 300

Waltz, M.C., 297
Watson, H.A., 155
Westman, H.P., 82
White, J.F., 133, 283
White, J.F., et. al., 182, 195
Wilkinson, E.J., et. al., 283

Young, L., 54, 193

Subject Index

ABCD matrices, 345
Analog phase modulator, 94
Analog phase shifter, 280
Attenuation equation, 21
    for multiple diodes, 160, 179
Attenuators, variable diode, 4, 41, 197, 287

Balanced modulator, 286
Bandwidth limit, 97
Bandwidth of matched switch, 208
Baseband switching, 29
Bias circuits, 45
    bandwidth, 45, 46
Bias lead, 49
Bias port leakage, 82
Bias T, 54, 57, 66
Binary 180° phase modulator, 91, 257, 286
Bode limit, 158, 191, 340
Bulk diode switches, 293

Capacitance compensation, 57
Coefficient of rf conductivity, 110-111, 309
Communications applications, 10
Conic transmission line, 100
Continuous phase shifter, 280
Cutoff frequency, 85
    calculation, PIN diode, 96
    limitation on phase shifters, 92-93, 95
    limitation on switching, 88, 100
    measurement, 300

DC block, 46
Digital phase shifter, 238
    hi-lo pass, 273
    loaded line, 263
    reflection, 248
    switched line, 238
    summary of, 279
Diode 1N263, 1
    D5151, 133
    HP 3001, 22, 39, 133, 142
    MC 7000, 257
    T7G, 133
    cartridge measurement, 305
    cutoff frequency (see cutoff)
    junction measurement, 305
    lifetime measurement, 311

measurement, 138-139, 297
  parasitics, 18
  resonance, f_R, 96, 100
Discriminator, phase, 287
Doubly balanced mixer/modulator, 285
Driver circuits, 361
Duplexers, 227
Dynamic Q of diode, 301, 302
  theory, 329

Error in phase shift, 239, 248, 259, 266, 276

Ferrite, 5
Filters for bias circuits, 69, 71
Filter theory, 6, 157, 193, 333

Harmonic generation, 107
Heat storage capacity of mesa, 111-112
Helical bias lead, 51
High speed bias circuit, 45, 63
Hi-Lo pass phase shifter, 273
Hot carrier diode limiter, 232
Hot switching, 116

Insertion loss limit, 86, 88, 90
  of phase shifters, 244, 251, 262, 272
  wideband, 157
Instrumentation applications, 11
Integrated circuits, 7, 104, 145
Intermod products, 107
Isolation of multiple diodes, 159, 164
Isolation of multiple throw switches, 189
Isolation limit, 88, 90

Junction measurement, 305
  theory, 323
  thermal properties, 113, 312

Kuroda's identity, 58, 60, 71, 339

Leveler, power, 229
Limit level, P_{LIM}, 223
Limiters, 4, 215
Linear phase shifter, 281
Loaded line phase shifter, 263
Low VSWR switch, 53, 197
Lumped circuit (IC) attenuator, 209
Lumped circuit switch, 178

Matched switch, 197

SUBJECT INDEX

Measurement of diodes, 297
Measurement of limiters, 232
Megawatt switching, 112-113, 225, 229
Microstrip switch design, 145
Millimeter wave switching, 90
Modes of switching, 27
Mode 1 switching, 29
Mode 2 switching, 31
Multiple diode switches, 157, 200, 202, 205
Multiple throw switches, 183, 201

Phase modulator, binary, 91
    linear, 94, 280
Phase shift equation, 22
Phase shifters, 4, 235
    analog, 280
    digital (see digital)
PIN diode theory, 48, 222
Power limit, 111-112
    phase shifter, 114, 246, 253, 260, 263, 272, 276
    switch, average, 108
    switch, CW, 109-110
    switch, peak, 106

Q, dynamic (see dynamic)
Q of diode, 85, 86, 90, 95-96

Radar applications, 8
Recovery time of limiters, 219
Reflection phase shifters, 248
Reflection switches, 39, 145, 183, 227
Reliability, 5, 317
Resonance of diode, f_R, 96, 100
Resonant switching, 31
Richards equation, 338
Ringing, 65
Rise-time, 65

Sampling gate, 290
Sampling oscilloscope, 2
Series switch, 29, 60
Shunt switch, 29, 55
Single sideband modulator, 247, 281, 287, 290
Smith chart, 357
Spike leakage, 225
Stripline limiter design, 131
Stripline switch design, 124, 142
Stub switch design, 143, 193
Stub switched phase shifter, 266
Suppression of rf leakage, 45, 82
    of switching transients, 46, 74, 78, 117

Switched line phase shifter, 238
Switching speed, 115

T, bias, 54, 57, 65
T junction correction, 140
Tchebyscheff coefficients, 334-335
Thermal resistance measurement, 312
Transfer switch, 184
Transient impedance measurement, 302
Transmission line equation, 354
    theory, 351
    model of diode, 36, 135
Tuned insertion loss (multiple diodes), 174
Tuned isolation (multiple diodes), 168

Vector generator, 287

Waveguide switching, 67, 141-142, 151
Wideband switching, 157

X-band limiter, 227
X-band switch, 67